THE ESSENTIAL DICTIONARY OF
HUMAN BIOLOGY

2ND EDITION

THE ESSENTIAL DICTIONARY OF HUMAN BIOLOGY

2ND EDITION

Terry Newton

Ashley Joyce

Australia • Brazil • Japan • Korea • Mexico • Singapore • Spain • United Kingdom • United States

The Essential Dictionary of Human Biology
2nd Edition
T J Newton
A P Joyce

Acquisitions editor: Libby Houston
Managing editor: Kathryn Fairfax
Production editor: Annette Sayers, Eleanna Raissis
Editor: Catherine Page
Art director: Astred Hicks
Proofreader: Nadia Billings
Text design: Cristina Neri
Cover design: Astred Hicks
Illustrator: Lorenzo Lucia
Reprint: Jess Lovell
Typeset in Minion by diacriTech

First published 1996 by McGraw Hill Australia.
Reprinted 2003 (twice), 2005, 2007, 2008 by McGraw Hill Australia.
Second edition published 2009 by McGraw Hill Australia.
This edition published 2010 by Cengage Learning Australia.

For product information and technology assistance,
in Australia call **1300 790 853**;
in New Zealand call **0800 449 725**

For permission to use material from this text or product, please email **aust.permissions@cengage.com**

ISBN 978 0 17 021375 2

Cengage Learning Australia
Level 7, 80 Dorcas Street
South Melbourne, Victoria Australia 3205

Cengage Learning New Zealand
Unit 4B Rosedale Office Park
331 Rosedale Road, Albany, North Shore 0632, NZ

For learning solutions, visit **cengage.com.au**

Printed in Australia by Ligare Pty Limited.
5 6 7 8 9 10 11 19 18 17 16 15

PREFACE

There has never been more interest in the biology of the human species than there is today. In secondary schools, whether it is offered as a discrete course or as an optional part of a biology course, human biology is a very popular choice. At the tertiary level, almost all Australian universities have courses in human biology that attract large numbers of students. Further education courses are now offered due to increasing public interest in the subject. This interest is fostered by constant media reports on new discoveries in the field of human biology as well as feature articles in newspapers and magazines and documentaries on television.

The Essential Dictionary of Human Biology has been compiled for the use of secondary students of human biology, tertiary students taking introductory courses in human biology and other interested people. The dictionary covers terms relating to human biology in its broadest sense; not just human physiology and anatomy, but also areas such as sociology, anthropology, demography, genetics and the origins and prehistory of the human species. Although many texts contain glossaries, the definitions given are necessarily brief, only refer to terms used in the text and do not include the cross-referencing that is possible in a dictionary.

There are thousands of scientific terms used in communicating information about our species, so in a work such as this it is very difficult to decide what to include and what to leave out. As our guiding principle, we have tried to include those terms that are likely to be encountered by students up to first year tertiary level, and technical terms that may be used in media reports or feature articles. We did not set out to comprehensively define all the major human diseases, but we have included diseases that are often used as examples in human biology courses. You will therefore find a large number of diseases described in the dictionary.

A number of people helped in the process of compiling and producing the first edition of this dictionary, and we are still very grateful to all those people. In particular, we wish to thank Dr Geoff Meyer of the Centre for Human Biology at the University of Western Australia for his support and for his critical comments on parts of the original manuscript. We also acknowledge the experts in various fields who reviewed the manuscript and provided feedback on errors, ambiguities and possible omissions. Other helpful comments on the first edition manuscript were provided by June Gouldthorp and Michael Jaques.

For this revised edition, we would like to thank Catherine Page for her tireless editing, particularly of the many new definitions, and her useful suggestions. Thanks are also due to Libby Houston Annette Sayers, and Eleanna Raissis of the Schools Division of McGraw-Hill Education, for their faith in the project and their encouragement during the preparation of the manuscript.

Terry Newton

Ashley Joyce

USING THIS DICTIONARY

In compiling the dictionary, ease of use has been a major consideration. The following principles have been adopted to make the book simple to use and to maximise its usefulness:

- To avoid the reader having to look up multiple entries to arrive at a satisfactory definition of a word, simple terms have been used in the definitions, with scientific terms in parentheses, where appropriate.
- Where there are two or more words with the same meaning, the most commonly used alternative has been defined in detail and the others are quoted as alternatives, with a prompt to refer to the main definition if necessary.
- Cross-referencing of words has been used so that a more complete view of how a word is used may be obtained. Every term printed in SMALL CAPITALS is defined elsewhere in the dictionary, and users should refer to these entries for further information. Similarly, following the '*see*' and '*see also*' references at the ends of entries will give readers a greater understanding of the terms.
- Entries in the dictionary may be in the singular or the plural form of a word, depending on the form that is most commonly used. Where appropriate, the other form of the word—e.g. papillae (singular *papilla*)—has been indicated in italic type.
- Italic type has also been used for alternative names, for scientific names, such as *Homo sapiens*, and for terms that have been defined elsewhere, but where no additional information is presented at the cross reference. In some definitions, italic type has been used to add emphasis to a particular term.
- Many terms have alternative spellings. The spelling most common in Australia has been used, but alternative spellings are indicated. As a general rule American spellings have not been quoted as alternatives.
- Diagrams and tables have been included where appropriate to enhance the definitions and to make the book more useful as a reference work. Where the reader is directed to a diagram, referring to that diagram will help clarify the meaning of the term.
- The pronunciation of many words varies between countries and even between regions within a country such as Australia. Since science is essentially an international endeavour, there is often no consensus about the correct pronunciation of a scientific term. For this reason a pronunciation guide for each term has not been included. However, for terms where the pronunciation differs considerably from the spelling of the word (such as *chancres*), we have indicated the usual pronunciation phonetically.

A

A band part of the microscopic structure of skeletal muscle; *see* sarcomere

abdomen the region between the diaphragm and the pelvis

abdominal cavity the body cavity below the diaphragm; contains the stomach, intestines, liver and other organs

abdominal thrust manoeuvre a first-aid procedure to help a person who is choking; a quick, upward thrust against the diaphragm is used to force air out of the lungs with sufficient force to eject any material lodged in the trachea; also called the *Heimlich manoeuvre*

abduction movement of a limb away from the mid-line (axis) of the body (e.g. raising the arms from the sides to a horizontal position); *see also* diagram of movements at joints, right

ABO blood group system a system of blood grouping based on the presence or absence of two ANTIGENS (antigen A and antigen B) on the red blood cells; persons with antigen A are classified as belonging to group A, those with antigen B as group B, those with both antigens as AB and those with neither as O; an individual with blood group O is known as a *universal donor*, because in an emergency persons belonging to A, B or AB groups can receive O blood provided it is transfused slowly; O blood has no A or B antigens, and so the recipient's blood will not react against it; an individual with blood group AB is known as a *universal recipient* (*receiver*) and is able to receive a transfusion of A, B, O or AB type blood in an emergency; persons of AB group are unable to produce antibodies against either the A or B antigen and so will not react against blood of a different group; *see also* antibody

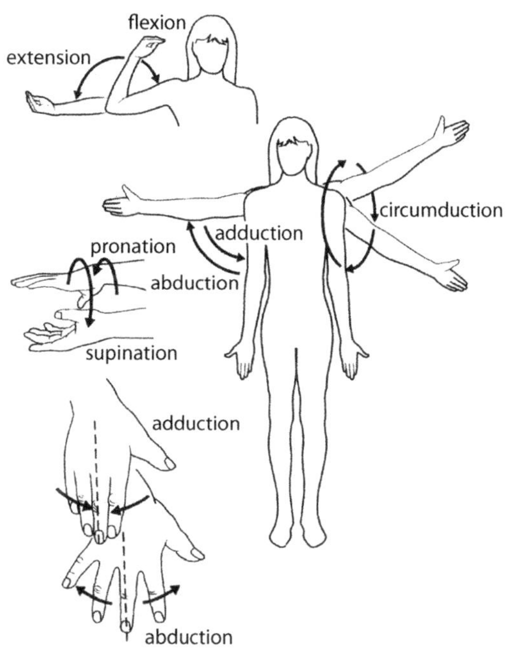

Movements at a joint

aborigines the original inhabitants of a country or region; the inhabitants of a region in prehistoric times; descendants of the original inhabitants of a country or region; also known as *indigenous people*; Aborigines are the original inhabitants of Australia

abortion the premature loss or removal of an embryo or foetus that is not capable of survival outside the uterus; a *spontaneous abortion*, or *miscarriage*, is a natural event and may be caused by a defective egg or sperm or some irregularity in the implantation of the embryo; an *induced abortion* is where one of various means is used to deliberately remove the embryo or foetus from the uterus; may be performed for medical reasons or as a means of birth control

abscess a collection of pus in an inflamed area (e.g. a boil or pimple); *see also* inflammation, ulcer

absolute age the actual age (in years) of a fossil or artefact; *see also* relative age, dating

absolute dating determining the actual age of material; *see* dating

absorption in general, the taking up of liquids by solids, or of gases by solids or liquids; in humans, the intake of fluids or other substances by cells of the skin or mucous membranes; also, the movement of materials into the blood or lymph from the alimentary canal

accelerator mass spectrometry (AMS) radiocarbon dating a technique that can be used to give radiocarbon dates for very small samples of material; *see* radiocarbon dating

accessory duct the smaller of two ducts from the pancreas that empties into the duodenum; the larger duct is the PANCREATIC DUCT

accessory ligament a ligament about a joint that is in addition to the ARTICULAR CAPSULE; one that strengthens or supports another

accessory organ one of several organs that aid in digestion but are not part of the alimentary canal (e.g. salivary glands, liver, gall bladder, pancreas); *see* diagram of digestive system (p. 77)

accessory sex organs structures that transport, protect and nourish the gametes after they leave the testes or ovaries; in males, the epididymis, vas deferens (ductus deferens), seminal vesicles, prostate gland, bulbo-urethral glands, scrotum and penis; in females, the uterine tubes, uterus, vagina and vulva; *see also* primary sex organs

acclimation a term used in some American textbooks meaning the same as ACCLIMATISATION

acclimatisation becoming accustomed to different conditions over a period of time (e.g. when a person who normally lives at sea-level goes to a high altitude, acclimatisation to the changed atmospheric conditions occurs over a period of weeks); also known as *somatogenetic adaptability*

accommodation a change in the curvature of the lens of the eye to focus light rays coming from close or distant objects; for near vision the ciliary muscle around the lens contracts and the lens bulges more; for distant vision the ciliary muscle relaxes and the lens becomes flatter: *see* diagram, opposite top

acetabulum (plural, *acetabulae*) the socket of the pelvis in which the head of the thigh bone (femur) fits; *see* diagram opposite bottom

acetylcholine a NEUROTRANSMITTER molecule

acetylcholinesterase inhibitor a drug that increases the level of the neurotransmitter acetylcholine in the brain; helps to stabilise the symptoms of diseases such as ALZHEIMER'S DISEASE and PARKINSON'S DISEASE

Achilles reflex when the ACHILLES TENDON is struck sharply, the contraction of the calf muscles results in the bending of the foot; also called *ankle reflex*

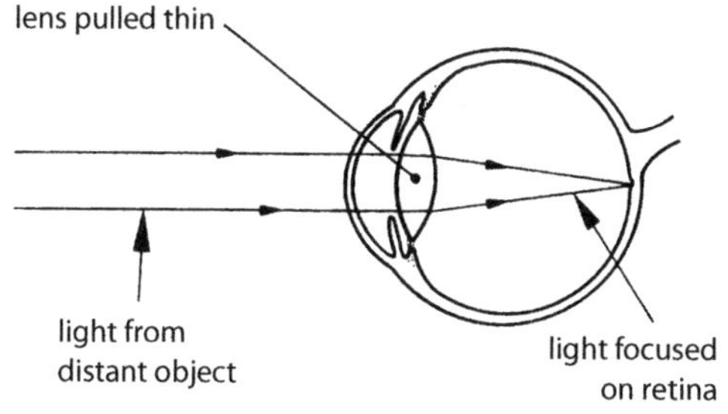

(a) Accommodated for distant vision

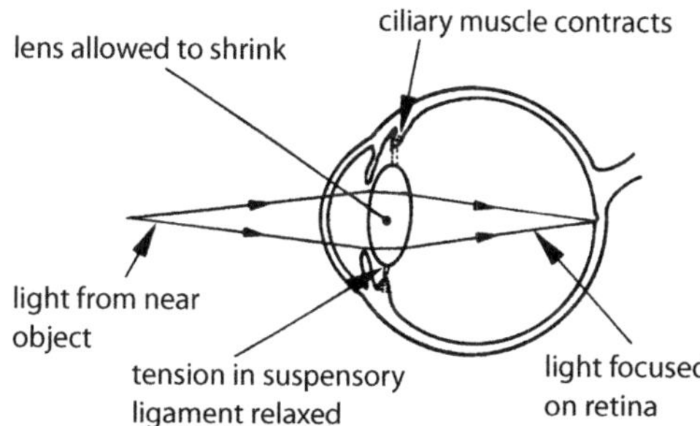

(b) Accommodated for near vision

Accommodation of the lens of the eye

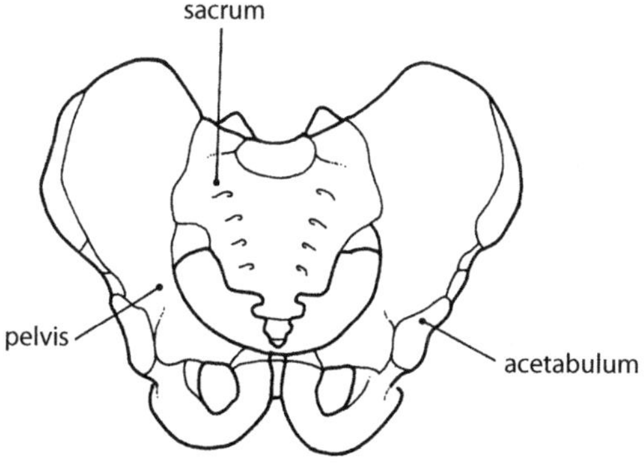

The pelvis, showing the acetabulum

Achilles tendon an alternative name for CALCANEAL TENDON

achondroplasia a form of dwarfism characterised by short limbs, a prominent head, intelligence within the normal range and a waddling gait; often due to the mutation of a single gene, which is dominant and carried on one of the non-sex chromosomes

achromatopsia an inherited form of total COLOUR-BLINDNESS

acid a substance that releases hydrogen ions (H^+) when dissolved in water; a proton donor; a substance that has a pH of less than 7; *see also* diagram of pH scale (p. 208)

acid deposition the falling of acid substances onto land or oceans; oxides of nitrogen and sulphur dioxide react with water vapour in the atmosphere to form acids; the acids may fall to earth as rain (*acid rain*), hail or snow; sometimes the acids are deposited with particles such as soot

acid rain a form of ACID DEPOSITION

acidosis a condition in which the blood has a higher hydrogen ion concentration than normal, resulting in decreased pH; respiratory acidosis is caused by an abnormally low breathing rate; metabolic acidosis may result from accumulation of acidic metabolic products or loss of bicarbonate ions from the body with diarrhoea; *see also* alkalosis

acne infection and inflammation of oil-producing glands in the skin (sebaceous glands) that usually develops at puberty because, under the influence of increased secretion of sex hormones, the oil glands grow in size and increase oil production

acquired characteristics
1. characteristics that are gained during a person's lifetime, produced by the environment or by use or disuse (e.g. a suntan or enlarged muscles in athletes)
2. in Jean Baptiste Lamarck's hypothesis, characteristics that are acquired in response to environmental demands and then passed on to offspring (e.g. the long neck of a giraffe would have been acquired by stretching to reach leaves at the tops of trees); experimental evidence has not been able to confirm this idea

acquired immune deficiency syndrome (AIDS) an extremely serious and often fatal disease caused by infection with human immunodeficiency virus (HIV), which damages the immune system so that the patient becomes susceptible to infections and to some forms of cancer; death usually occurs due to a secondary infection or to cancer; the disease is transferred through the exchange of body fluids, for example through sexual intercourse or through infected blood (e.g. sharing hypodermic needles)

acquired immunity IMMUNITY gained during life as a result of exposure to an antigen

acquired reflex a reaction that has been learned in response to a STIMULUS

acromegaly a condition caused by oversecretion of growth hormone during adulthood, characterised by thickened bones and enlargement of other tissues

acrosome a dense capsule in the head of a sperm containing enzymes that aid the penetration of the sperm into an egg

acrosome reaction the release of enzymes from the acrosome in the head of the sperm; the enzymes help dissolve cells and intercellular material covering the egg, allowing the sperm to penetrate and fertilise the egg

actin a protein of which myofilaments of muscle are made (*see* muscle fibres); also the substance of which the cytoskeleton of cells is composed

actinomycetes filamentous bacteria; some produce antibiotics and some are pathogenic in humans

action potential a change in the electrical charge (potential difference) on a nerve or muscle cell membrane due to sodium ions moving into the cell at a particular place on the membrane; the movement of sodium ions reverses the voltage across the membrane, and this change is transmitted along the cell membrane of a nerve cell or a muscle fibre, and is then called a *nerve impulse*; *see also* depolarisation membrane potential diagram (p. 73)

activation energy energy required to begin a chemical reaction; enzymes work by decreasing the activation energy for reactions in living organisms

active immunity IMMUNITY produced by the body manufacturing antibodies against a foreign antigen

active process a process that occurs only if there is an input of energy (e.g. active transport of materials across a cell membrane)

active site the part of an ENZYME molecule that combines with one of the molecules with which the enzyme reacts (the substrate); *see* diagram opposite top

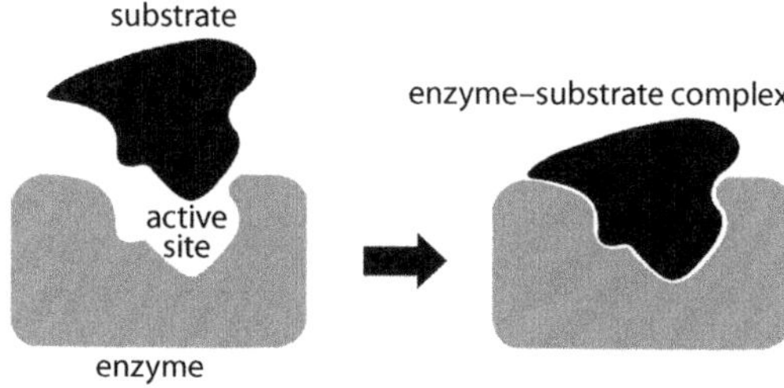

Active site of an enzyme

active transport the use of energy from cellular respiration to move substances, usually ions, across the cell membrane

acuity the resolving power of the eye; the clarity of vision

acute disease a disease that rapidly reaches a peak and has severe symptoms

adaptation **1.** a particular structure, physiological process or form of behaviour that makes an organism better able to survive and reproduce in a particular environment (e.g. the short, rounded body shape of Eskimos, with a small surface area to volume ratio, is considered to be an adaptation to a cold environment) **2.** the decrease in perception of a sensation over time while the stimulus is still present (e.g. the sense of smell adapts rapidly) **3.** in vision, the adjustment of the pupil of the eye to variations in the amount of light

adaptive radiation the formation of many new species following the availability of new environments or the development of a new adaptation

addiction the physical dependence upon a substance such as alcohol or other drugs, characterised by tolerance of increasingly larger doses, craving for the drug, and withdrawal symptoms when use of the substance is discontinued

Addison's disease a disorder caused by undersecretion of hormones from the adrenal cortex, especially glucocorticoids and mineralocorticoids; characterised by muscular weakness, mental lethargy, weight loss, low blood pressure and dehydration

adduction the movement of a limb towards the mid-line of the body (e.g. movement of the arms from a horizontal position to beside the body); *see also* diagram showing movements at joints (p. 1)

adenine one of the four bases that form part of the DNA and RNA molecule; *see* deoxyribonucleic acid, ribonucleic acid

adenohypophysis the anterior lobe of the PITUITARY GLAND

adenoid an enlarged tonsil at the back of the throat; *see* tonsils

adenosine a substance that functions as a HORMONE and a NEUROTRANSMITTER; also forms part of the ADENOSINE DIPHOSPHATE and ADENOSINE TRIPHOSPHATE molecules

adenosine diphosphate (ADP) the substance formed when the end phosphate group is removed from a molecule of adenosine triphosphate, composed of adenosine and two phosphate groups; *see* diagram next page

adenosine monophosphate (AMP) a molecule consisting of an adenosine group with one phosphate; the basic structure for ADP and ATP molecules; *see also* cyclic adenosine monophosphate

adenosine triphosphate (ATP) a compound consisting of adenine and a ribose sugar to which are attached three phosphate groups; used as an energy source

adenine
adenosine
phosphates
ribose
adenosine diphosphate (ADP)
adenosine triphosphate (ATP)

Structure of ADP and ATP

by cells; when a phosphate group is lost from the molecule, adenosine diphosphate (ADP) is formed with the release of energy, which can be used for cell activities; during respiration ADP is converted to ATP, thus ATP transfers energy from reactions that release energy to those that require energy, *see* diagram above

adipose tissue fat storage tissue; a type of connective tissue; *see* diagram below

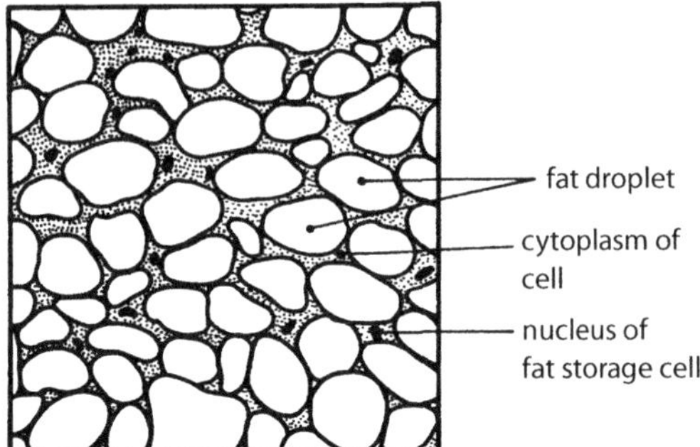

Adipose tissue

adjuvant any substance that increases the power of an ANTIGEN to stimulate ANTIBODY production

adolescence the period of transition from childhood to adulthood

adolescent growth spurt the increase in the rate of body growth during the adolescent years, which occurs as a result of hormonal changes; in girls usually between 11 and 14 years of age; in boys between 13 and 16

ADP abbreviation for ADENOSINE DIPHOSPHATE

adrenal cortex the outer portion of the ADRENAL GLAND

adrenal glands two pyramid-shaped endocrine glands, each located above a kidney; also called the *suprarenal glands*; each is, in effect, two separate glands; the inner portion, the *adrenal medulla*, secretes the hormones adrenaline and noradrenaline; the outer part, the *adrenal cortex*, produces hormones such as aldosterone that regulate sodium and potassium metabolism and others such as cortisol that affect carbohydrate metabolism; *see* diagram next page

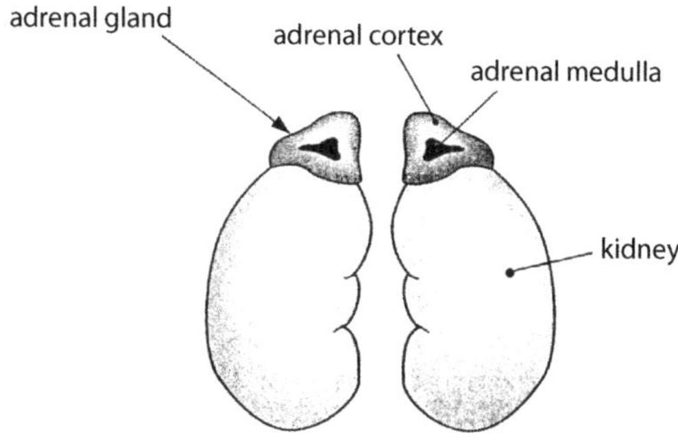

Adrenal glands

adrenal medulla the inner portion of the ADRENAL GLAND, controlled by the sympathetic division of the autonomic nervous system and may therefore be considered a part of that system

adrenaline a hormone secreted by the adrenal medulla that prepares the body for fight-or-flight responses and has a similar action to stimulation by the sympathetic division of the autonomic nervous system; it also carries nervous impulses across the junction (synapse) of some adjacent nerve cells; also called *epinephrine*; *see also* noradrenaline, table of effects of sympathetic stimulation (p. 186), table of hormones (p. 128–30)

adrenergic fibre a nerve fibre that releases noradrenaline to act as the NEUROTRANSMITTER at a synapse; *see also* cholinergic fibre

adrenocorticotropic hormone (ACTH) a hormone secreted by the anterior lobe of the pituitary that controls the production and release of certain hormones from the adrenal cortex, especially cortisol; secretion of ACTH is controlled by a releasing factor produced by the hypothalamus; also spelt adrenocorticotrophic, *see* table of hormones (p. 128–30)

adrenocorticotropin an alternative name for ADRENOCORTICOTROPIC HORMONE, also spelt adrenocorticotrophic

adult-onset diabetes an alternative name for *type 2 diabetes*; *see* diabetes mellitus

adult stem cells *see* stem cells

aerobic respiration respiration requiring oxygen; complete oxidation of glucose to form carbon dioxide and water; after the anaerobic phase of respiration produces pyruvic acid from glucose, breakdown continues in the mitochondria through the KREBS CYCLE and the electron transport chain, eventually producing carbon dioxide and water; the complete breakdown of one molecule of glucose can release enough energy to build up 38 molecules of ATP—two from the anaerobic phase and up to 36 from the aerobic phase; *see also* glycolysis, electron transport system

aetiology the study of the causes of disease

afferent a term referring to conducting or carrying into an organ; used in reference to nerves and blood vessels (e.g. an afferent blood vessel carries blood *into* an organ); *see also* efferent

afferent arteriole the blood vessel that enters an organ; specifically, the arteriole that carries blood into the glomerulus of a kidney nephron; *see also* efferent arteriole, diagram of kidney nephron (p. 183)

afferent division the part of the peripheral NERVOUS SYSTEM that carries impulses *into* the brain and spinal cord

afferent neuron a neuron that carries an impulse towards the central nervous system; also called a *sensory neuron*

African origin model a model of human evolution that states that archaic *Homo sapiens* arose and evolved into anatomically modern *Homo sapiens* in Africa and then spread out through the rest of the Old World, replacing or perhaps interbreeding with local populations of hominins; also known as the *out of Africa hypothesis*; *see also* multiregional evolution model

afterbirth the placenta and remains of the umbilical cord, amnion and chorion that are expelled from the uterus shortly after the birth of a baby

afterimage the persistence of a sensation even though the stimulus has been removed (e.g. if one looks at a bright light then closes the eyes, the image of the light remains)

agar a carbohydrate derived from seaweeds that forms a gel (jelly) when mixed with water and allowed to solidify, used as a medium for growing cultures of micro-organisms

age profile an alternative name for POPULATION PYRAMID

age–sex structure a measure of the composition of a population in terms of males and females at different ages

age structure of a population, the number of individuals in each age-group

ageing the progressive decline in the effective functioning of individual cells and whole organs that occurs over time, leading to a gradual decline in the efficiency of the body's homeostatic mechanisms

agglutination the clumping together of microorganisms or of blood cells due to a reaction between substances on the surfaces of the cells (agglutinogens) and antibodies in the surroundings; occurs when blood of incompatible groups is mixed

agglutinin a type of antibody; *see* agglutinogen

agglutinogen an antigen on the surface of a cell (especially a red blood cell) that induces formation of antibodies known as *agglutinins*, which cause the cells to bind together in clumps (agglutination); the type of agglutinogens a person has is inherited; *see also* antibody

aggressive behaviour behaviour directed by one animal towards another, including threats and attacks, often associated with mating and defence of territory

agonist often referred to as the *prime mover*, the muscle that is mainly responsible for a particular movement and opposes the movement of another muscle (the antagonist); when a bone is moved at a joint, as the agonist contracts the antagonist relaxes; if agonist and antagonist contract together, the joint is locked; *see* diagram below

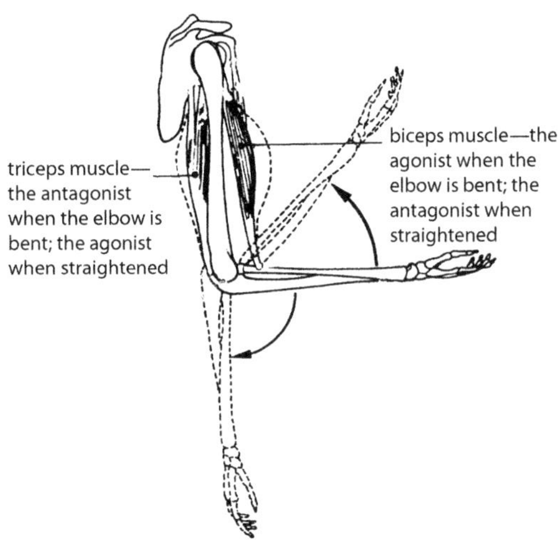

Muscles functioning as agonists and antagonists

agranular endoplasmic reticulum ENDOPLASMIC RETICULUM with no ribosomes attached to it

agranulocyte a type of white blood cell; *see* leucocytes

AID ARTIFICIAL INSEMINATION by donor

AIDS an abbreviation for ACQUIRED IMMUNE DEFICIENCY SYNDROME

AIDS-related complex the former term for PERSISTENT GENERALISED LYMPHADENOPATHY

alanine one of the 20 amino acids that are common in proteins; not essential in the human diet; *see also* list of amino acids (p. 11)

alarm reaction an alternative name for FIGHT-OR-FLIGHT RESPONSE

albinism an inherited condition caused by a recessive allele carried on a non-sex chromosome; the individual, known as an *albino*, is unable to produce the pigment melanin, which results in white skin, white hair and pink eyes (due to the reflection from blood vessels in the eye)

albino an individual lacking pigmentation; *see* albinism

albumins a group of proteins found in virtually all animal tissues, soluble in water and the smallest but most abundant of the plasma proteins; they make up 55% of the protein in the blood plasma and contribute to the blood's viscosity; they also help to regulate blood volume because they do not diffuse through the capillary walls, so that water tends to be drawn into the capillaries from the tissue fluid (by osmosis)

albuminuria the presence of albumin in the urine

alcohol **1.** a group of chemical compounds in which an alkyl group (CH_3—, CH_3CH_2—, $CH_3CH_2CH_2$— etc.) is joined to an OH group **2.** the colourless active ingredient in alcoholic drinks; ethanol, also called ethyl alcohol (C_2H_5OH); when consumed, acts as a depressant on the nervous system and impairs the performance of skilled tasks

aldosterone one of the mineralocorticoid hormones secreted by the adrenal cortex; it acts on the kidney, causing it to retain sodium ions and excrete potassium ions; it also maintains the sodium ion content of the body and therefore controls fluid and electrolyte balance; *see* table of hormones (p. 128)

alimentary canal the tube through which food passes, extending from the mouth to the anus; comprises mouth, oesophagus, stomach, intestines; also called the *alimentary tract*, *gut*, *digestive tract* or *gastrointestinal tract*; *see* diagram of the digestive system (p. 77)

alkali an alternative name for BASE

alkalosis a condition in which the blood has a lower hydrogen ion concentration than normal, resulting in an increased pH; respiratory alkalosis is the result of overbreathing (HYPERVENTILATION); metabolic alkalosis may result from excessive intake of alkaline drugs or excessive vomiting of the acidic contents of the stomach; *see also* acidosis

all-or-none law a law or principle that states that certain tissues respond in the same way to a stimulus regardless of the strength of the stimulus; in muscles, an individual fibre contracts to its maximum extent or not at all; in nerve cells, if a stimulus is strong enough to trigger a nerve impulse, the impulse is always the same even though the size of the stimulus may vary

all-or-none response a response that is of a constant size regardless of the strength of the stimulus; with respect to nerve cells, a nerve impulse is transmitted at full strength or not at all; with respect to muscle fibres, there is maximum contraction or none at all, because individual fibres cannot partially contract

allantois one of the FOETAL MEMBRANES

allele any alternative form of a gene that occurs at a given point (the locus) in a chromosome; a gene can have a number of different forms (multiple alleles), but each individual normally has only two alleles of each gene; *see also* homozygous, heterozygous

allele frequency a measure of how often each allele of a gene occurs in a population

Allen's rule the tendency for mammals in cold climates to have shorter and bulkier limbs, allowing less loss of body heat, whereas mammals in hot climates tend to have long, slender limbs, allowing greater loss of body heat; *see also* Bergmann's principle

allergic response the reaction to a substance that is normally harmless or that would not cause an IMMUNE RESPONSE in most people; frequent symptoms are itching and inflammation; tissue injury may occur; in severe cases the response may endanger the person's life; sometimes called an *allergic reaction*

allergy an overreaction of the body's immune response to minute traces of foreign matter (antigens); the response may appear as a rash, itching, swelling or breathing difficulty

alpha cells cells in the islets of the PANCREAS that secrete the hormone glucagon

alpha-foetoprotein protein found in the foetus; abnormally large amounts can indicate SPINA BIFIDA or other NEURAL TUBE DEFECTS

alternative medicine any healing technique that is not part of conventional medicine; examples are naturopathy, homeopathy, herbalism, aromatherapy

altitude sickness an illness experienced by people who climb or are transported to high altitudes, caused by the reduced amount of oxygen available in the air; symptoms are breathlessness, nausea, headache and weakness; symptoms begin at about 3000 metres and are definite at 5000 metres or above; the body slowly acclimatises over a period of about 3 weeks; also called *mountain sickness*

altruistic behaviour self-sacrificing behaviour; having an unselfish regard for others; there is debate about whether animals other than humans can show genuine altruistic behaviour; also called *altruism*

alveolar air air that is in the ALVEOLI of the lungs

alveolar duct a branch of a bronchiole around which alveoli and alveolar sacs are arranged; *see* diagram of respiratory system (p. 236)

alveolar sac a group of alveoli sharing a common air passage; *see* diagram of respiratory system (p. 236)

alveolus (plural *alveoli*) **1.** A microscopic air sac in the lung in which exchange of gases between the air and the blood occur; efficient gas exchange can occur because the walls of the alveoli are very thin and the total surface area of the alveoli is very large (in humans about 70 m^2); *see* diagram of respiratory system (p. 236) **2.** The milk-secreting part of a mammary gland

Alzheimer's disease the most common form of DEMENTIA; usually diagnosed in people over the age of 65; an early symptom is memory loss, which may be followed by confusion, mood swings, aggression and withdrawal

amenorrhoea the absence of MENSTRUATION, which can be caused by endocrine disorders, changes in body weight or strenuous sports training; also occurs during PREGNANCY, during BREASTFEEDING and after MENOPAUSE

amine an ORGANIC COMPOUND containing nitrogen; the term was derived from the word AMMONIA

amino acid a type of organic acid; the unit from which proteins are formed; there are 10 amino acids that must be included in the human diet because they cannot be made in the body: arginine, histidine, isoleucine, leucine, lysine, methionine, phenylalanine, threonine, tryptophan and valine; *see* list of amino acids, opposite *see* diagram top right

ammonia chemical formula NH_3; produced by DEAMINATION and converted by the liver to UREA

amnesia a lack or loss of memory

amniocentesis a technique for determining sex or detecting chromosomal abnormalities in a foetus; the mother's abdominal wall and the foetal membranes are punctured so that a small amount of the amniotic fluid surrounding the foetus can be removed and foetal cells in the fluid can be examined; *see* diagram next page

Table 1 *The 20 amino acids commonly found in proteins**

	Abbreviations	
Name	*Three letters*	*One letter*
alanine	Ala	A
arginine	**Arg**	**R**
asparagine	Asn	N
aspartic acid	Asp	D
cysteine	Cys	C
glutamic acid	Glu	E
glutamine	Gln	Q
glycine	Gly	G
histidine	**His**	**H**
isoleucine	**Ile**	**I**
leucine	**Leu**	**L**
lysine	**Lys**	**K**
methionine	**Met**	**M**
phenylalanine	**Phe**	**F**
proline	Pro	P
serine	Ser	S
threonine	**Thr**	**T**
tryptophan	**Trp**	**W**
tyrosine	Tyr	Y
valine	**Val**	**V**

**Those shown in bold type are essential in the human diet.*

Amniocentesis

amnion one of the FOETAL MEMBRANES

amniotic cavity the cavity enclosed by the amnion; *see* foetal membranes

amniotic fluid fluid in the amniotic cavity; *see* foetal membranes

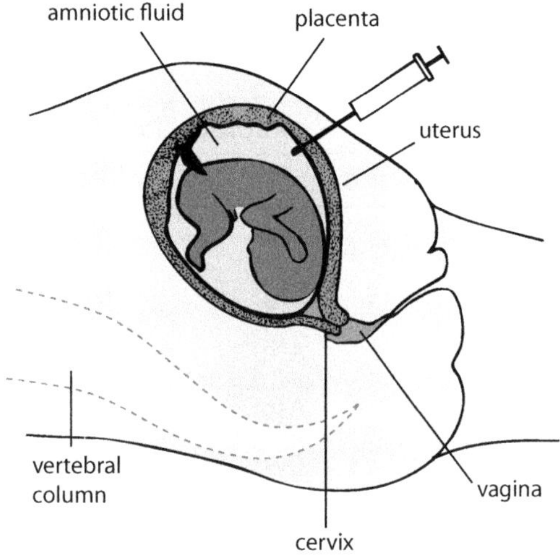

amoebic dysentery inflammation of the intestines caused by infection with a single-celled organism, *Entamoeba hystolitica*

amoebic meningitis inflammation of the MENINGES of the brain caused by amoebae, single-celled organisms; amoebae live in warm water and may enter through a swimmer's nose

amphetamine a drug used medically to treat depression, obesity and other disorders by stimulating the central nervous system; prolonged use can cause tolerance and/or psychological dependence; excessive use over long periods can cause mental illness

amplification *see* gene amplification

ampulla a sac-like swelling of a canal; in the inner ear, the swelling at one end of each of the SEMICIRCULAR CANALS

amygdala part of the BASAL GANGLIA and the LIMBIC SYSTEM of the brain; concerned with emotions; also called the *amygdaloid nucleus*

amygdaloid nucleus alternative name for AMYGDALA

amylase an enzyme that breaks down starch into sugar; *see also* salivary amylase, pancreatic amylase

anabolic processes processes in which small molecules are built up into large ones; *see* metabolism

anabolic steroids also known as *anabolic androgenic steroids* (AAS); STEROIDS (e.g. the male sex hormones) that stimulate growth of tissues, especially muscles

anabolism the joining of small molecules to make larger molecules; *see* metabolism

anaemia a condition in which there is a reduced number of red cells in the blood or a reduced amount of haemoglobin; the blood therefore has a reduced oxygen-carrying capacity; *macrocytic anaemia*, characterised by abnormally large red blood cells, is caused by a deficiency of folic acid; *pernicious anaemia* is insufficient production of red blood cells due to lack of vitamin B_{12}; *aplastic anaemia* is lack of red blood cells due to inhibition or destruction of the red bone marrow; *sickle-cell anaemia* is an inherited disease caused by a recessive allele carried on a non-sex chromosome; it causes early death in persons inheriting two alleles for sickle-cell anaemia, but persons receiving only one allele suffer from *sickle-cell trait* (abnormal haemoglobin that causes the red blood cells to become sickle-shaped when oxygen concentration is low), which provides some protection against becoming infected with malaria; *thalassaemia* is also an inherited form of anaemia, resulting from defects in the formation of haemoglobin and controlled by recessive alleles; *thalassaemia major* is the homozygous condition and always results in illness; *thalassaemia minor* is the heterozygous condition and may not result in symptoms; the thalassaemia allele is more common in people of Mediterranean origin

anaerobic respiration respiration in the absence of oxygen; the breakdown of glucose to pyruvic acid and then lactic acid with the release of enough energy to form two ATP molecules from each molecule of glucose

anaesthesia a partial or total loss of feeling or sensation, usually the loss of pain sensation; a drug used to produce anaesthesia is called an *anaesthetic*

anaesthetic a drug that causes total or partial loss of feeling; usually taken to mean loss of pain sensation (e.g. chloroform, ether)

anagenesis a pattern of SPECIATION

anal sphincter a ring of muscle around the anus

analgesia absence of the normal sense of pain

analgesic a drug used mainly for the relief of pain (e.g. morphine, aspirin)

analogous trait a physical trait (characteristic) that has a similar function in two species but is performed by a different structure (e.g. the wings of a bird and those of a flying insect both perform the same function but have different structures)

anamnestic response the accelerated, more intense production of antibodies that takes place upon a second or later exposure to an antigen; *see also* memory cell

anaphase the third phase of MITOSIS and of each of the two divisions of MEIOSIS

anatomical position the accepted position of the body when making descriptions of the anatomy of the body; the body is erect, facing the observer, the arms are at the sides and the palms of the hands are facing forward; *see* diagram below

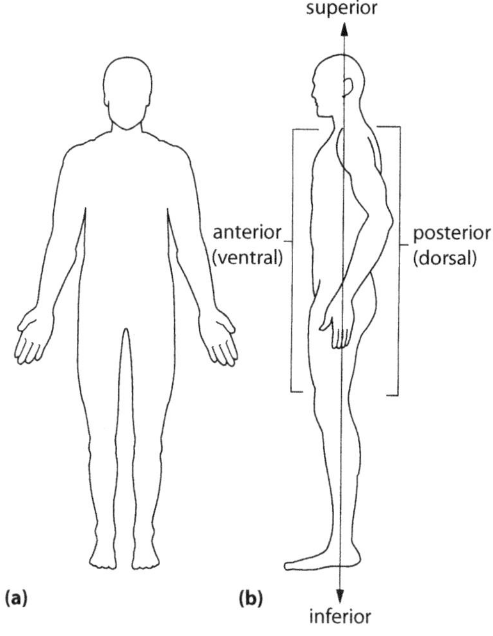

(a) Anatomical position; (b) terms used to describe areas of the body

anatomically modern *Homo sapiens* the modern form of the human species; *see Homo sapiens* for details

anatomy the structure, or the study of the structure, of an organism and the relationship of its parts to each other

andro- prefix meaning male (e.g. androgen is a general term for male sex hormones)

androgen a general name for a male sex hormone that stimulates development and maintenance of male secondary sex characteristics (e.g. testosterone, androsterone)

androgen insensitivity syndrome a genetic disorder carried on the X CHROMOSOME that makes a FOETUS with the chromosomal complement for a male unresponsive to ANDROGENS; individuals are born looking externally like females; internally there is a short blind-pouch VAGINA but no UTERUS, OVIDUCTS or OVARIES

andropause male MENOPAUSE; a gradual reduction in levels of male sex hormones that can occur in men from the age of 50 onwards

anencephaly a gross defect in which the majority of the brain does not form during foetal development and results in the child being stillborn or dying within minutes or hours of birth; one of a group of conditions known as neural tube defects; risk of such defects can be reduced by increased consumption of the vitamin folic acid before and during pregnancy

aneuploidy a condition in which there is an abnormal number of chromosomes in a person's cells; either one too many or one fewer than normal; *see also* trisomy, monosomy

aneurism a balloon-like swelling of a blood vessel due to thinning or weakening of the wall that, if untreated, grows larger and eventually bursts; causes include atherosclerosis, syphilis and inborn defects of blood vessels; also spelt *aneurysm*

angina a common name for ANGINA PECTORIS

angina pectoris a pain in the chest caused by reduced flow of blood to the heart muscle; does not necessarily involve heart or blood vessel disease; often referred to as *angina*

angioplasty a surgical procedure that is used to widen blocked blood vessels and restore normal blood flow, frequently involving the use of balloons or stents, *see* balloon angioplasty

angiotensin a substance that regulates aldosterone secretion; also a powerful vasoconstrictor; *see* renin

angiotensinogen a protein in blood plasma that can be converted into angiotensin; *see* renin

angle of convergence alternative name for CARRYING ANGLE

anion a negatively charged ion (e.g. Cl^-)

annealing an alternative name for CHEMICAL HYBRIDISATION

anorexia loss of appetite or desire for food

anorexia nervosa a nervous disorder characterised by an extreme loss of appetite that leads to emaciation, malnutrition and, in some cases, death

anoxia a lack of oxygen in the tissues

antagonist a muscle that works in opposition to another muscle (the agonist); when a bone is moved at a joint, the agonist contracts while the antagonist relaxes; if agonist and antagonist contract together, the joint is locked; *see* agonist diagram (p. 8)

antagonistic muscle pair two muscles that move the same structure, such as a limb, in opposite directions (e.g. the biceps in the upper arm pulls on the bones of the forearm to bend the elbow, and the triceps pulls on the same bones to straighten the elbow); *see* agonist diagram (p. 8)

ante- a prefix meaning before (e.g. antenatal—before the birth of a baby)

anterior situated near or towards the front, or head, end of an animal (e.g. the eyes are at the anterior end of the animal, the front limbs are anterior to the rear limbs)

anterior root one of the two roots that link a spinal nerve to the spinal cord; *see* root

anterior tibialis the common name for the muscle that bends the foot at the ankle when walking so that the weight of the body falls on the heel; also known as the *tibialis anterior*; *see* diagram below

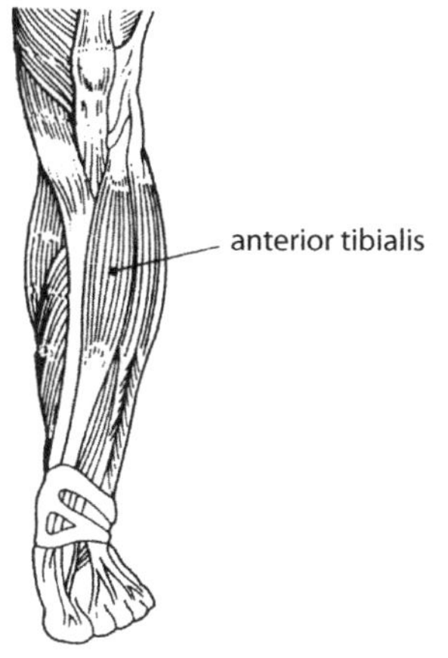

Anterior tibialis

anthrax a serious disease in humans and animals (usually cattle and sheep); caused by infection with the bacterium *Bacillis anthracis*

anthropo- a prefix that means human (e.g. an anthropometer measures parts of the human body)

anthropocentrism the belief that humans are the central beings of the universe; a view that puts human interests above all else and views and interprets all things in terms of human beliefs and values (e.g. an anthropocentric view of environmental concerns considers only the impact of environmental change on humans and human activities, an opposite view to that of DEEP ECOLOGY)

anthropogenic produced or caused by humans in earlier classifications

anthropoid in earlier classifications, a member of the suborder Anthropoidea, within the order Primates; considered the more biologically advanced of the two primate suborders; the monkeys, apes and humans; monkeys, apes and humans now classified as simiiformes; *see* classification of primates (p. 219)

Anthropoidea in classical classification, one of the two suborders of the order Primates; animals in this suborder are called ANTHROPOIDS; a cladistic classification separates the primates into suborders Strepsirrhini and Haplorrhini, with Anthropoidea as a further subdivision of Haplorrhini; *see* strepsirrhines and haplorrhines

anthropologist a scientist who specialises in the study of ANTHROPOLOGY

anthropology the study of the biological and cultural variation of the human species, including the origin of humans; *biological anthropology* (or *physical anthropology*) is a branch of anthropology that focuses on the biological evolution of humans and human ancestors, the relationship of humans to other organisms, and the patterns of biological variation within and among human populations; *palaeoanthropology* is an area of biological anthropology that studies the fossil remains of human ancestors; *cultural anthropology* studies social patterns among human societies; cultural behaviour is learned and socially transmitted rather than instinctive and transmitted by inheritance; *linguistic anthropology* is an area of cultural anthropology that focuses on the nature of language and its relationship to culture; *anthropological archaeology* studies cultural variation in prehistoric (and some historic) populations through an analysis of the cultural remains

anthropometer an instrument used for measuring parts of the human body

anthropometry measurement of the human body; study of the variation among humans by measurement

anthropomorphism giving human form or personality to non-human organisms (e.g. the idea that one's pets have human feelings rather than instinctive behaviour)

anthrozoology study of the interaction between humans and animals

anti- a prefix that means against, or opposed to (e.g. anticoagulant—a chemical that opposes or prevents blood clotting)

antibiotic a chemical that is able to inhibit the growth of, or kill, micro-organisms (e.g. penicillin, streptomycin)

antibody a specific protein produced in response to a substance foreign to the body (ANTIGEN) that reacts with the antigen to neutralise or destroy it; all antibodies are globular proteins (immunoglobulins) and are produced by lymphocytes called B cells; if the antigen is a toxin (e.g. spider venom or bacterial products), the antibody is called an *antitoxin*; if the antigen is on the surface of a cell, the antibody is called an *agglutinin*

antibody-mediated immunity an alternative name for HUMORAL IMMUNITY

anticoagulant a substance that delays, or prevents, the clotting of blood; HEPARIN is an anticoagulant that occurs naturally in the lungs and liver; warfarin is given to patients who may be prone to blood clotting in the blood vessels (thrombosis)

anticodon a sequence of three bases in a transfer RNA molecule that binds to a complementary sequence of three bases (the codon) in a messenger RNA molecule, to specify a particular amino acid during protein synthesis; *see also* codon

antidepressant a drug used to relieve depression

antidiuretic a substance that inhibits the formation of urine

antidiuretic hormone (ADH) a hormone produced by the hypothalamus, and released by the posterior lobe of the pituitary gland, that stimulates water reabsorption in the distal and collecting tubules of the kidney; it reduces urine production and can also raise blood pressure by causing constriction of small arteries; also known as *vasopressin*; *see* table of hormones (p. 128), neurosecretion

antigen any foreign substance (usually a protein or carbohydrate) capable of causing formation of antibodies when introduced into the tissues; the ANTIBODY produced is specific to the particular antigen

antigen–antibody complex a compound formed when an ANTIBODY combines with an ANTIGEN

antihistamine a drug used to counteract the effects of histamine; used particularly to relieve the symptoms of an allergy

anti-inflammatory a substance that prevents or reduces INFLAMMATION; aspirin and ibuprofen are anti-inflammatory drugs

antiseptic the general term for a substance applied to the body to destroy harmful micro-organisms, generally in the form of a solution or cream; a substance used to prevent infection in a wound

antiserum a liquid containing antibodies or antitoxins; *see* serum

antitoxin an ANTIBODY produced against a toxin

anus the opening at the end of the alimentary canal; the outlet of the rectum

anvil one of the bones in the middle ear; *see* auditory ossicles

anxiety neurosis a minor mental illness marked by feelings of excessive apprehension or anxiety, which may be treated by prescription of tranquillisers

aorta the large artery arising from the left ventricle of the heart that distributes blood to the body; *see also* diagram of major arteries (p. 20)

aortic body a group of cells attached to the walls of the aorta near the heart that are sensitive to changes in the concentrations of oxygen and carbon dioxide in the blood and to the blood's pH level; decreasing pH, rising carbon dioxide concentration, and perhaps decreasing oxygen concentration, trigger the cells to send messages to the respiratory centre in the brain so that breathing rate increases

ape a tail-less primate; a member of the superfamily Hominoidea; in classical classifications, the *lesser apes* belong to the family Hylobatidae and include the gibbons and siamangs; the *great apes*, members of the family Pongidae, are chimpanzees, gorillas and orang-utans; more recent classifications separate the orang-utans into a subfamily Ponginae and humans, chimps and gorillas into a subfamily Homininae

aplastic anaemia ANAEMIA resulting from inhibition or destruction of the red bone marrow

apneustic area a part of the brain involved in the regulation of breathing; *see* respiratory centre

apnoea a temporary absence of breathing that may be brought about by stimuli such as sudden, sharp pain, or sudden exposure to cold (such as a cold shower); *see also* sleep apnoea

apoptosis programmed cell death; a sequence of events that culminates in the death of a cell; may be due to ageing of, or injury to, the cell

appendicitis inflammation of the appendix

appendicular skeleton that part of the skeleton that is made up of the bones of the upper and lower limbs, and the bones of the shoulders and pelvis

appendix a short, tube-like sac attached to the caecum of the large intestine; also known as the *vermiform appendix*; *see also* diagram of digestive system (p. 77)

aquaculture food production in water (e.g. fish farming); may be called *mariculture* if the organisms involved are sea-living (marine)

aqueous humour the watery fluid filling the front cavity of the eyeball, between the cornea and the lens; *see also* vitreous humour, diagram of eye (p. 94)

arachnoid one of the membranes covering the central nervous system; *see* meninges

arboreal living in trees (e.g. many monkeys, the gibbons and the orang-utan are arboreal)

arboreal adaptation a modification of an organism that enables it to live in trees (e.g. the grasping hands and feet of primates)

arbovirus a virus that is carried by arthropods such as mosquitoes, ticks and lice and is transmitted to humans by the bites of those animals (e.g. the yellow fever virus carried by mosquitoes)

Archaean an alternative name for the ARCHAEOZOIC era

Archaeozoic one of the major subdivisions of geological time; the 2nd era, 3800–2500 million years ago; fossils from this era are very rare, and the oldest known are from north-west Australia; also known as the *Archaean era*; also referred to as an EON in some time scales; *see* geological time scale (p. 111)

archaeology the study of culture through the remains left by societies, often through excavation to uncover cultural remains (e.g. study of stone tools, pottery or remains of dwellings); archaeology is subdivided into two sub-branches, historical archaeology and prehistorical archaeology

archaic having the characteristics of an earlier period; in human biology, describes fossils with primitive features (e.g. archaic *Homo sapiens*)

archaic *Homo sapiens* an early variant of *HOMO SAPIENS*

arches curved arrangement of bones in the human foot, which help to absorb many of the shocks that occur during walking and running; the *transverse arch* runs from side to side and is formed by the tarsal bones and the rear parts of the 5 metatarsals; the *longitudinal arch* runs from front to back and is formed by the arrangement of the tarsal and metatarsal bones; the longitudinal arch is present in the human foot but not in that of other primates and is one of the characteristics that enables humans to stand and walk erect; it helps the foot to support the weight of the body and to provide leverage while walking; *see* diagram of arches top right

Ardipithecus a genus of hominid named in 1995, created because close examination of the fossils, previously classified as AUSTRALOPITHECUS RAMIDUS, showed them to be sufficiently different from previous finds to justify placing them in a new genus

areola **1.** the pigmented region surrounding the nipple of the breast **2.** any tiny space in a tissue

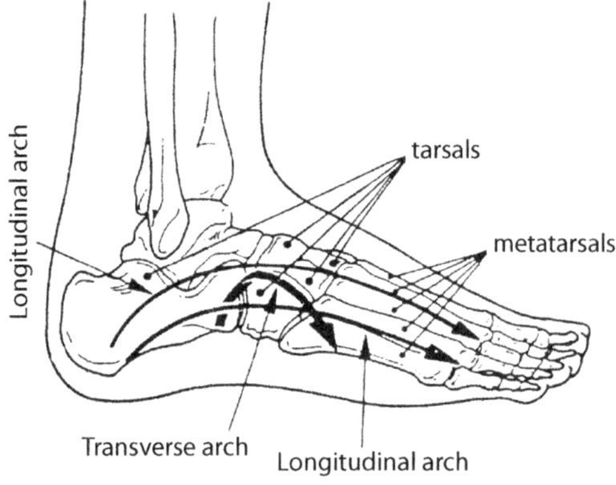

Arches of the foot

areolar tissue a connective tissue made up of loosely woven fibres that supports and connects other tissues; also called *loose connective tissue*; *see* diagram below

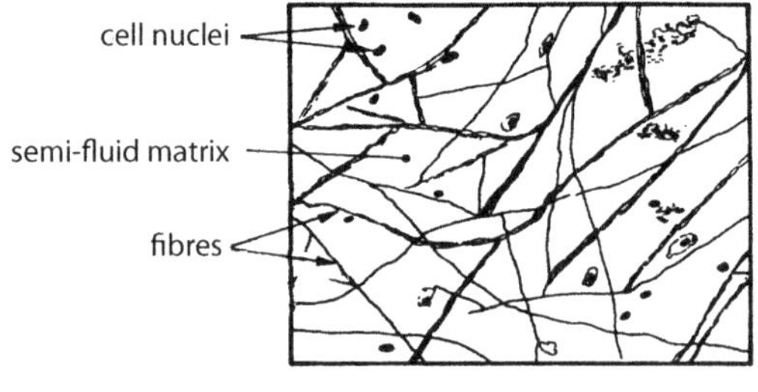

Areolar tissue

arginine one of the 20 amino acids that are common in proteins; essential in the human diet; *see also* list of amino acids (p. 11)

arithmetic mean the sum of a number of values divided by the number of values; commonly called the *average*

arrector pili smooth muscles attached to hairs, contraction of which makes the hairs become more vertical, resulting in 'goose bumps'

ART *see* assisted reproductive technology

artefact **1.** an object made or modified by humans (e.g. stone tools, carvings, cave paintings) **2.** any feature or structure formed during preparation of tissue for microscopic examination, possibly leading to misinterpretation of the specimen; may be spelt *artifact*

arteriole a very small artery (diameter less than about 0.5 mm); smooth muscle in the wall can change the diameter of the arteriole so that it plays a major role in regulating flow of blood through the capillaries it supplies; *see also* vasoconstriction, vasodilation

arteriosclerosis a degenerative change in the walls of arteries involving loss of elasticity of the artery walls

artery a blood vessel that carries blood away from the heart; *see* diagram of major arteries, next page

arthritis a general term referring to inflammation of the joints; the most common form is *osteoarthritis*, a degenerative disease that restricts movement at the joint; involves degeneration of the articular cartilage and the development of bony spurs from the exposed ends of the bones at the joint; the spurs restrict movement and cause pain; *rheumatoid arthritis* is a severe form of arthritis involving swelling, pain and loss of function due to inflammation; it affects the synovial membrane, causing it to thicken with accumulation of synovial fluid, resulting in severe pain; as the disease progresses the membrane produces an abnormal tissue called *pannus* that sticks to the articular cartilage and greatly restricts movement at the joint

articular capsule the sleeve-like structure surrounding a synovial joint, consisting of a fibrous capsule and a synovial membrane; *see* diagram of structure of synovial joint (p. 264)

articular cartilage the cartilage that covers the articulating surfaces of the bones forming a joint; *see* diagram of structure of synovial joint (p. 264)

articular disc an alternative name for MENISCUS

articulation an alternative name for JOINT

artificial immunity IMMUNITY produced by giving a person an antigen or antibodies

artificial insemination introduction of semen into the vagina or uterus using a syringe, used in some cases of infertility in humans; semen may come from husband or from a donor, in which case the process is known as *donor insemination* (*DI*) or *artificial insemination by donor* (*AID*)

artificial pacemaker an electronic device, usually implanted within the body, that takes over the function of the heart's natural PACEMAKER; a device that uses electrical impulses to regulate the heart's natural rhythm

arytenoid cartilages a pair of cartilages in the LARYNX

ascending limb in the kidney, the arm of the LOOP OF HENLE leading back to the cortex; *see also* diagram of nephron (p. 183)

ascending tract a bundle of sensory fibres in the central nervous system; *see* tract

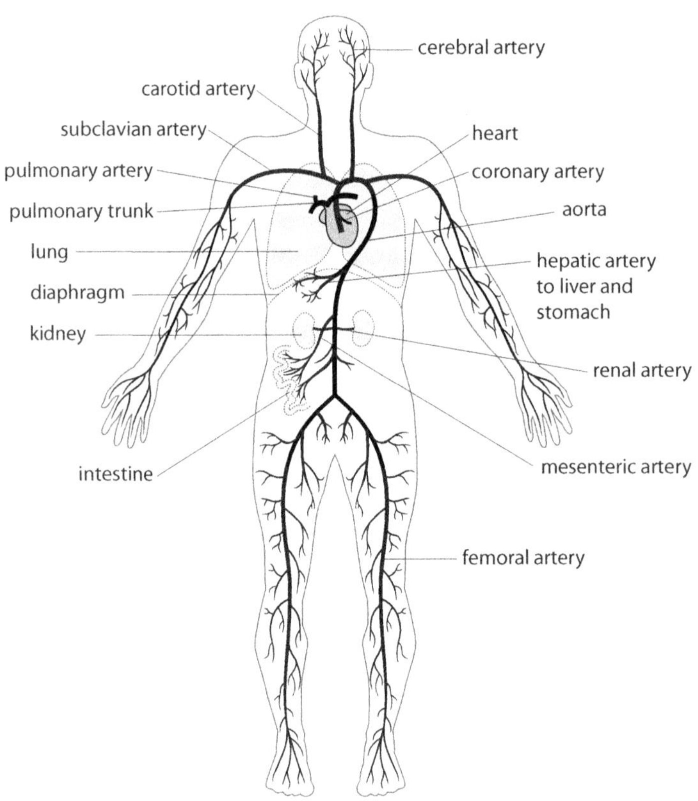

The major arteries

ascorbic acid a water-soluble vitamin, rapidly destroyed by heat and essential for many metabolic reactions, the best known of which is the production of the protein collagen in the formation of connective tissue; deficiency causes scurvy (poor wound healing, fragile blood vessel walls and defective bone growth); good sources include citrus fruits, green vegetables and tomatoes; also called *vitamin C*

asexual reproduction the type of reproduction in which new individuals are produced without any union of sex cells

asparagine one of the 20 amino acids that are common in proteins; not essential in the human diet; *see also* list of amino acids (p. 11)

aspartate an alternative name for ASPARTIC ACID

aspartic acid one of the 20 amino acids that are common in proteins; not essential in the human diet; also called *aspartate*; *see also* list of amino acids (p. 11)

asphyxia loss of consciousness resulting from deficiency of the oxygen supply to the blood; suffocation

aspirate to remove by suction (e.g. the removal of blood or pus by syringe)

assimilation **1.** the conversion of absorbed food into the complex compounds required by the body **2.** the process whereby individuals, or groups, who migrate to a country gradually acquire the basic attitudes, habits and mode of life of the adopted country

assisted reproductive technology (ART) any procedure used to achieve pregnancy by artificial means

association area a part of the cerebral cortex of the brain concerned with intellectual and emotional processes such as memory, reasoning, judgment and personality; separate from motor and sensory areas of the cerebral cortex

association neuron an alternative name for CONNECTOR NEURON

assortative mating mating that is not random because the choice of partner is influenced by that person's inherited characteristics; that is, mating between individuals who are biologically similar for some characteristics (e.g. humans tend to marry within their own ethnic group)

aster a star-shaped structure surrounding the centrioles during the nuclear division of a cell, comprised of a bundle of microtubules produced from the central CENTROSOME

asthma an allergic reaction involving breathing difficulty; smooth muscles in the small air passages contract and mucus is secreted so that air movement is restricted

astigmatism an irregular curvature of the cornea or lens of the eye; curvature in the horizontal plane may differ from that in the vertical plane so that light rays entering the eye in different planes are focused at different points, causing images to be partly out of focus; can be corrected by spectacles with lenses that have more curvature in one plane than another

asymptomatic refers to the stage of a disease when no symptoms are evident (e.g. the early stages of HIV/AIDS infection)

atheroma a fatty deposit that may occur on the internal walls of arteries; may obstruct the flow of blood

atherosclerosis a gradual degenerative change of the walls of the arteries that many people experience as they get older, characterised by deposits in the artery walls (often of lipid materials) that restrict blood flow through the vessels and can promote blood clotting

athlete's foot a common name for TINEA

atlas the first cervical vertebra, which articulates with the occipital bone of the skull and the axis (the second cervical vertebra)

atom the basic unit of matter that makes up an element; the smallest particle into which an element can be broken down and still retain the properties of that element; consists of a nucleus and electrons

atomic mass the mass of an atom compared with a carbon atom, which has been given a mass of 12; approximately equal to the combined number of protons and neutrons in the nucleus of an atom; hydrogen has an atomic mass of 1, oxygen 16 and uranium 238; also called *atomic weight*

atomic nucleus the central core of an atom, containing protons and neutrons; frequently referred to as *nucleus*

atomic number the number of protons in the nucleus of an atom; each element has a unique atomic number

ATP an abbreviation for ADENOSINE TRIPHOSPHATE

atrial fibrillation occurs when the atria of the heart lose their natural rhythm and contract very rapidly; *see* fibrillation

atrial systole contraction of the atria of the heart; *see* cardiac cycle

atrioventricular node (AV node) a compact mass of modified cardiac muscle cells between the heart's atria and ventricles that relays electrical impulses to the ventricles; *see also* pacemaker and diagram (p. 200)

atrioventricular valves (AV valves) flaps of tissue between the atrium and ventricle of the heart allowing blood flow in one direction only, from atrium to ventricle; the flaps of fibrous tissue grow out of the walls of the heart and, at their free edges, are held in place by cords (chordae tendineae) joined to the heart muscle; the AV valve on the right side has three flaps and is called the *tricuspid valve*; that on the left has two flaps and is called the *bicuspid valve* (formerly known as the *mitral valve*); *see* diagram of heart (p. 121)

atrioventricular bundle a bundle of specialised muscle fibres that conduct impulses from the atrioventricular node to the ventricles; *see* pacemaker

atrium a chamber of the heart that receives blood from the veins; the heart has two atria, one receiving blood that has passed through the lungs and one receiving blood from the rest of the body; formerly called the *auricle*; *see* diagram of heart (p. 121)

atrophy the wasting away or decrease in size of a part of the body due to failure, nutritional problems or lack of use

attenuation the process of weakening or reducing the virulence of a micro-organism so that it no longer causes disease but will still stimulate antibody formation; the weakened organism can then be used in a vaccine

auditory concerning the sense of hearing

auditory canal a tube making up part of the outer ear, leading from the pinna to the eardrum; also called the *external auditory meatus*; *see* diagram of ear (p. 82)

auditory nerve a common name for the 8th cranial nerve; more correctly called the *vestibulocochlear nerve*; *see* table of cranial nerves (p. 65)

auditory ossicles three small bones that extend across the middle ear and transfer vibrations of the eardrum to the oval window of the inner ear; the *malleus* (*hammer*) is attached to the eardrum and passes vibrations to the *incus* (*anvil*), which in turn passes them to the *stapes* (*stirrup*), which fits into the oval window; *see* diagram of ear (p. 82)

auditory tube an alternative name for EUSTACHIAN TUBE

auricle **1.** the pinna; the part of the outer ear on the outside of the head **2.** formerly used to describe one of the atria, a receiving chamber of the heart

auscultation examining the body by listening to sounds within the body, usually with a stethoscope

australopithecine a general term used to refer to any species in the genus *Australopithecus*; represented by fossil remains that have been found only in Africa; sometimes referred to as australopiths

Australopithecus a genus of fossil hominid that lived in Africa between 4 and 1 million years BP, possessing a relatively small brain and human-like hands and feet; walking bipedally; *see also* species of *Australopithecus* by name

Australopithecus aethiopicus an early form of East African australopithecine; more primitive in appearance than *Australopithecus boisei*; classified by some authorities as *Paranthropus aethiopicus*

Australopithecus afarensis a species of australopithecine that lived in eastern Africa between 4 and 3 million years BP; although its frame was upright and bipedal, the teeth and postcranial skeleton showed a number of primitive and ape-like features; fossils found so far suggest strong sexual dimorphism; the first specimen found was nicknamed 'Lucy' after the popular Beatles' song 'Lucy in the Sky with Diamonds'

Australopithecus africanus a species of australopithecine that lived in southern Africa between 4 and 3 million years BP; first described by the Australian-born South African anatomist, Raymond Dart, in *Nature* early in 1925, when he suggested the name *Australopithecus africanus*, 'the southern ape of Africa'; the skull and teeth of *A. africanus* show more similarities to the genus HOMO than any of the other species of australopithecines

Australopithecus anamensis one of the most primitive species of australopithecine yet discovered; fossils were found at two sites near Lake Turkana in Kenya and the discovery announced in 1995; evidence suggests that this species of hominid strode upright at least 4 million years ago; the fossils are complete upper and lower jaws, teeth from several individuals, a piece of skull, arm bones and a leg bone; taken together, the fossils push the emergence of bipedalism 500 000 years earlier than indicated by any other data; pieces of the right tibia (shinbone) are very human-like, while the top portion of the tibia is thick, in order to support the extra weight of walking on two limbs instead of 4; the knob on top of the tibia is concave—rather than convex, as it is in apes—creating the more stable knee joint needed for balance; the fossils were found in areas that were once densely wooded, suggesting that these hominids developed bipedal locomotion in the relative safety of the forests; the most dramatic difference between *Australopithecus anamensis* and all later hominids is the skull; the tooth rows in the jaw are parallel, like those of a chimp or gorilla, and the teeth are relatively large and covered in thick enamel, indicating that the diet included nuts and hard fruits

Australopithecus bahrelghazali a species of australopithecine named after fossils were unearthed in Chad, Africa, in January 1995, 2500 kilometres from the sites where fossils of other australopithecine species had been discovered, making it the first australopithecine to be found west of Africa's Great Rift Valley; named after Bahr-El-Ghazal, the site where it was found; the fossil was dated at 3.5 to 3 million years old and possessed ape-like as well as more evolved features

Australopithecus boisei the most robust species of australopithecine; believed to have lived in eastern Africa between 2.5 and 1 million years BP; characterised by extremely large back teeth and a large supporting facial and cranial structure, indicating large chewing muscles—hence its nickname 'Nutcracker Man'; *A. boisei* and *A. robustus* have a larger brain than the other australopithecines; formerly called

Zinjanthropus; now classified by some authorities into a separate genus of robust australopithecines—*PARANTHROPUS*

Australopithecus crassidens a robust australopithecine represented by fossils from Swartkrans in southern Africa; not usually regarded as a separate species but rather included in the species *AUSTRALOPITHECUS ROBUSTUS*

Australopithecus garhi a species of gracile australopithecine represented by fossils from the Afar Depression in Ethiopia; the DENTAL ARCADE is U-shaped; PREMOLAR and MOLAR teeth are larger than in any other GRACILE form of AUSTRALOPITHECINE

Australopithecus ramidus the most primitive species of australopithecine yet discovered, believed to have lived in eastern Africa around 4.4 million years BP; fossil evidence suggests an upright creature that was still a forest dweller but it is uncertain whether it was fully bipedal; in 1995 the discoverer of the fossils, Tim White, classified it into a new genus, *Ardipithecus*, because it was considered to be significantly different from the australopithecines; now known as *Ardipithecus ramidus*

Australopithecus robustus a robust species of australopithecine that lived in southern Africa between 2 and 1 million years BP; discovered by Robert Broom and originally named *PARANTHROPUS ROBUSTUS*; characterised by large back teeth (although not as large as those possessed by *AUSTRALOPITHECUS BOISEI*); *A. robustus* and *A. boisei* have a larger brain than the other australopithecines; some anthropologists do not recognise *A. robustus* as a distinct species and include it with *A. boisei*; some authorities are now classifying the australopithecines into two genera—Australopithecus for the gracile forms and *PARANTHROPUS* for the robust forms, in which case *A. robustus* would be called *Paranthropus robustus*

autoclave a very strong closed vessel in which steam under pressure reaches a temperature of around 115°C (hot enough to kill most bacterial spores), enabling the contents to be sterilised; used in the preparation of materials for the culture of micro-organisms; a pressure cooker can be used as a small autoclave

autocrine a mode of hormone action in which a hormone binds to receptors on, and affects the function of, the type of cell that produced it; secretion of a substance, such as a growth factor, that stimulates the secretory cell itself to perform some function

autocrine hormone a hormone that brings about a response by interacting with receptors on the surface of the cell that secretes it; the response is usually cell division

autogenous originating from a person's own body; an autogenous transplant occurs where tissue from one part of the body is transplanted to another part

autoimmune disease a disease in which antibodies are produced that attack the person's own body; *see* self

autoimmune response the production of antibodies or effector T cells that attack a person's own tissue antigens; *see* self

autologous transfusion a TRANSFUSION where the patient's own blood is used

autonomic division an alternative name for AUTONOMIC NERVOUS SYSTEM

autonomic nervous system the part of the efferent division of the peripheral NERVOUS SYSTEM that carries nerve impulses from the brain and spinal cord to internal organs (smooth and cardiac muscle, and glands); *see* table of effects (p. 186)

autopsy the examination of the body after death, often to try to establish cause of death

autoradiography the technique of producing an image on an X-RAY film by means of radiation released by the subject being photographed; e.g. in diagnosing disease, autoradiographs may be made after the subject has been injected with or has ingested some radioactive material

autosomal inheritance when genetic information is passed from one generation to the next on the non-sex chromosomes (autosomes)

autosome a non-sex chromosome; of the human set of 46 chromosomes, 44 are autosomes (the other two are SEX CHROMOSOMES)

AV node an abbreviation for ATRIOVENTRICULAR NODE; *see* pacemaker

AV valves an abbreviation for ATRIOVENTRICULAR VALVES

average the sum of a set of numbers divided by the number in the set; also called the *arithmetic mean*

axial skeleton that part of the skeleton that lies around the central axis of the body; the skull, vertebral column, ribs and sternum

axilla the armpit; the small hollow beneath the arm where it joins the body at the shoulders

axis **1.** the second cervical vertebra, which articulates with the atlas (the first cervical vertebra) **2.** the imaginary line about which a joint or structure moves or is symmetrical

axon an extension from the body of a nerve cell (neuron); each nerve cell has a single axon that carries nerve impulses away from the cell body; some axons have side branches called *collaterals*; axons vary in length from a few millimetres (in the brain) to up to a metre or more (from the spinal cord to a toe); *see also* dendrite, diagram of nerve cell (p. 187)

axon collateral *see* axon

Azoic an alternative name for the HADEAN era

AZT an alternative name for ZIDOVUDINE

B

B cell a lymphocyte that, during an immune response, develops either into a plasma cell or into a memory cell; the body has a huge number of different B cells, each capable of responding to a specific antigen; when activated by an antigen, the B cells either become plasma cells that secrete the appropriate antibody, or memory cells that can bring about a more rapid response should the same antigen occur in the future; also called a *B lymphocyte*

B lymphocyte *see* B cell

bacillus (plural *bacilli*) a bacterium with rod-shaped cells; *see* bacteria

backbone a common name for the VERTEBRAL COLUMN

bacteria (singular *bacterium*) very small single-celled organisms (just visible with a light microscope); do not have a nucleus, but genetic material is free in the cytoplasm; reproduce asexually by cell division, but genetic material may be transferred between cells; some are free living, but some are parasitic; many are pathogenic, causing diseases in plants and animals, including humans; many are useful, such as those that break down dead organic matter and those used in the production of cheese, yoghurt and other foods; classified according to the shape of the cell; *cocci* (singular *coccus*) have spherical cells; some species occur singly (monococcus), some in pairs (diplococci), some in chains (streptococci) and some in clusters (staphylococci); *bacilli* (singular *bacillus*) have rod-shaped cells (e.g. the bacteria that cause anthrax and diphtheria); *spirilla* (singular *spirillum*) have cells that are twisted, spiral or corkscrew in shape; *spirochaetae* (singular *spirochaete*) are cells in the form of a tight spiral that move by twisting the cell (e.g. the bacterium that causes syphilis); *vibrios* have a cell that is curved, often like a comma (e.g. *Vibrio cholerae*, the bacterium that causes cholera); *see* diagram below

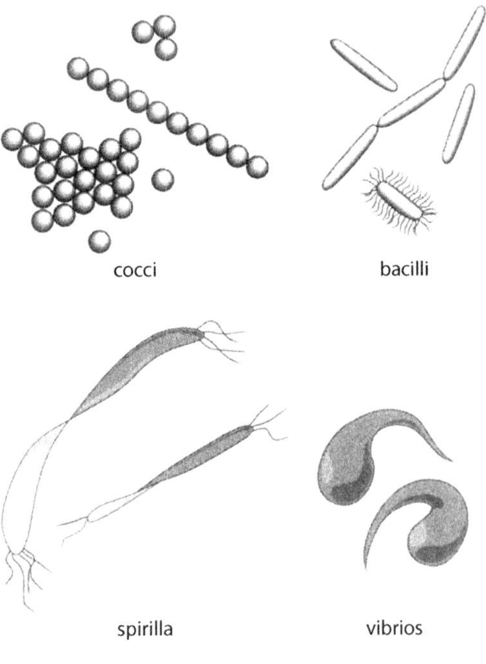

Bacteria

bacterial culture a sample of BACTERIA that is allowed to grow in STERILE conditions after being rubbed onto AGAR contained in a PETRI DISH

bacteriology the field of science that deals with the study of bacteria

bacteriophage a virus that infects bacterial cells; also known as a *phage*

balance in general, a state of equilibrium; specifically, the maintenance of a stable orientation to the environment; organs involved in balance are the SEMICIRCULAR CANALS, the UTRICLE and the SACCULE of the inner ear

balanced diet a diet that contains all ESSENTIAL NUTRIENTS in appropriate amounts and the correct amount of energy for body size and level of activity

ball-and-socket joint a type of SYNOVIAL JOINT

balloon angioplasty a surgical procedure in which a CATHETER equipped with a tiny balloon is inserted into an artery that has been narrowed, usually by the accumulation of fatty deposits; once in place, the balloon is inflated to widen the artery so that blood flows freely

bar graph a graph in which the length of a horizontal rectangular bar represents a particular quantity; *see* diagram this page; *see also* column graph

barbiturate sedative drugs that were once used to relieve anxiety and to promote sleep, now largely replaced by the safer benzodiazepine drugs (the minor tranquillisers)

baroreceptors receptors for blood pressure; *see* pressoreceptors

Barr body the inactive *X*-chromosome in female cells, visible as a dark-staining body in the nucleus; *see also* inactive-*X* hypothesis

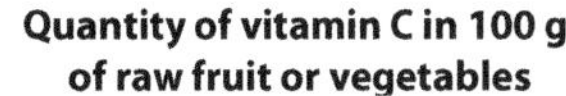

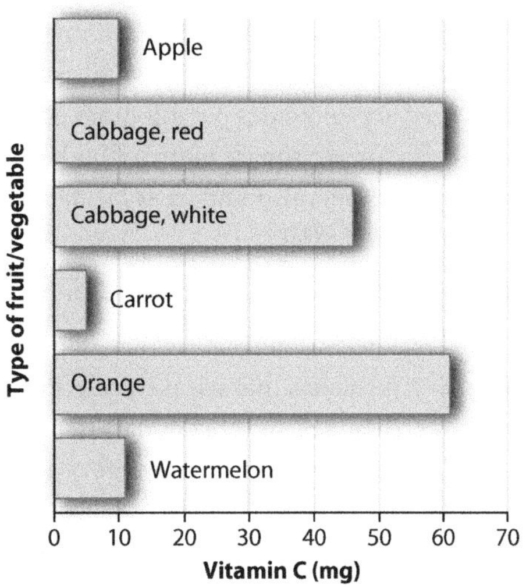

Bar graph

Bartholin's glands a pair of glands on either side of the vaginal opening that produce mucus to give lubrication during sexual intercourse; also called *greater vestibular glands*

basal ganglia (singular *basal ganglion*) regions of grey matter inside each cerebral hemisphere that control large unconscious movements of the skeletal muscles (e.g. arm swinging while walking) and regulation of the muscle tone required for particular body movements; damage to basal ganglia can result in muscle tremors and involuntary contractions of skeletal muscles; also called *basal nuclei* or *cerebral nuclei*

basal metabolic level (BML) an alternative term for BASAL METABOLIC RATE

basal metabolic rate (BMR) the rate of chemical reactions when a person is at rest; measured under standard conditions, which include a comfortable and constant air temperature, the subject at rest but awake, the subject having fasted for at least

12 hours before the test and having done no exercise for 30 minutes before the test; measured as heat produced per unit time and expressed in terms of energy utilisation; units used are kilojoules per kilogram of body mass (or per square metre of body surface) per hour; also known as *basal metabolic level (BML)*

basal nuclei an alternative name for BASAL GANGLIA

base **1.** a substance that releases hydroxyl ions when dissolved in water and has a pH greater than 7; the chemical opposite of an acid; *see also* pH scale (p. 208) **2.** a substance joined to a sugar and a phosphate to make a NUCLEOTIDE; the nucleotides of DNA have the bases adenine, thymine, cytosine and guanine; those of RNA have uracil instead of thymine; *see also* deoxyribonucleic acid

base pairs pairs of nucleotide bases on opposite strands of a DNA molecule; the base adenine always pairs with thymine, and guanine always pairs with cytosine; the length of a piece of DNA is often described in terms of the number of base pairs it has; *see also* deoxyribonucleic acid

base pairing the formation of chemical bonds between complementary bases in the two chains of nucleotides that make up a DNA molecule (adenine always pairs with thymine, cytosine always pairs with guanine); also, the pairing that takes place between part of the DNA molecule and messenger RNA during transcription; *see also* deoxyribonucleic acid, protein synthesis

base triplet a sequence of three nucleotides in a DNA molecule that is the code for one amino acid; *see also* deoxyribonucleic acid

basement membrane a thin layer of extracellular material that can be seen with a light microscope; occurs between an epithelium and the underlying connective tissue; consists of non-cellular protein-rich and polysaccharide-rich layers

basicranium the bones at the base of the CRANIUM

basilar membrane a membrane in the cochlea of the inner ear that separates the cochlear duct from the scala tympani, and upon which the organ of Corti lies

basophil a type of white blood cell involved in inflammation that releases an anticlotting agent at the site of an injury and produces the substance histamine, which delays the spread of invading micro-organisms

behaviour any observable action or activity carried out by an animal

behaviour modification a procedure that uses the principles of CONDITIONING to change behaviour; rewards are given for desirable behaviour, and punishment may be imposed for undesirable behaviour; may be used to modify behaviours such as smoking, drinking, overeating, or disruptive behaviour in children; may be conducted by the person whose behaviour is to be modified or by another person

behavioural genetics the study of the inheritance of behavioural characteristics; *see also* behaviour

behavioural response the change in BEHAVIOUR of an individual as a result of a stimulus (e.g. putting on warmer clothes in response to cold conditions)

belly **1.** the prominent, fleshy part of a skeletal muscle; *see* diagram opposite **2.** in everyday use, the abdomen

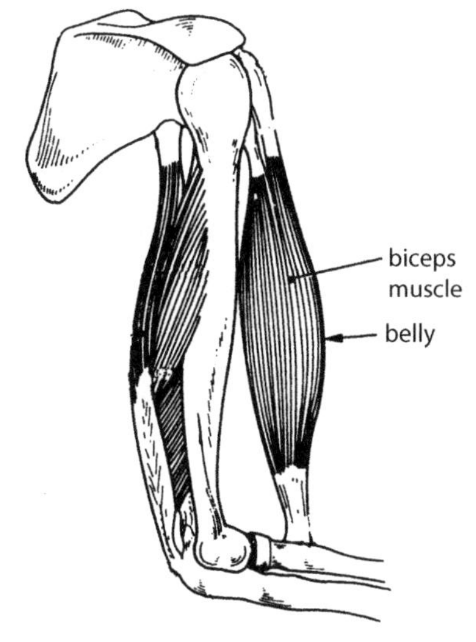

The belly of a muscle

Benedict's test a chemical test used to detect the presence of reducing sugar, particularly glucose, in food samples, body fluids or other solutions; Benedict's reagent, blue in colour, is a solution of copper sulphate, sodium citrate and sodium carbonate in water; when heated with a solution containing reducing sugar, copper I oxide is formed and the colour changes to greenish-yellow, orange or brick-red, depending on the concentration of sugar; *see also* Fehling's test

benign tumour a tumour that does not spread to cause secondary growths; *see* neoplasm

benzedrine a form of AMPHETAMINE; a CENTRAL NERVOUS SYSTEM stimulant that increases energy and decreases appetite; used to treat some forms of depression; originally used to treat CONGESTION of the nasal and bronchial passages

Bergmann's principle the tendency among mammals of similar shape for the larger mammal to lose heat less rapidly than the smaller mammal; also, the tendency among mammals for individuals within a species or closely related species to be progressively larger the further they live from the equator; *see also* Allen's rule

Beri-beri a disorder caused by deficiency of vitamin B_1, characterised by digestive disturbances and skeletal muscle paralysis; *see also* table of vitamins (p. 289)

beta cells cells in the islets of the PANCREAS that secrete the hormone insulin

beta-endorphin one of a group of chemicals (neuropeptides) found in the brain; *see* endorphins

bi- a prefix meaning two (e.g. biceps—a muscle with two points of origin)

bicarbonate ion HCO_3^-; a negatively charged ion important in maintaining the pH of body fluids; *see also* ion

biceps **1.** biceps brachii, the large muscle making up the front part of the upper arm—the muscle that contracts to bend the arm at the elbow; one end is attached to the shoulder blade (scapula), the other to the radius in the forearm; antagonistic to the triceps brachii *see* diagram top left **2.** biceps femoris, one of the hamstring muscles in the thigh

bicuspid valve the atrioventricular valve on the left side of the heart; *see* atrioventricular (AV) valves

bilateral symmetry a body plan in which the right and left sides of the body are mirror images of each other; a characteristic of vertebrates

bilayer a structure consisting of two layers of molecules; in particular, the CELL MEMBRANE, which consists of a bilayer of PHOSPHOLIPID molecules

bile a bitter, alkaline, yellow secretion of the liver, stored in the gall bladder and released into the small intestine; assists in the digestion and absorption of fats; may be called gall; *see also* bile salts, bile pigments

bile pigments two pigments—*bilirubin* and *biliverdin*—that result from haemoglobin breakdown in the liver and are excreted as waste in the bile; bilirubin is red and biliverdin is green; the bile pigments contribute to the colour of the faeces

bile salts salts in the bile that assist in breaking fats into tiny droplets (emulsification) ready for chemical digestion

biliary calculi (singular *biliary calculus*) an alternative name for GALLSTONES

bilirubin one of the BILE PIGMENTS

biliverdin one of the BILE PIGMENTS

bilophodont having two cross ridges on the molar teeth (e.g. in monkeys the molars have four cusps, with the front two cusps and the back two cusps each connected by a ridge)

binocular vision where both eyes can see an object at the same time due to overlapping fields of view; each eye sees a slightly different view of an object, so that when the brain superimposes the images we see in three dimensions (depth as well as height and width); assists judgment of distances; this three-dimensional effect is called *stereoscopic vision*; *see* diagram, top right

binomial nomenclature the system of naming organisms using the generic (genus) and specific (species) names to describe a species; the words used are always Latin or Greek, or made to sound like Latin or Greek; the generic name is always written with an initial capital letter and the specific name always with an initial small letter; the scientific name is printed in italics or underlined (e.g. the human species is *Homo sapiens*)

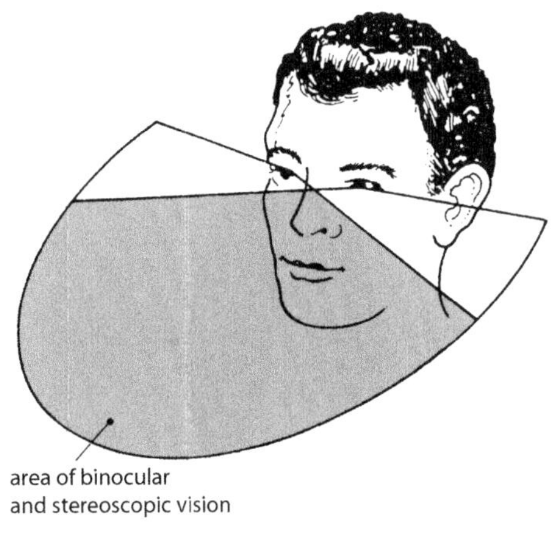

Binocular and stereoscopic vision

bio- a prefix meaning life or living organisms (e.g. biology—the study of life)

biochemistry the study of chemical substances and chemical reactions occurring in living things

biocultural approach a method of investigating human evolution and behaviour that looks at the interaction between biology and culture

biodegradable refers to a substance that is capable of being broken down by the action of living organisms, usually bacteria

biodiversity the number, relative abundance and genetic diversity of species in a given area; a global treaty to preserve world biodiversity was signed by almost all nations in 1992

biofeedback a process in which an individual receives constant signals about various internal body functions; monitoring devices can be used to provide information about heart rate, blood pressure, muscle tension, electrical activity in the brain and other functions; with the provision of suitable biofeedback, some people can learn to consciously control functions that are normally involuntary; biofeedback has been shown to be useful in the treatment of conditions like migraine and tension headaches

biogeochemical cycles cycling of elements and compounds from the environment, to organisms, and then back to the environment; examples are the carbon, oxygen, nitrogen, phosphorus and water cycles

biogeography the study of the geographical distribution of species

biological anthropology a branch of ANTHROPOLOGY that focuses on the biological evolution of humans and human ancestors

biological clock an internal mechanism that produces CIRCADIAN RHYTHMS

biological control the control of pests, or disease-causing organisms, using natural predators to control the size of pest populations; two of the best known Australian examples are the use of the *Cactoblastis* moth to control prickly pear and the *Myxoma* virus to control rabbits; release of sterilised males of pest species may also be described as biological control; when females in the wild population mate with the sterile males no offspring result, so the population is reduced

biological distance a measure of the genetic differences between populations (e.g. Scandinavians are more biologically distant from Africans than from people of the Mediterranean)

biology the scientific study of life; from the Greek *bios*, meaning life

bionic an artificial replacement for a body part that has normal biological capability and performance, enhanced by the use of electronic or mechanical components (e.g. bionic limbs)

bionic ear *see* cochlear implant

biopsy the removal of a sample of cells, tissue or fluid from the living body for examination, usually with a microscope and/or biochemical analysis

biosphere that part of the earth's land, water and atmosphere that is occupied by living organisms

biotechnology the use of biological processes for industrial and other purposes; some such uses are very old (e.g. the use of yeasts in making wine), while others are more recent (e.g. the use of genetically manipulated bacteria to produce human growth hormone)

biotic potential the maximum growth rate of a population under ideal conditions; the estimated rate of increase in the population of a species that would occur if there were no deaths due to predation, parasitism, disease or other such factors; it is never achieved in nature because of the effects of NATURAL SELECTION

biotin a water-soluble vitamin made by bacteria in the gut; dietary sources include liver, kidney, egg yolk and yeast; deficiency causes muscle pain, fatigue and poor appetite; sometimes known as *vitamin H*

bipedal locomotion walking upright on two legs; also called *bipedalism*

bipedalism *see* bipedal locomotion

bipolar disorder a mental disorder involving mood swings from very excited to depressed; formerly called *manic depression*

bipolar neuron a NEURON with two extensions—one AXON and one DENDRITE—arising from opposite sides of the CELL BODY; these are SENSORY NEURONS, such as those found in the RETINA of the eye; *see also* multipolar neuron

bird flu the common name for *avian influenza*; caused by a virus that normally infects birds, but on rare occasions has infected humans

birth canal the passage formed by the dilated cervix and vagina through which the foetus travels at birth

birth control the regulation of fertility through the prevention of conception or implantation; *see also* contraceptive

birth defect any defect or disease that is present at birth; also known as *congenital disorder*

birth rate of a population, the number of people born in a year, for every 1000 people in the population at the middle of that year

Biuret test a chemical test for proteins or polypeptides; dilute copper sulphate is added to the test solution, which is then made alkaline by the addition of sodium hydroxide; if proteins or polypeptides are present, a violet or purple precipitate of copper hydroxide is produced; the stronger the colour, the greater the concentration of protein/polypeptide

bivalent any pair of homologous chromosomes lying alongside each other during prophase I of MEIOSIS

bladder a common name for URINARY BLADDER

blastocyst a hollow ball of cells formed during early embryonic development, consisting of an internal cavity, an outer layer of cells and, at one side, a mass of cells (the *inner cell mass*); the outer layer of cells becomes part of the membranes making up the foetal portion of the placenta, and the inner cell mass becomes the embryo; at the blastocyst stage, about 7 or 8 days after fertilisation, the blastocyst implants itself in the lining of the uterus; *see also* implantation

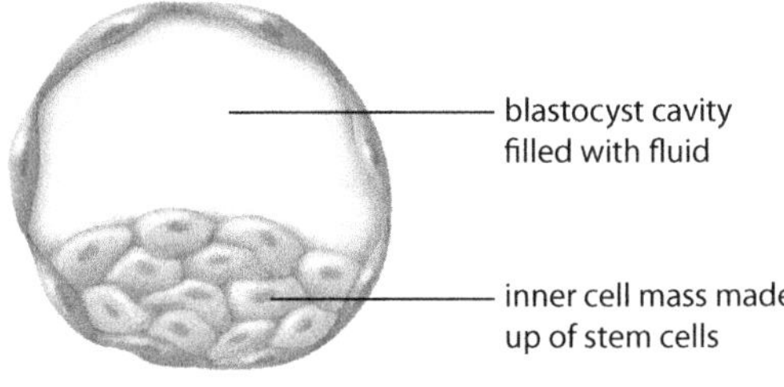

A blastocyst

blind spot the point on the retina where the optic nerve leaves the eyeball; contains no light-sensitive nerve endings; also called the *optic disc*; *see also* diagram of eye (p. 94)

blood a fluid tissue that circulates through the heart, arteries, veins and capillaries; contains blood cells suspended in a fluid matrix, the PLASMA; the main transport medium of the body; *see also* red blood cell, leucocyte, platelet

blood–brain barrier a barrier between the blood in the capillaries of the brain and the cells in the brain that function as connective tissue; a selective barrier that protects the brain from some harmful substances; the blood–brain barrier is formed by the walls of the brain capillaries

blood clotting the process by which blood coagulates to reduce bleeding from a wound; damaged blood vessels and platelets secrete the enzyme *thromboplastin*, which, in the presence of calcium ions, causes *prothrombin* to change into *thrombin*; the enzyme thrombin causes *fibrinogen* to change into *fibrin* threads, which trap blood cells and form the clot; over time the mass of fibrin threads contracts, forming a scab—a process known as CLOT RETRACTION

blood coagulation an alternative term for BLOOD CLOTTING

blood count the number of red cells or white cells in a given volume of blood, usually a cubic millimetre; normal red cell count ranges from 4.8 million per cubic millimetre in females to 5.4 million per cubic millimetre in males; normal white cell count for males and females is 5000–9000 per cubic millimetre

blood doping increasing the number of red blood cells in the body in order to improve athletic performance; done by blood transfusion or by using the hormone ERYTHROPOIETIN, which stimulates red cell production

blood flow in general, the flow of blood through the circulatory system; in particular, the amount of blood flowing through an organ in a given time

blood glucose the amount of glucose in the blood plasma; normally averages 90 mg/100 mL of blood; held fairly constant by the action of the hormones glucagon and insulin

blood group system a classification of blood based on the presence, or absence, of certain antigens on a person's red blood cells, or on the response of a person's blood to the introduction of particular antibodies; there are at least 300 blood group systems, but the two most important are the ABO system and the Rh system; also known as *blood type*; *see also* ABO blood group system, Rh blood group

blood poisoning an alternative name for SEPTICAEMIA

blood pressure the pressure of the blood on the walls of the blood vessels, especially the arteries; measured using an instrument called a sphygmomanometer; units of measurement are millimetres of mercury (mm Hg), i.e. the height of a column of mercury that the pressure could support; *systolic blood pressure* is the pressure of the blood on the arterial walls while the ventricles are contracting, the highest pressure in the large arteries (in young adult females, about 110 mm Hg; in young adult males, about 120 mm Hg); *diastolic blood pressure* is the pressure of blood on the arterial walls while the ventricles are relaxed, the lowest blood pressure in the large arteries (in young adult females, about 70 mm Hg; in young adult males, about 80 mm Hg)

blood screening checking blood for evidence of viruses or other contaminants that may cause disease

blood sugar an alternative term for BLOOD GLUCOSE

blood typing determining a person's blood group; *see also* blood group system

blue baby a condition usually due to the failure of the FORAMEN OVALE or the DUCTUS ARTERIOSUS to close properly after birth

blunt ends ends of DNA produced by a STRAIGHT CUT of a sequence of nucleotide bases

body cavity a space within the body that has particular boundaries and contains certain internal organs (e.g. the abdominal cavity contains many of the organs of the digestive system, the thoracic cavity contains organs such as the heart, lungs and trachea)

body mass index (BMI) a ratio of a person's weight to height, calculated as:

$$\frac{\text{weight (kg)}}{\text{height} \times \text{height (m)}}$$

used as a measure of whether a person is overweight or underweight; *see* body mass index interpretation table below

Table 2 *Interpreting body mass index*

Condition	*BMI*
Underweight	less than 18.50
Acceptable weight	18.50–24.99
Overweight	25.00–29.99
Obesity	30.00–39.99
Extreme obesity	40.0 or more

Bohr effect a property of haemoglobin; the oxygen-carrying capacity of haemoglobin varies with pH; in more acidic environments the oxygen-carrying capacity of haemoglobin is reduced; thus, when muscles are exercising vigorously and there is a high concentration of dissolved carbon dioxide (producing carbonic acid), the haemoglobin in the blood releases more oxygen; also known as the *Bohr shift*

bolus a soft, rounded mass of chewed food that is swallowed after being shaped by the tongue and moistened by saliva

bomb calorimeter a CALORIMETER in which organic material is burned; the amount of heat released is measured to determine the ENERGY VALUES of foods

bond the attractive force that holds atoms together in molecules; also called a *chemical bond*

bonding **1.** the joining together of atoms or molecules with chemical bonds **2.** the development of unconditional love and warmth between a parent and a baby, for which the contact between baby and parents in the period immediately following birth is very important

bone **1.** one of the units that makes up the skeleton; *see also* diagram of skeleton (p. 249) **2.** a hard connective tissue that forms most of the skeleton; composed of a matrix of collagen fibres hardened by calcium phosphate; may be compact bone or cancellous bone; *compact bone* is bone tissue that contains few spaces; the bone has concentric rings (lamellae) of hard material that fit tightly together; occurs in a layer over spongy bone; protects and supports the LONG BONES; also called *dense bone*; *see* diagram of microscopic structure of compact bone below; *cancellous bone* is made up of an irregular network of thin plates of bone; contains many large spaces; makes up most of the head of long bones and most of the bone tissue in short, flat and irregular bones; also called *spongy bone*; *see* diagrams below

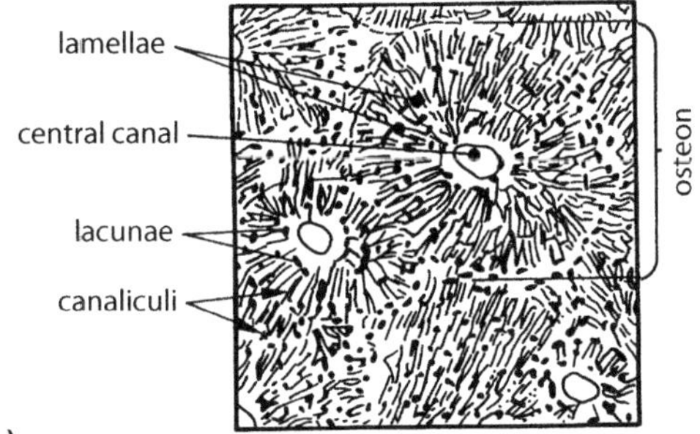

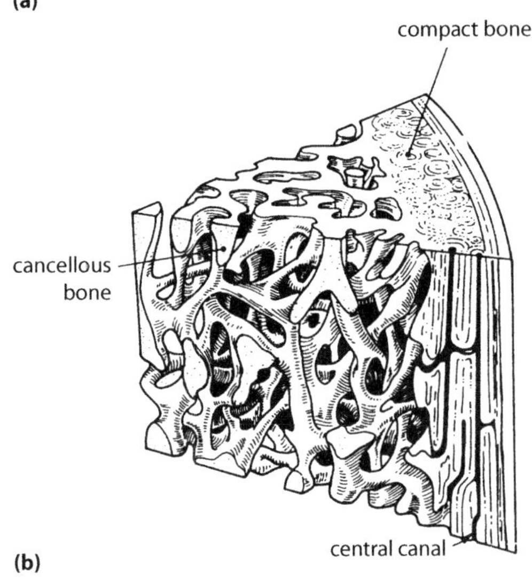

Microscopic structure of (a) dense bone; (b) cancellous bone

bonobo the pigmy CHIMPANZEE, *Pan paniscus*

bony labyrinth part of the inner ear; *see* labyrinth

botulism food poisoning caused by a bacterium that may infect preserved food, especially canned meat and vegetables; rare, but can be fatal

bowel cancer a cancerous growth in the large intestine—colon, rectum or appendix; cancer of the small intestine is called *small bowel cancer*

Bowman's capsule an alternative name for GLOMERULAR CAPSULE

BP abbreviation for *before present*; the internationally accepted form of designating past dates; the present has been set arbitrarily at the year 1950, thus a date of 100 000 years BP means 100 000 years before the year 1950

brachialis the muscle that lies beneath the biceps of the upper arm and is involved in bending the forearm

brachiation a specialised form of locomotion through trees in which the animal moves from branch to branch by hand-over-hand branch swinging; gibbons and siamangs are true brachiators, because this is their major method of locomotion; chimpanzees are semibrachiators, as they move on all four limbs on the ground and brachiate in the trees; *see* diagram below

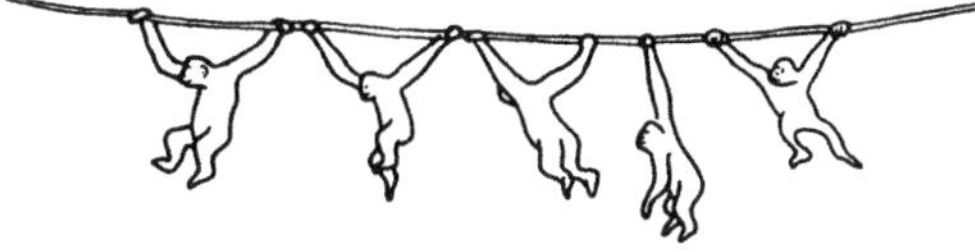

Brachiation

brachycephalic having a short, broad head; *see* cephalic index

brain the enlarged part of the central nervous system that is located in the cranial cavity of the skull; *see* diagram below; *see also* parts of the brain by name

brain electrical activity mapping (BEAM) an alternative name for an ELECTROENCEPHALOGRAPH (EEG)

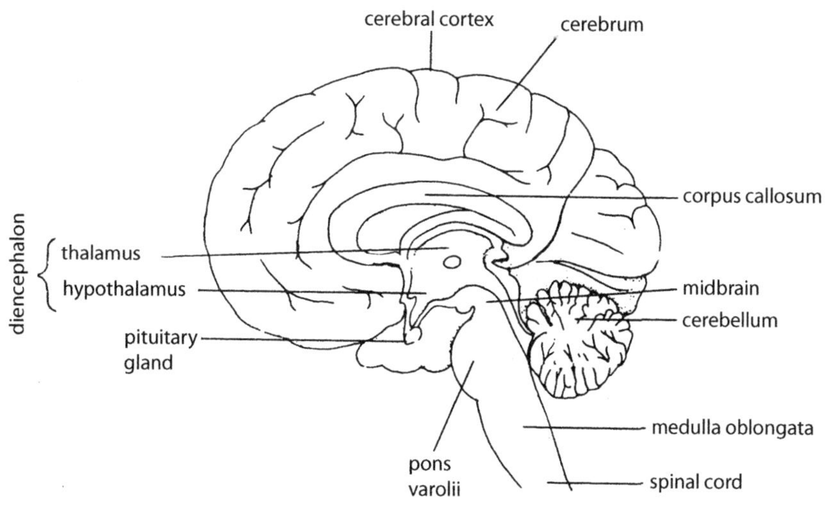

A section through the brain

brain stem the portion of the brain immediately above the spinal cord, which relays messages to and from the spinal cord; consists of the medulla, pons and midbrain; *see* diagram of brain previous page

brain waves electrical activity in the brain resulting from the millions of nerve impulses (action potentials) that occur at any one time; *see also* electroencephalogram

branching evolution a pattern of SPECIATION

breast milk milk produced by the mammary glands; contains antibodies, produced by the mother, which gives the developing infant some protection against infections

breastfeeding suckling of a baby at the mother's breast (as opposed to bottle-feeding); the act of breastfeeding may help to build a bond between child and mother

breathing taking air into, and out of, the lungs; *see also* inspiration, expiration

breech birth when the foetus lies with its buttocks in the lower part of the uterus instead of the usual position with the head over the opening of the uterus; the baby is born feet first

Bright's disease an alternative name for GLOMERULONEPHRITIS

broad ligament a wide fold of the PERITONEUM that is attached on each side of the uterus; the upper edge of the broad ligament encloses the uterine tubes

Broca's area an area of the frontal lobe of the cerebral cortex of the brain (in 95% of people, the left frontal lobe) responsible for translating thoughts into speech; concerned with co-ordination of the muscles involved in word formation and co-ordination of breathing while speaking; also called the *motor speech area*; *see* diagram of motor areas of the cerebrum (p. 48)

bronchi (singular *bronchus*) branches of the trachea; the two primary bronchi take air from the trachea (windpipe) to each lung; secondary bronchi take air to each lobe of the lung; tertiary bronchi divide into bronchioles; bronchi have supporting cartilage in their walls; *see also* diagram of respiratory system (p. 236)

bronchial asthma a condition characterised by smooth muscle spasms in the bronchi, resulting in wheezing and difficulty in breathing; frequently an allergic reaction

bronchiole a very small air tube in the lung; a small branch from a tertiary bronchus that leads to the alveoli in the lungs; its walls contain no cartilage, only smooth muscle; *see* diagram of respiratory system (p. 236)

bronchitis inflammation of the bronchi in the lungs

brow ridge the ridge of bone above the eye sockets of the skull; brow ridges are very prominent in *Homo erectus* and archaic *Homo sapiens*; also called the *supraorbital torus*

bruise internal bleeding and clotting into tissue spaces near the body surface, seen as a dark blue patch at the body surface

brush border the surface of cells covered with microvilli (e.g. the surface of cells lining the outside of the villi of the small intestine); *see* diagram (p. 287)

buccal relating to the mouth (e.g. the mouth cavity is also called the buccal cavity, and buccal epithelium lines the inside of the mouth)

buffer a substance that resists a change in pH when an acid or a base is added to a chemical solution

buffer system the mechanism that maintains the pH of body fluids at a relatively constant level; most buffer systems consist of a weak acid and a weak base; they prevent big changes to pH of body fluids by changing strong acids and bases into weak acids and bases

bulbo-urethral gland in males, one of a pair of small yellow glands on either side of the urethra and below the prostate gland that secrete a clear fluid into the urethra; the fluid, an alkaline component of semen, protects sperm by neutralising the acid of the urethra; also known as *Cowper's gland*; *see also* diagram of the male reproductive system (p. 235)

bulimia an eating disorder that may result from stress, depression or fear of being overweight, characterised by periods of uncontrollable overeating followed by forced vomiting or overdoses of laxatives

bulk transport an alternative name for VESICULAR TRANSPORT

bundle of His an alternative name for *atrioventricular bundle*; *see* pacemaker

bunion a painful swelling and overgrowth of bone on the foot; the deformity usually affects the big toe, the tip of which is forced towards the other toes, producing inflammation and thickening of the bursa, bone spurs and calluses

Burkitt's lymphoma a type of cancer that results from the translocation of the long arms of chromosomes 8 and 14 in one type of white blood cell; the cancer can grow into an enormous jaw tumour in a matter of days; treatment with anti-cancer drugs is effective in rapidly reducing the tumour; usually not inherited, as the translocation normally occurs in a body cell rather than an egg or sperm cell; common among Central African people; has links with the Epstein–Barr virus

bursa (plural *bursae*) a small sac or cavity filled with fluid that reduces friction at synovial joints; *see also* diagram of structure of synovial joint (p. 264)

bursitis the inflammation of the bursae surrounding a joint

buttocks two fleshy masses on the rear portion of the lower trunk, formed by the gluteal muscles

C

C phase the part of the CELL CYCLE when the cytoplasm divides (cytokinesis)

caecum part of the large intestine; a pouch where the small intestine joins the large intestine; *see also* diagram of digestive system (p. 77)

caesarean section a procedure in which a foetus is delivered through a cut made in the abdominal wall and uterus of the mother

caffeine a white, bitter-tasting, crystalline substance, first isolated from coffee in 1820 and probably the world's most popular drug; a stimulant that occurs in coffee, tea, chocolate and cola drinks; physiological effects include increasing blood pressure, increasing metabolic rate, inhibiting glucose metabolism and increasing urine production

Cainozoic era the 6th and most recent of the major subdivisions of geological time; from 65 million years ago to the present; also known as the 'Age of Mammals'; the first primates appeared during the Cainozoic era; also known as the *Cenozoic era*; *see* geological time scale (p. 111)

calcaneal tendon the tendon that attaches the calf (gastrocnemius) muscles to the heel bone; frequently referred to as the *Achilles tendon*; may also be called *calcaneus tendon* or *calcanean tendon*; *see* diagram, top right

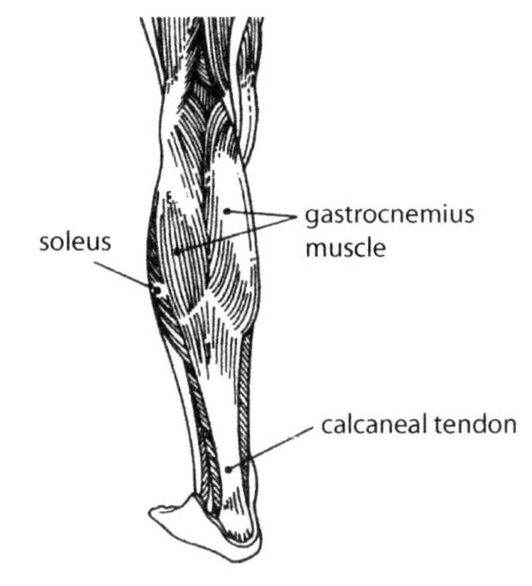

Calcaneal tendon

calcaneus the heel bone; the longest of the bones that make up the ankle; *see also* diagram of skeleton (p. 249)

calcitonin a hormone produced by particular cells in the thyroid gland, involved in the homeostasis of blood calcium and phosphate levels; lowers the amount of calcium and phosphate in the blood by increasing uptake of calcium by bones and by inhibiting bone breakdown; has an opposite effect to parathyroid hormone, which causes release of calcium and phosphate into the blood

callus **1.** the growth of new bone tissue in and around a bone fracture, which is ultimately replaced by mature bone

2. an acquired, localised thickening, frequently of the skin at a point of friction

calorie a unit of energy; the amount of heat required to raise the temperature of 1 g of water through 1°C; one *large calorie,* the unit used to describe the energy value of food, is equal to 1000 small calories; in Australia, energy values are now described in terms of joules and kilojoules

calorimeter an instrument used for measuring quantities of heat; the name comes from the former unit of heat, the calorie, but heat is now measured in joules; *see also* bomb calorimeter

calyx (plural *calyces*) a division of the renal pelvis; *see* kidney

Cambrian period one of the 7 periods of geological time in the Palaeozoic era, between the Ediacaran and Ordovician periods, about 542–488 million years ago; *see* geological time scale (p. 111)

canaliculus (plural *canaliculi*) a small canal between cells; in the matrix of bone, canaliculi join the spaces that contain bone cells; *see also* diagram of microscopic structure of bone (p. 34)

cancellous bone bone containing many large spaces; *see* bone

cancer a malignant growth characterised by uncontrolled cell division and capable of spreading to other body parts; *see also* neoplasm

canine a sharp, pointed tooth; *see* teeth

cannabis the Indian hemp plant, *Cannabis sativa*, or the drug derived from that plant; *marijuana* is prepared from the dried plant material and *hashish* is resin scraped from the plant; effects of cannabis on the user are variable depending on the amount used, the situation and the mood, expectation and motivation of the user; common effects are relaxation, enhanced appreciation of sound and colour and talkativeness; the main ingredient affecting behaviour is tetrahydrocannabinol (THC)

capacitation a change that sperm undergo in the female reproductive tract that enables them to fertilise an egg; sperm are not normally able to fertilise an egg until they have been in the female reproductive tract for several hours; the change is not well understood but is thought to involve a change to the sperm membrane so that they can penetrate the egg

capillarity the tendency of a liquid substance to move upward against the pull of gravity through a narrow space

capillary a microscopic blood vessel that links arterioles and venules and allows exchange of materials between the blood and the body cells

capsule **1.** (of bacteria) a protective layer that occurs outside the cell wall of some species; many disease-causing bacteria have a capsule **2.** (of a SYNOVIAL JOINT) *see* articular capsule

carbaminohaemoglobin a molecule resulting from a combination of carbon dioxide and HAEMOGLOBIN

carbohydrate an organic compound composed of carbon, hydrogen and oxygen; includes sugars, starches, cellulose and glycogen; carbohydrates can be divided into three main groups on the basis of the size of the molecule; *monosaccharides* (*simple sugars*) contain 3–7 carbon atoms—the most common are 6-carbon (hexose) sugars (e.g. glucose, fructose, galactose, mannose) and 5-carbon (pentose) sugars (e.g. ribose, deoxyribose); *disaccharides* consist of two simple sugar molecules chemically combined—sucrose (table sugar), lactose and maltose are disaccharides; *polysaccharides* are composed of three or more simple sugar molecules chemically combined (e.g. starch, glycogen, cellulose)

carbon-14 a naturally occurring, radioactive form of carbon that gradually decays to form nitrogen; has a half-life of 5730 years; used in radiocarbon dating; *see* diagram below

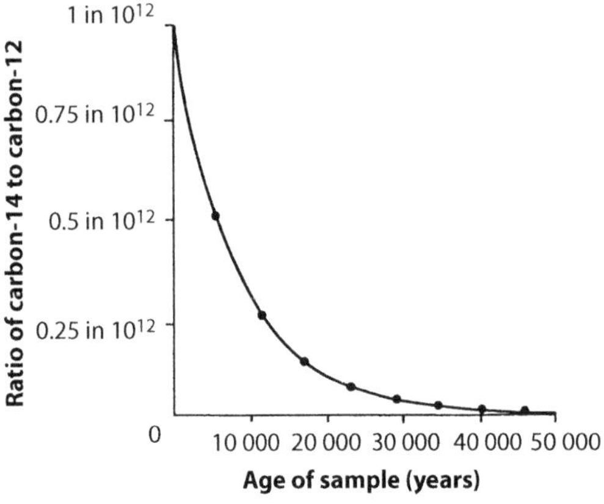

Rate of decay of carbon-14 to nitrogen

carbon-14 dating a method of dating fossils using decay of carbon-14; *see* radiocarbon dating

carbon monoxide poisoning poisoning that results from increased levels of carbon monoxide in the blood and its ability to more readily combine with haemoglobin than oxygen; tissues are therefore starved of oxygen

carbonic acid H_2CO_3; the acid formed when carbon dioxide dissolves in water; *see also* carbonic anhydrase

carbonic anhydrase an enzyme that facilitates the combination of carbon dioxide with water to form carbonic acid; although not essential for the reaction to occur, the enzyme occurs in red blood cells to greatly speed up the reaction; the carbonic acid then dissociates into hydrogen ions and bicarbonate ions; about 72% of carbon dioxide in blood is transported as bicarbonate ions

Carboniferous period the period of geological time in the Palaeozoic era between the Devonian and Permian periods; lasted from 359 to 299 million years ago; *see* geological time scale (p. 111)

carboxyhaemoglobin a combination between carbon monoxide and haemoglobin, which is much more stable than that between oxygen and haemoglobin; carboxyhaemoglobin in the blood reduces the oxygen-carrying capacity and high concentrations cause carbon monoxide poisoning

carboxypeptidase an enzyme involved in protein digestion, formed from the inactive procarboxypeptidase that occurs in pancreatic juice

carcinogen any cancer-causing agent (e.g. X-rays, ultraviolet radiation, tobacco tar, organic solvents); also known as *carcinogenic agents*

carcinoma a malignant tumour of epithelial tissue (e.g. cancers of the skin, lung, alimentary canal, brain, liver and glands); *see also* neoplasm

cardiac relating to the heart

cardiac arrest a complete stoppage of the heart

cardiac centre the part of the brain that regulates heartbeat; located in the medulla oblongata; also called the *cardiovascular regulating centre*; consists of a group of nerve cells called the *cardioacceleratory* *centre*, which sends impulses to speed up the heart and strengthen the beat, and a *cardioinhibitory centre*, which decreases heart rate and strength of contraction

cardiac cycle the cycle of events that occurs in one complete heartbeat; takes about 0.8 seconds; *systole* is contraction of heart muscle and *diastole* is relaxation; a complete cycle consists of systole and diastole of both atria and systole and diastole of both ventricles; *atrial systole* is the contraction of the two atria pushing blood into the ventricles and lasts about 0.1 seconds; *ventricular systole* (contraction of the ventricles) follows atrial systole, and is when blood is pushed into the arteries; this lasts about 0.3 seconds; diastole of both atria and ventricles occurs for the remaining 0.4 seconds

cardiac muscle the muscle that forms the wall of the heart; *see* muscular tissue

cardiac output the volume of blood pumped by one ventricle of the heart (usually measured from the left ventricle) in 1 minute; the volume of blood pumped by a ventricle for each beat (stroke volume) multiplied by the number of beats per minute (heart rate); cardiac output = stroke volume × heart rate per minute; the average for resting adults is about 5.25 L/min; also known as the *cardiac minute output*

cardiac sphincter a circular muscle between the stomach and oesophagus that relaxes during swallowing to allow food to pass easily from the oesophagus into the stomach; also called the *lower oesophageal sphincter*

cardioacceleratory centre the part of the brain that increases speed and strength of heartbeat; *see* cardiac centre

cardioinhibitory centre the part of the brain that decreases speed and strength of heartbeat; *see* cardiac centre

cardiology the study of the heart and the diseases associated with it

cardiopulmonary resuscitation (CPR) a technique used to restore heartbeat and breathing to a dying person; *see* resuscitation

cardiovascular disease disease of the heart and/or blood vessels

cardiovascular regulating centre alternative name for CARDIAC CENTRE

cardiovascular system the transport system of the body; *see* circulatory system

caries the gradual loss of minerals from the enamel and dentine of a tooth; can also mean decay of bone; if referring to the teeth, can also be called *dental caries* or *tooth decay*

carnivorous describes an animal that feeds on other animals

carotene a yellow pigment that occurs, along with MELANIN, in the skin of people of Asian origin; *see also* melanin

carotid artery the artery supplying blood to the head and neck; *see also* diagram of major arteries (p. 20)

carotid body a group of cells within the walls of the carotid arteries that is sensitive to changes in the concentrations of oxygen and carbon dioxide and pH of the blood; it sends nerve impulses to the vasomotor centre, which regulates blood pressure

carotid sinus a widening of the carotid artery at the point where it branches into two; contains receptors that monitor blood pressure

carpal tunnel syndrome a painful disorder of the wrist and hand caused by pressure on the nerve that passes between the carpal bones in the wrist

carpals the 8 small bones that make up the wrist; they are arranged in two rows with 4 bones in each row; collectively the bones are known as the *carpus*; *see also* diagram of skeleton (p. 249)

carpus the bones of the wrist; *see also* carpals

carrier **1.** an individual who has a recessive allele that is masked by the presence of a dominant allele so that the recessive character is not seen in the individual's appearance (e.g. a person may carry a recessive allele for albinism but is not an albino) **2.** a person infected with a transmissible disease who does not show symptoms of the disease but is able to transmit the disease to others **3.** a substance, often a protein, that transports another substance (e.g. carrier molecules in the membranes of cells may transport materials across the membrane)

carrier proteins *see* carrier (definition 3)

carrier-mediated transport transport of ions or molecules across a cell membrane by special protein carriers; includes both FACILITATED DIFFUSION and ACTIVE TRANSPORT

carrying angle the arrangement of the thigh bones (femurs) to form an angle to the vertical; it occurs in hominids because of the broad pelvis and enables them to walk upright without rolling from side to side; *see* diagram top right

carrying capacity the maximum population size that is capable of being supported in a given environment

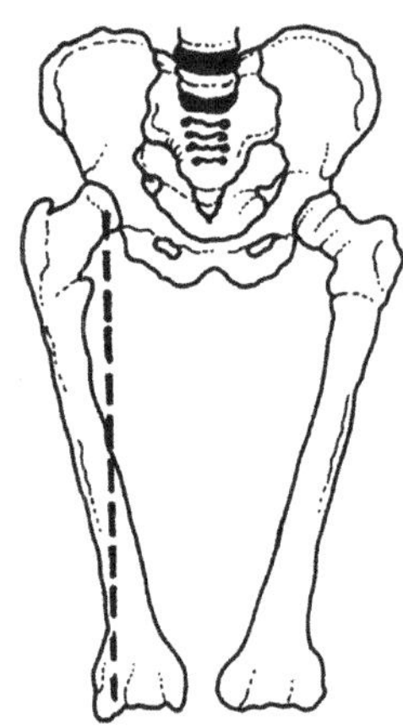

Carrying angle

cartilage a connective tissue that is hard but flexible and makes up part of the skeletal system; commonly known as gristle; *hyaline cartilage* gives flexibility and support—it is found at joints over the ends of LONG BONES, at the ends of the ribs and forming the nose, larynx, trachea and bronchi; *fibrocartilage* has bundles of fibres giving strength and rigidity—it is found in the discs between the vertebrae, in the cartilage of the knee, at the point where the two hip bones fuse at the front of the pelvic girdle and in certain places where tendons attach to bone; *elastic cartilage* has a network of elastic fibres providing strength and maintaining the shape of certain organs such as the larynx, epiglottis, the external part of the ear and the Eustachian tubes

cartilaginous joint a slightly movable JOINT

case study an in-depth study of a person or group, frequently done in order to make generalisations about a larger group or about society as a whole; a form of QUALITATIVE research

casein the main protein found in milk and milk products

castration the removal of the testes

CAT scan a method of TOMOGRAPHY

catabolic processes *see* catabolism

catabolism part of METABOLISM involving the breaking down of large molecules into smaller molecules

catalase an enzyme that breaks down hydrogen peroxide (H_2O_2) into oxygen and water; found in most plant and animal cells

catalyst a substance that speeds up a chemical reaction without being changed in the reaction

cataract the complete or partial clouding of the lens of the eye or its capsule so that it becomes opaque; it may be treated by removal of the lens; ultraviolet light in sunlight contributes to cataract formation

catarrhine refers to members of a subgroup of the primates called the CATARRHINI

Catarrhini in the classical classification of the order primates, a subgroup of the order which is classified according to the position of the nostrils (close together and downward directed, in contrast to the Platyrrhini which have nostrils further apart); includes all the Old World monkeys, apes and humans; members of this group are known as *catarrhines*; some taxonomists prefer to divide the primates into strepsirrhines (lemurs and lorises) and haplorrhines (tarsiers, monkeys, apes and humans); *see* diagram below *see also* Platyrrhini, diagram of simplified classification of living primates (p. 219)

catastrophism the hypothesis that patterns of evolutionary change in organisms that are observed in the fossil record can be explained by repeated catastrophes, which are followed by repopulation from other areas by different organisms

(a) A Catarrhine monkey; (b) A Platyrrhine monkey

catecholamines a group of chemicals including ADRENALINE, NORADRENALINE and DOPAMINE

cathartic a drug or medicinal preparation that relieves constipation by stimulating muscular activity of the large intestine and promoting flow of liquid to the bowels, resulting in defecation; strong cathartics are referred to as *purgatives*, and mild forms are called *laxatives*

catheter a narrow, hollow tube that can be inserted into a body cavity or into a blood vessel and is used to remove fluids (such as urine and blood) or for the introduction of diagnostic substances or medication

cation a positively charged ion (e.g. H^+)

Ceboidea in traditional classification, one of the three superfamilies of anthropoids, including all the New World monkeys (such as marmosets, squirrel monkeys and spider monkeys); more recent classifications place New World monkeys in the parvorder Platyrrhini; *see also* classification of primates (p. 219)

cell the basic structural and functional unit of a living organism; all cells have a similar basic structure but details differ according to the function of the cell; *see* cell components by name, eukaryote, prokaryote, diagram of cell structure, page below

cell body the part of a nerve cell (neuron) containing the nucleus; an axon and dendrites extend from the cell body; *see* diagram of nerve cell (p. 187)

cell cycle the continuous cycle of growth and division of cells; the stages through which cells go when actively dividing—the three stages of interphase, the four phases of mitosis and cytoplasmic division; the *G1 phase* (first gap) is the first part of interphase; many cells remain in the G1 phase until they die; actively growing cells enter the *S phase* (synthesis), during which the DNA is duplicated; the *G2 phase* (second gap) follows, during which other cell molecules are synthesised; this is followed by the *M phase* (mitosis), when the nucleus divides, and then the *C phase* (cytokinesis), when the cytoplasm divides; *see* diagram of cell cycle, opposite

cell division the process by which a cell reproduces itself; consists of a division of the nucleus and a division of the cytoplasm; *see also* meiosis, mitosis

cell membrane a membrane that forms the external boundary of a cell; also known as the *plasma membrane*; made up of a bilayer (two layers) of PHOSPHOLIPID molecules; *see also* diagram of cell structure below

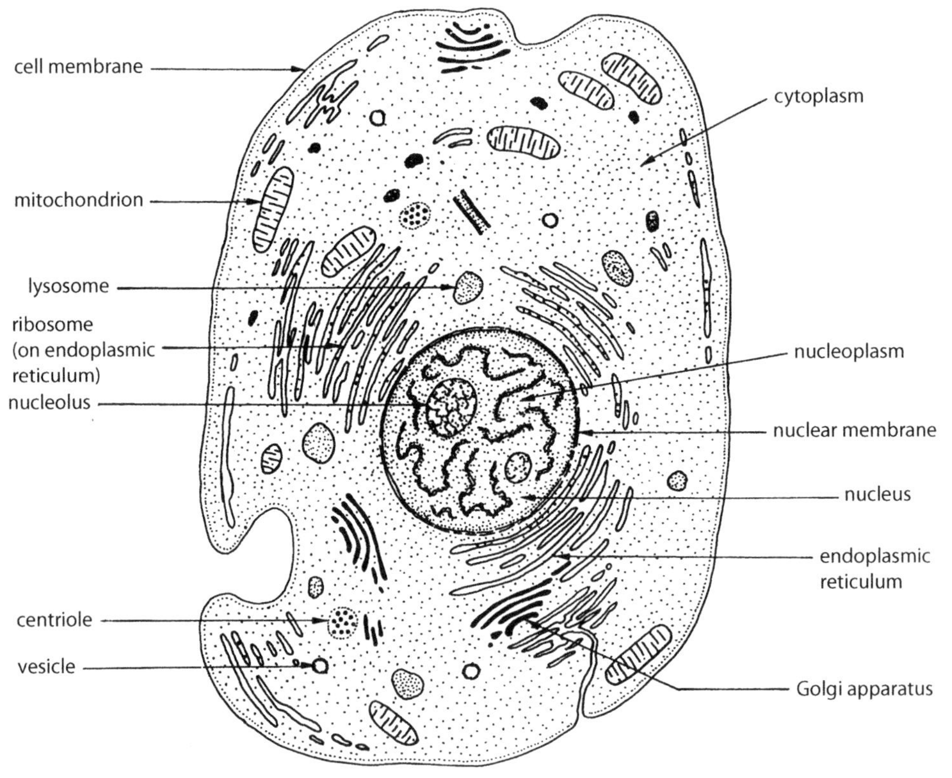

Structure of a cell

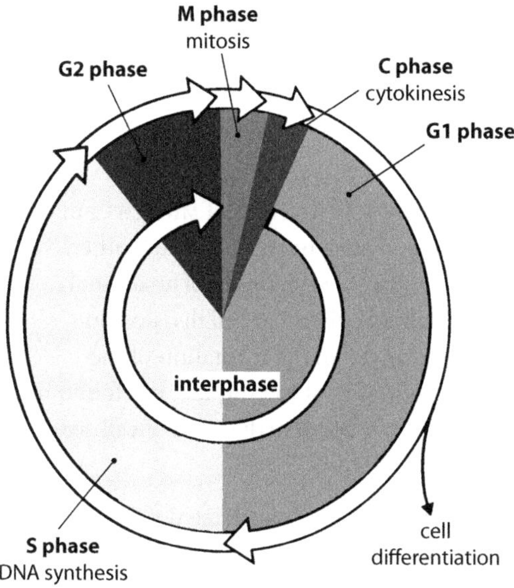

The cell cycle

cell organelle a structure within a cell; *see* organelle

cell theory the proposition that states that all living organisms are composed of cells; cells are the basic living units within organisms, and the chemical reactions of life occur within cells; all cells arise from preexisting cells

cell wall a fairly rigid structure that surrounds the cell membrane of plant and bacterial cells; made of cellulose but bacterial cell walls also contain peptidoglycan

cell-based therapy a treatment in which stem cells are induced to make replacement cells for damaged or diseased tissues

cell-identity marker a biochemical characteristic that distinguishes between different CELL types; every cell has a different pattern of chemicals on its surface, allowing cells to recognise each other (e.g. a sperm cell recognises an egg by the egg's specific cell-identity markers)

cell-mediated immunity another name for CELLULAR IMMUNITY

cell-replacement therapy a treatment which replaces damaged cells with healthy ones

cellular immunity that part of the immune response in which special lymphocytes (T cells) attach to foreign agents (antigens) to destroy them; it is particularly effective against infection by fungi, cancer cells, transplants of foreign tissue, parasites and viral infections inside cells; also called *cell-mediated immunity*; *see also* humoral immunity

cellular respiration the chemical reactions that make energy available for the cell; the breakdown of organic molecules, taken in as food, to release energy; also called *tissue respiration* or sometimes *internal respiration*; *see also* aerobic respiration, anaerobic respiration

cellulose the main constituent of the CELL WALL in green plants; an insoluble polysaccharide composed of straight chains of glucose molecules that contributes to the roughage (fibre) in the human diet

cementum a tissue, impregnated with calcium carbonate, that holds a tooth in its socket; also known as cement; *see also* diagram of tooth (p. 267)

Cenozoic era an alternative name for the CAINOZOIC ERA

central canal **1.** a hollow that runs through the centre of the spinal cord, is filled with cerebrospinal fluid and is connected to the medulla of the brain; *see* diagram next page **2.** in bone, a circular channel running through the centre of a Haversian system that contains connective tissue and blood vessels; also called a HAVERSIAN CANAL

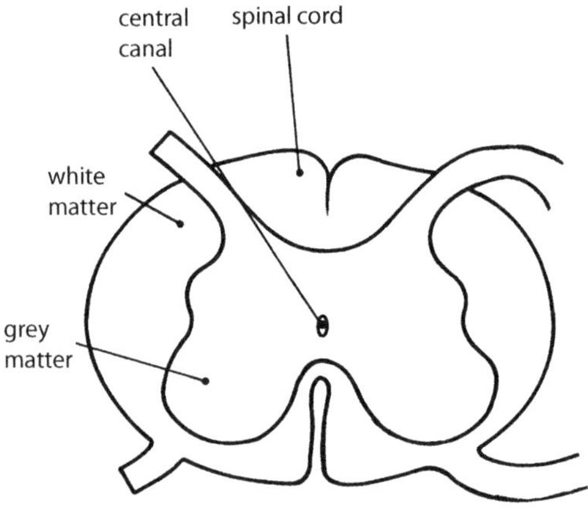

Central canal within the spinal cord

central chemoreceptor a CHEMORECEPTOR located in the MEDULLA OBLONGATA

central nervous system (CNS) the brain and spinal cord; *see* nervous system

central thermoreceptor a THERMORECEPTOR located in the HYPOTHALAMUS

centre of gravity (CG) the point where the entire weight of an object is in balance; in humans, the centre of gravity is in the pelvic region, between the VERTEBRAL COLUMN and the front of the abdomen, about four centimetres below the navel

centrifugation separating substances or structures of different densities by spinning them rapidly (e.g. red blood cells, white blood cells and plasma can be separated by centrifugation)

centrifuge a machine that spins very fast and is used to separate solids from liquids and liquids of differing densities; used to separate blood samples into plasma, red cells and white cells so that the proportions of each can be determined; an *ultracentrifuge* spins at very high speeds (up to 60 000 revolutions per minute) and can be used to separate the various types of organelles from a tissue preparation or various sizes of large molecules

centrioles two rings of MICROTUBULES, arranged at right angles to each other, within the CENTROSOME of an animal cell; they replicate at cell division and are involved in the formation of the microtubules of the spindle; not found in plant cells; *see also* diagram of cell structure, page 44 bottom

centromere a constricted portion of a chromosome that joins the two chromatids when a chromosome duplicates; the point where a chromosome attaches to a spindle fibre during cell division; *see also* diagram of chromatid (p. 52) and diagrams of phases of mitosis (p. 172)

centrosome an area located near the nucleus of a cell from which microtubules arise; within the centrosome of an animal cell is a pair of centrioles; also called the *microtubule-organising centre*

centrum the thick, strong, weight-bearing part of a vertebra; *see* diagram, below

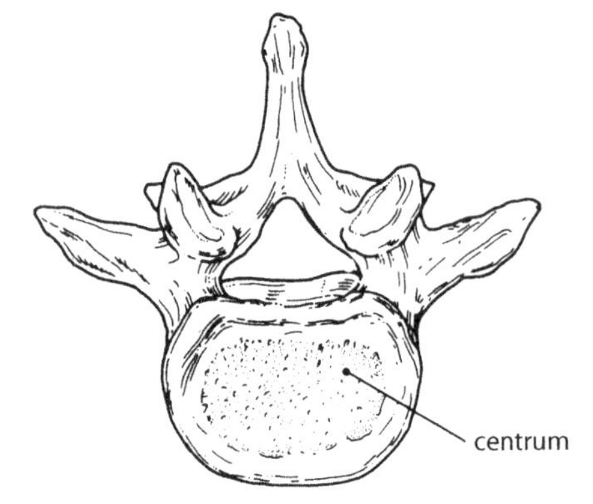

Centrum of vertebra

cephalic relating to the head

cephalic index a measure of head shape calculated by dividing maximum head width by head length (from front to back); an index of 0.80 or more is classified as *brachycephalic* (short head); 0.79 to 0.76 is *mesocephalic*; 0.75 or less is *dolichocephalic* (long head)

cephalocaudal development from head to foot; refers to the development of muscle co-ordination in infants; movements of the head develop before movement of the hands, and hand co-ordination is achieved before leg and foot co-ordination

cephalosporin a type of antibiotic obtained from fungi; similar to PENICILLIN in chemical structure and mode of action

Cercopithecoidea traditionally classified as one of the three superfamilies of anthropoids; more recent classifications include cercopithecoids as one of the two superfamilies of the parvorder Catarrhini; includes all the Old World monkeys such as langurs, macaques and baboons; characterised by ISCHIAL CALLOSITIES, two premolars on each side of each jaw, distinctive cusp patterns on the molar teeth, and usually considerable SEXUAL DIMORPHISM; *see also* classification diagram of living primates (p. 219)

cerebellar peduncle a bundle of nerve fibres that connects the cerebellum to the brain stem; there are three pairs of such fibres

cerebellum the part of the brain behind and below the cerebrum concerned with co-ordination of skeletal muscles for movement, balance and posture; *see also* diagram of brain (p. 35)

cerebral artery the artery that carries blood to the brain; *see also* diagram of major arteries (p. 20)

cerebral cortex the outer layer (grey matter) of the cerebrum of the brain, consisting of nerve cell bodies; to increase the surface area, it has folds called *convolutions*, with grooves or fissures between the folds; has numerous and complex functions including interpretation of sensory impulses, conscious control of muscular movement, and emotional and intellectual processes

cerebral haemorrhage bleeding from one of the blood vessels in the brain; one of the causes of stroke

cerebral hemisphere one of the two halves of the cerebrum; the left and right cerebral hemispheres are almost completely separated by a deep fissure, the *longitudinal fissure*, but are connected internally by a bundle of nerve fibres called the *corpus callosum*

cerebral nuclei an alternative name for BASAL GANGLIA

cerebral palsy a group of disorders caused by damage to the areas of the brain that send messages to the muscles (motor areas); the damage may occur before birth, during birth or during infancy; causes include rubella during the first 3 months of pregnancy, exposure to radiation before birth, oxygen deficiency during birth and accumulation of fluid on the brain during infancy

cerebral peduncle one of a pair of bundles of nerve fibres that conduct impulses between the cerebral cortex and the pons and spinal cord; the main connection between the upper parts of the brain and lower parts of the brain and spinal cord; located on the lower surface of the midbrain

cerebral thrombosis a blood clot in one of the blood vessels in the brain; one of the causes of stroke; *see also* thrombosis

cerebrospinal fluid (CSF) a clear, watery fluid produced in the cavities of the brain, containing glucose, proteins, urea, salts and some lymphocytes; fills the brain cavities and the central canal of the spinal cord and surrounds the brain and spinal cord; is a shock-absorbing medium to protect the brain and spinal cord from injury; delivers nutrients and removes waste and toxic substances from the nervous tissue of the central nervous system

cerebrovascular accident (CVA) the medical term for STROKE

cerebrum the largest part of the brain, made up of left and right hemispheres; co-ordinates and processes input from receptors, controls responses by muscles and glands, and is associated with memory, learning, thinking and intelligence; *see* diagram below; *see also* cerebral cortex, diagram of brain (p. 35)

cerumen ear wax, secreted by special glands near the opening of the ear canal

cervical refers to the neck, or the neck-like region of an organ or structure (e.g. cervical vertebrae—vertebrae in the neck); *see also* cervix

cervical cap a thin rubber cap that is fitted across the cervix of the female before sexual intercourse and prevents conception by preventing sperm from entering the uterus; *see* diagram opposite top

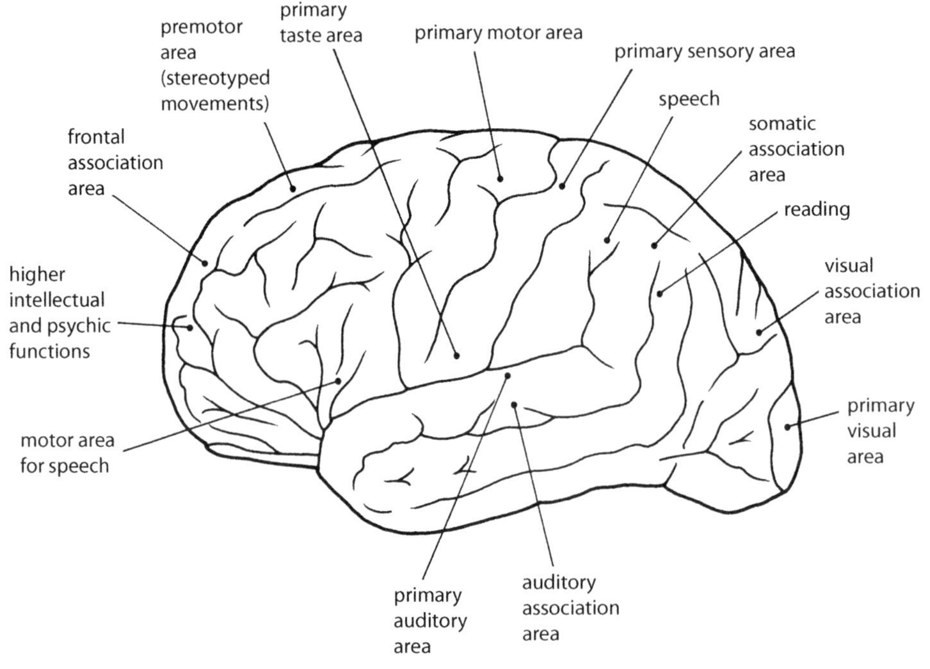

Functional areas of the cerebrum

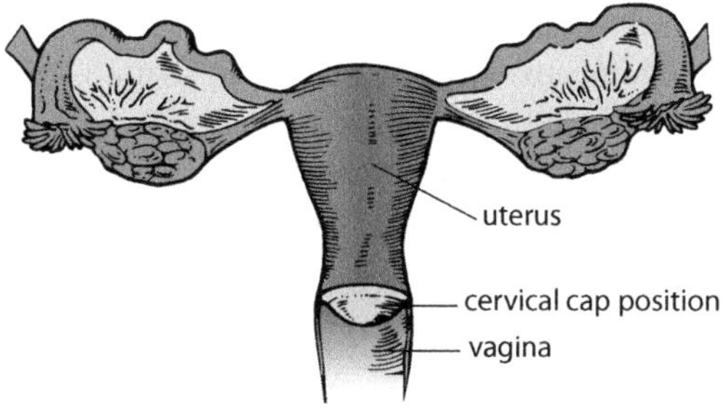

The position of the cervical cap in the uterus

cervical dysplasia a change in the shape, growth and number of cells in the cervix of the uterus that if mild may return to normal but if severe may progress to cancer; can be detected by a PAPANICOLAOU TEST

cervical vertebrae the bones of the spinal column located in the neck region; *see* vertebral column

cervix the neck or neck-like part; in particular, the neck of the uterus, which leads into the vagina; *see also* diagram of female reproductive system (p. 235)

chancres the initial stage of a syphilis infection; small sores that appear on the penis, labia of the vagina or other parts of the body; pronounced 'shankers'

channel protein a protein in the cell membrane that has a pore through which materials can pass between the extracellular fluid and the cytoplasm

characteristic any structural or functional feature of an organism that is determined by INHERITANCE; may also be called a *character*; although genetically determined, some characteristics are modified by the environment of the organism (e.g. human skin colour); *trait* is sometimes used to mean the same as characteristic, but in other cases, characteristic may mean the type of feature (e.g. eye colour), and trait may mean the form of the characteristic (e.g. blue eyes)

Chargaff's rules rules stating that in the DNA of any cell, the amount of cytosine will be equal to the amount of guanine, and the amount of adenine equal to the amount of thymine

cheek teeth premolars and molars

cheekbone an alternative name for ZYGOMATIC BONE

chemical bond the attractive force that holds atoms together in molecules; breaking a chemical bond results in the release of energy; also simply known as *bond*

chemical digestion the chemical breakdown of food into small molecules that can be readily absorbed into the capillaries of the blood and lymph

chemical element a substance made up of atoms that are all of the same type; *see* element

chemical hybridisation a technique for comparing the degree of similarity between two single strands of DNA and/or RNA; the single strands are mixed and those bases that are able to pair do so, creating a hybrid molecule; the greater the amount of pairing, the more closely related are the two nucleic acids; used to determine the evolutionary relationships between related species; also called *molecular hybridisation* or *annealing*

chemical reaction the combination, or the breaking apart, of atoms in which chemical bonds are produced or broken and new products with different properties are formed

chemistry the study of the properties and structure of matter and of changes in the composition and structure of matter

chemoreceptor a receptor that detects the presence of chemicals; chemoreceptors are located on or near the AORTIC BODY and the CAROTID BODY, and in the MEDULLA OBLONGATA

chemotherapy the treatment of an illness or a disease by the use of chemicals; treatment of infections by antibiotics is a form of chemotherapy, but in common usage the term is often taken to mean a method of treating cancer

chest cavity the body cavity enclosed by the rib cage and the diaphragm; contains the heart and lungs; also known as the *thoracic cavity*

chiasma (plural *chiasmata*) **1.** the crossing or intersection of two structures (e.g. the optic nerves cross over at the optic chiasma) **2.** the place where crossing over occurs between chromatids (*see* diagram p. 52; p. 66) **3.** the cross shape that is formed when chromatids cross over (*see* diagram p. 53; p. 66)

chicken pox a highly contagious disease caused by the *Herpes zoster* virus; itchy, fluid-filled blisters develop on the skin

chief cells cells in the gastric glands of the stomach that produce pepsinogen (which is converted into the enzyme pepsin)

childhood the period of human development from age 2 to about age 12

chimera an organism that contains a mixture of cells that are genetically different; may result from fertilisation of an egg by more than one sperm or the fusion of two zygotes; can also be produced by genetic engineering

chimpanzee one of the anthropoid apes; there are two species, the common chimpanzee, *Pan troglodytes*, and the lesser known pygmy chimpanzee, *Pan paniscus*; chimpanzees are considered by many scientists to be the most intelligent of the apes; they live naturally in equatorial Africa, are smaller than gorillas, have larger ears and tend to spend more time in trees

chiropractor a person who diagnoses and treats problems related to the skeletal and muscular systems, particularly disorders of the spine; using a form of 'natural' therapy with manipulation rather than drugs or surgery

Chlamydia one of the most common sexually transmitted diseases, which is transmitted by a bacterium during sexual intercourse; in males, causes inflammation of the urethra accompanied by a yellow discharge from the penis and pain when passing urine; in females, can cause inflammation of the uterus and uterine tubes (*pelvic inflammatory disease, PID*); often referred to as *non-specific genital infection* (*NSG*)

chloride shift the diffusion of chloride ions from the plasma into the red blood cells and bicarbonate ions from the red blood cells into the plasma that maintains the ionic balance between red blood cells and the plasma

chlorinated hydrocarbons hydrocarbon compounds in which some hydrogen atoms have been exchanged for chlorine atoms, including the pesticides Chlordane, Dieldrin and DDT and the PCBs used in electrical installations; highly poisonous to organisms and not biodegradable

chlorofluorocarbons (CFCs) gases containing chlorine, fluorine and carbon that are not natural but produced entirely by industrial processes; once used as the propellant gas in aerosol spray cans; also used as a refrigerant, but that use now being phased out; partly responsible for destruction of ozone in the earth's upper atmosphere and thus an increase in the amount of ultraviolet radiation reaching the earth's surface; also contribute to the greenhouse effect

cholecystokinin a hormone released from the lining of the duodenum when fat and acid are present; causes contraction of the gall bladder and release of bile; also stimulates the pancreas to secrete enzymes that digest fat and protein

cholera an infectious disease due to a specific bacterium; individuals suffering from the disease have severe diarrhoea, vomiting and cramps; frequently in epidemic proportions, often fatal

cholesterol $C_{27}H_{46}O$; a compound, classified as a lipid, that is found in animal fats and in cell membranes; used by the body to make bile salts and steroid hormones; high levels in the blood are considered to be a risk factor for cardiovascular disease

cholinergic fibre a nerve fibre that releases acetylcholine to act as the NEUROTRANSMITTER at a synapse; *see also* adrenergic fibre

chondrin a firm matrix of protein–carbohydrate complex found in CARTILAGE; translucent, and bluish-white in colour

chondroblast a cell that forms the fibres and matrix of cartilage

chondrocyte a mature cartilage cell

chordae tendineae cords that join the flaps (cusps) of the ATRIOVENTRICULAR VALVES to the heart wall, holding the cusps in place and preventing them from turning inside out; sometimes known as *tendinous cords; see also* diagram of the heart (p. 121)

Chordata a phylum consisting of organisms that possess a NOTOCHORD at some period during their life; mammals are members of this phylum; in mammals the notochord has been modified to form a bony backbone or vertebral column

chorion one of the FOETAL MEMBRANES

chorionic biopsy an alternative name for CHORIONIC VILLUS SAMPLING (CVS)

chorionic villi (singular *chorionic villus*) finger-like projections of the chorion that are part of the placenta and grow into the lining of the uterus; they contain blood capillaries from the developing foetus and continue growing until they are bathed by the mother's blood; they bring foetal and maternal blood into close proximity (although they do not normally mix); *see* diagram of developing foetal membranes (p. 101)

chorionic villus sampling (CVS) a technique by which foetal cells are removed from the chorion and examined for indications of possible defects; testing can take place at 8–12 weeks of pregnancy; also referred to as *chorionic biopsy*; *see* diagram, below

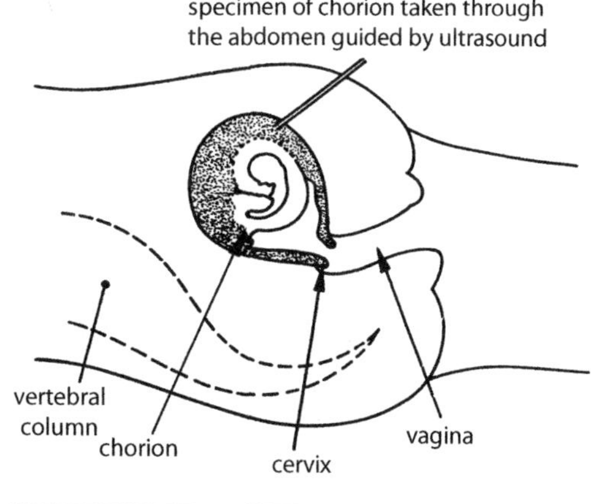

Chorionic villus sampling

choroid one of the layers of the wall of the eyeball, inside the sclera; contains many blood vessels, which supply nutrients to the retina, and a large amount of dark pigment that absorbs light rays, thus preventing reflection within the eyeball; at the front of the eyeball the choroid becomes the ciliary body that surrounds the lens; *see also* diagram of eye (p. 94)

chromatid one of a pair of duplicated chromosomes that are joined together by a small spherical body called the centromere; when chromosomes first become visible at the beginning of cell division they are in the form of pairs of chromatids; as cell division proceeds, the centromeres divide and each chromatid is then a chromosome; *see also* mitosis, meiosis, crossing over, diagram below

chromatin the thread-like mass of genetic material (mainly DNA) that is present in the nucleus of a cell that is not undergoing cell division; during cell division the chromatin shortens and thickens into the rod-shaped chromosomes; *see* diagram of chromatid below

chromatography a method of separating chemical substances by allowing them to flow through an absorbent material; separation occurs because smaller molecules move faster than larger ones

chromosomal mutation a change to the structure and/or number of chromosomes in offspring; the characteristics normally produced by that chromosome are changed or destroyed; a *translocation* is a chromosomal mutation in which chromosomes break and exchange pieces—part of one chromosome becomes attached to a different chromosome; a *duplication* results in part of a chromosome being duplicated so that there are extra copies of some genes; an *inversion* is where a segment of chromosome remains in the same position but is turned around end-to-end; *see also* mutation

chromosome a rod-like structure that appears in the nucleus at the beginning of cell division; contains DNA, which makes up the genes of the individual; humans have 46 chromosomes in each cell nucleus, 23 coming originally from the egg and 23 from the sperm; each individual thus has two of each type of chromosome, known as HOMOLOGOUS CHROMOSOMES

chromosome theory of heredity the theory that genes are located on the chromosomes and that each gene occupies a specific location on a chromosome

chromosome translocation a type of CHROMOSOMAL MUTATION in which chromosomes break and exchange pieces

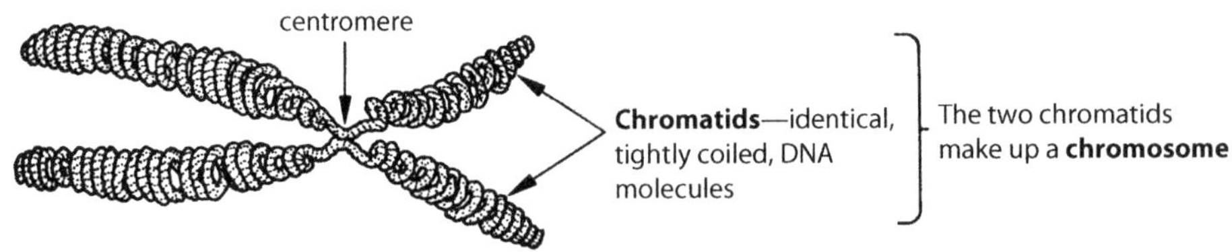

A duplicated chromosome in the form of a pair of chromatids

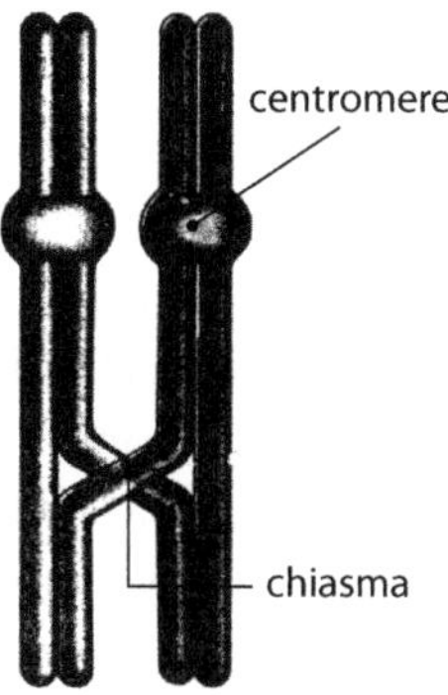

Chiasma occurs when two chromatids cross over

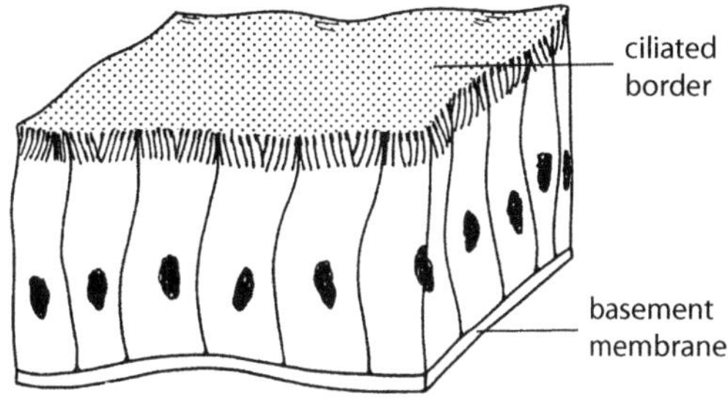

A tissue with cilia

chronic disease a disease that persists for a long time without being cured, or one that recurs frequently (e.g. in some countries, many diseases are chronic because of lack of medical facilities)

chronometric dating an alternative name for ABSOLUTE DATING; *see* dating

chyle a milky fluid consisting of fat globules in lymph; occurs in the lymph capillaries of the small intestine during absorption of digested food that contained fat

chylomicron a very small lipid droplet, coated with protein, that is absorbed into the lymph system from the small intestine; consists mainly of triglycerides

chyme the partly-digested, semifluid material that is passed from the stomach to the small intestine during digestion

cilia (singular *cilium*) hair-like processes projecting from a cell that beat rhythmically to move the whole cell or to move material across the cell surface; occur on cells lining the windpipe (trachea) and the uterine tubes; *see* diagram, top right

ciliary body a structure around the iris of the eye that contains ciliary muscles that change the shape of the lens; has folds or *ciliary processes* that secrete aqueous humour; *see also* diagram of eye (p. 94)

ciliary muscle part of the ciliary body of the eye; an involuntary muscle that alters the shape of the lens of the eye, enabling it to focus on near or distant objects; *see also* diagram of eye (p. 94)

ciliary process part of the CILIARY BODY of the eye

cilium singular of CILIA

circadian rhythms regular rhythms of activity, determined by internal mechanisms, that occur in cycles of approximately 24 hours (e.g. sleep patterns); circadian rhythms persist even in the absence of external cues such as day and night; the internal mechanism that produces such rhythms is called a *biological clock*; a *diurnal rhythm* is also a 24 hour rhythm, but is based on a regular pattern of light and dark

circular muscle smooth muscle with fibres arranged in a circle around an organ such as the stomach or the intestine

circulating hormone a HORMONE that is carried by the blood to act on target cells or organs that are remote from the site of production

circulation the movement of blood through the heart and blood vessels

circulation time the time required for blood to pass from the right atrium of the heart, through the lungs to the left ventricle, through the blood vessels of the foot and back to the right atrium; usually about 1 minute

circulatory system the transporting system of the body, consisting of heart, blood, blood vessels, lymph and lymph vessels; may also be called the *cardiovascular system*

circumcision the removal of the fold of skin (foreskin) over the tip of the penis; if carried out, is usually done 3–4 days after birth; in the Jewish faith it is a special ceremony performed on the 8th day after birth

circumduction the circular movement of a body part, especially at a movable joint where the distant end of a bone moves in a circle while the end at the joint remains relatively stable, so that the bone describes a conical shape (e.g. swinging the arm at the shoulder like a windmill); *see also* diagram (p. 1)

circumvallate papillae (singular *papilla*) the largest of the projections on the upper surface of the tongue that contain the taste buds; arranged in a V-shape at the back of the tongue; *see also* papillae

cirrhosis chronic inflammation of the liver in which liver cells are destroyed and replaced by fibrous connective tissue; may be caused by *hepatitis*, *alcoholism* or *parasites*

cisterna (plural *cysternae*) a space enclosed by the membranes of the ENDOPLASMIC RETICULUM or GOLGI APPARATUS of a cell

citric acid cycle an alternative name for KREBS CYCLE

cladistics a method of classification that groups together organisms with a common ancestry; the classification produced is like a family tree, but the main concern is the branching of the tree and not the amount of similarity of, or difference between, the groups; evolutionary classifications are based on degrees of difference and are in opposition to cladistics

cladogenesis a pattern of SPECIATION

cladogram the branching diagram used to represent relationships between groups of organisms; the cladogram represents the time at which groups diverged from one another in the past but ignores the degree of divergence; *see also* cladistics

class a category in the classification of organisms; a subdivision of a phylum that is further divided into orders; humans belong to the class Mammalia

classical conditioning a learning situation in which the organism learns to produce a new response to a stimulus with which it is familiar; *see* conditioning

classification placing things into groups; biological classification is used by biologists to group species with similar characteristics together

clavicle the collarbone; part of the shoulder girdle (pectoral girdle); a long slender bone with a double curve that acts as a connection between the breastbone (sternum) and the shoulder blade (scapula); *see also* diagram of skeleton (p. 249)

cleavage the process whereby the fertilised egg divides into a number of smaller cells; involves a series of mitotic divisions that result in an increased number of progressively smaller cells, so that the overall size of the embryo remains the same as the original fertilised egg

cleft lip a condition involving a split in the upper lip, often associated with CLEFT PALATE

cleft palate a condition in which the bones of the hard part of the roof of the mouth (the palate) do not unite before birth; *cleft lip*, a split in the upper lip, is often associated with cleft palate

climacteric the period during which a female undergoes a series of hormonal changes including the cessation of menstruation; thought to be caused by the ovaries becoming unable to respond to hormonal messages and therefore secreting less oestrogens; sometimes called the 'change of life'; may also be used to refer to reduced activity of the testes in males

climax an alternative name for ORGASM

clinal map a geographic representation of biological variation similar to a weather map in appearance; populations with similar patterns of variation are connected by a line

clinal zone the region over which a gradual change in the characteristics of a species occurs; *see also* cline

cline a gradual change in the characteristics of a species from one place to another; also, a gradual change in allele frequencies over distance (e.g. as one travels across Siberia from Asia to Europe there is a gradual change in the features of the indigenous people from Asian to European)

clinical trial a series of studies that use volunteers to investigate specific health questions; frequently used to evaluate the safety and effectiveness of new drugs and treatments by observing their effects on a large number of people

clitoris an organ of the female that contains erectile tissue, blood vessels and nerves; the equivalent of the male penis; *see also* diagram of female reproductive system (p. 235)

clone a group of cells derived from a single parent cell by mitosis; all the cells of a clone have the same genetic characteristics

cloning making an exact copy of something; *molecular cloning* produces exact copies of a fragment of DNA; *cellular cloning* produces a population of genetically identical cells; *cloning an organism* produces a new individual that is genetically identical to another organism; in *therapeutic cloning* an embryo is cloned so that stem cells can be harvested for therapeutic use; in *recombinant DNA technology*, cloning refers to the production of large numbers of vectors that carry the genetic information to make a specific protein

clot a semisolid mass of coagulated blood, consisting of a mesh of fibrin threads with trapped blood cells and platelets; usually occurs at the site of a wound, where it prevents blood loss

clot retraction the shrinking of a blood clot due to contraction of the fibrin threads; the retraction pulls the edges of damaged blood vessels closer together, reducing the chance of further bleeding; a clear fluid, the serum, is squeezed out of the retracting clot; *see also* blood clotting

clotting an abbreviation for BLOOD CLOTTING

clotting factor any substance that is essential for the normal clotting of blood (e.g. prothrombin activator, calcium ions and platelet phospholipids); also known as *coagulation factors*

coagulation an alternative term for BLOOD CLOTTING

cobalamin one of the B complex of vitamins; water soluble; the only B vitamin not found in vegetables; made by bacteria in the gut but may be obtained from meat, milk, eggs and cheese; deficiency alters DNA synthesis and retards cell division, which is particularly evident in the formation of abnormally large blood cells; deficiency causes pernicious anaemia; also called *cyanocobalamin* or *vitamin* B_{12}

cocaine $C_{17}H_{21}NO_4$; a bitter crystalline substance obtained from the leaves of the coca plant; a stimulant drug that affects the central nervous system by speeding up the activity of certain pathways in the brain; often used illicitly; may be used medicinally as a local anaesthetic

coccus (plural *cocci*) a bacterial cell with a spherical shape; *see* bacteria

coccyx the fused vertebrae at the end of the VERTEBRAL COLUMN

cochlea a spiral canal forming part of the inner ear; contains sensory cells involved in detecting sound and analysing pitch; resembles a snail shell; made up of three separate channels—the lower channel (filled with perilymph) is the *scala tympani*, the upper channel (also filled with perilymph) is the *scala vestibuli*, and a membranous channel between these (filled with endolymph) is the *scala media* or *cochlear duct*; *see* diagram below; *see also* diagram of ear (p. 82)

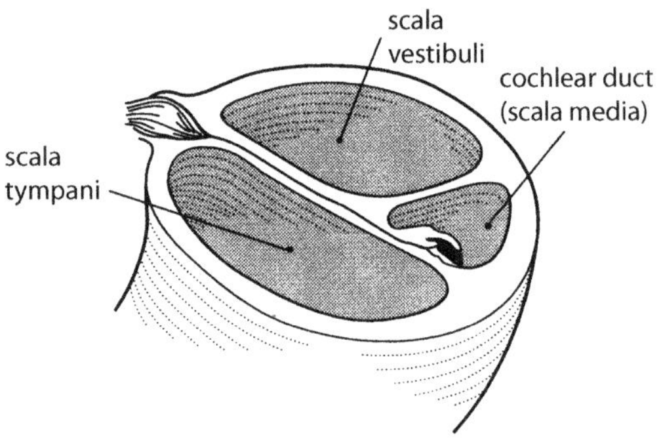

A section through the cochlea, showing the channels

cochlear duct an alternative name for the scala media; *see* cochlea

cochlear implant a device that is surgically implanted into the COCHLEA, enabling individuals with certain types of deafness to hear; also called a bionic ear

codeine a drug obtained from the resin of the seed pod of the OPIUM poppy

coding strand one of the strands of DNA (the other is called the TEMPLATE STRAND); the coding strand has the same base sequence as the MESSENGER RNA

codominance when contrasting alleles both affect the appearance of an individual and neither is dominant over the other; an individual with both alleles (the heterozygote) shows a blend of the characteristics controlled by each allele; e.g. in the ABO blood group system, the allele *A* (which determines production of antigen A) is codominant to the allele *B* (which determines production of antigen B), so that a person with the alleles *AB* can make both antigen A and antigen B

codon a sequence of three bases in a MESSENGER RNA molecule; the codon is the code for a particular amino acid; *see also* anticodon

coeliac disease chronic inflammation of the intestine due to an inability to tolerate gluten (gluten is a protein found in wheat, rye, oats and barley)

coelom the main body cavity of many animals, formed between layers of mesoderm in which internal organs are suspended; in the 2nd week of development of the human embryo, the middle layer of cells (the mesoderm) splits into two layers, one covering the ectoderm and one covering the endoderm; the space between these is the coelom; pronounced 'see-lome'

co-enzyme a non-protein substance that is associated with and activates an enzyme; many are derived from vitamins, e.g. nicotinamide adenine dinucleotide (NAD) is derived from the B vitamin niacin

co-evolution where two or more species of organisms have evolved together, resulting in a very close interrelationship with mutual adaptations for living together (e.g. between a parasite and host such as a tapeworm and its mammal host)

co-factor an additional chemical component, frequently a metal ion, required by certain enzymes in order to function; affects the activity of the enzyme in question but does not take part in and is unchanged by the chemical reaction taking place

cognitive learning altering the meaning of previous learning through the accumulation of new information and experience; sometimes called *latent learning* because it may not be put to immediate use, such as with the information gained from reading a book; *modelling* (learning by imitation) is also a form of cognitive learning

cohesion the tendency for like molecules to stick to each other, especially the attraction between the molecules of a liquid

coitus an alternative name for SEXUAL INTERCOURSE

coitus interruptus a method of birth control that depends on the withdrawal of the penis from the vagina just before ejaculation; semen is not deposited in the vagina and fertilisation cannot occur

cold sores *see* herpes

colitis inflammation of the colon and rectum

collagen a fibrous PROTEIN found in all types of connective tissue, especially tendons, bones, cartilage and ligaments; comprises about 40% of the body's protein

collecting tubule a vessel that receives urine from several kidney nephrons and passes it through the renal pyramids into the renal pelvis; also called a *collecting duct*; *see also* diagram of kidney nephron (p. 183)

colon part of the large intestine; consists of ascending, transverse and descending portions; *see also* diagram of digestive system (p. 77)

colostomy using surgery to form a new opening from the colon to the body surface; the new opening serves as an ANUS; carried out when the rectum has to be removed due to cancer or other disease

colostrum a watery, yellowish-white fluid with a high content of the mother's antibodies secreted from the breast a few days prior to and after childbirth, before the true milk is secreted

colour-blindness any deviation in the normal perception of colours; results from the lack of one or more of the photopigments in the cones of the retina of the eye; red–green colour blindness is an inherited sex-linked condition (the gene is carried on the *X*-chromosome) and is therefore much more common in males than females

column graph a graph in which the length of a vertical rectangular bar represents a particular quantity; *see also* bar graph

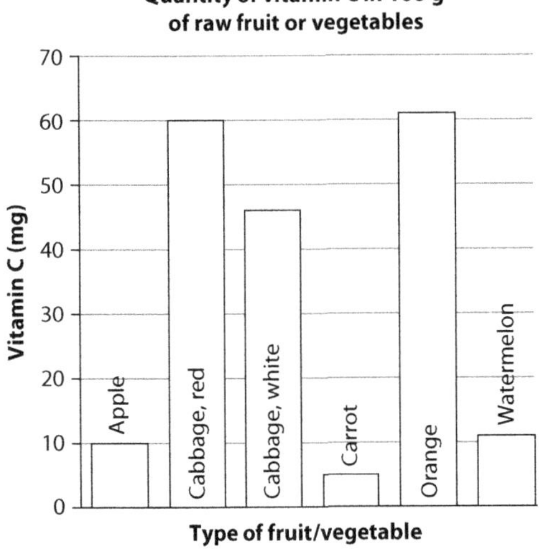

Column graph

columnar epithelium EPITHELIAL TISSUE with column-shaped cells

coma a state of unconsciousness from which a person cannot be aroused; *see also* sleep

combined pill a contraceptive pill containing substances similar to the female hormones oestrogen and progesterone; daily use prevents release of mature eggs from the ovary

commensalism the relationship between two species in which one species benefits and the other suffers no apparent harm or benefit; in humans, many micro-organisms live on or in the body in such a relationship

comminuted fracture a broken bone in which bone is splintered; *see* fracture

common bile duct a tube formed by the union of the duct from the liver (common hepatic duct) and the duct from the gall bladder (cystic duct); empties bile into the small intestine

communicable disease an alternative name for TRANSMISSIBLE DISEASE

compact bone a dense form of bone; *see* bone

comparative approach a method used in ANTHROPOLOGY to identify common and unique behaviours or characteristics in different populations; also known as *comparative studies*

competition individuals striving for the same resource, which may be in limited supply (e.g. food or shelter); may be between members of the same species (intraspecific) or between members of different species (interspecific)

complement a group of 11 proteins that occur in blood serum and are involved in the destruction of invading micro-organisms by initiating reactions that cause holes to develop in the membranes of the micro-organisms, attracting phagocytes, promoting phagocytosis and causing release of histamine, which provokes the inflammatory response; also known as *complement proteins* and *complement system*

complementary base pairing in nucleic acids, the bonding of adenine with thymine, uracil with thymine or guanine with cytosine; it holds the two strands of DNA together and different parts of RNA molecules in specific shapes and is fundamental to genetic replication, genetic expression and genetic recombination; *see* diagram of structure of DNA (p. 72)

complementary medicine alternative treatments used in conjunction with conventional medicine (e.g. using herbal supplements while undergoing radiotherapy for cancer); *see also* alternative medicine

complete fracture a break across an entire bone; *see* fracture

complete protein a PROTEIN containing all essential AMINO ACIDS

complex carbohydrate an alternative name for POLYSACCHARIDE

compound a substance made up of the atoms of two or more different elements chemically combined (e.g. water, H_2O, is a compound made up of two atoms of hydrogen and one of oxygen)

compound fracture a FRACTURE in which the broken bone protrudes through the skin

computed tomography (CT) an X-ray technique that provides a cross-sectional picture of any area of the body; *see* tomography

concentration a measure of the ratio of solute to solvent in a solution

concentration gradient a difference in concentration of a solution, often between the inside and outside of a cell; also called *diffusion gradient*

conception an alternative name for FERTILISATION in humans

conchae (singular *concha*) shell-shaped structures; refers particularly to scroll-shaped bones in the nasal cavity that are covered with mucous membrane and increase the surface area inside the cavity; they create turbulent air flow for efficient circulation, filtration and temperature conditioning of inhaled air; also refers to the hollow portion of the external ear not contained in the head

concussion injury to the brain from a blow to the head with no visible bruising but possible abrupt, temporary loss of consciousness

condenser a microscope lens between the light source and the specimen; concentrates light through the specimen

conditioning the simplest form of learning; learning associations between stimuli and responses so that a reflex action is changed by experience; in *classical conditioning* the original response is carried out to a new stimulus; an association is established between a new stimulus and a response (e.g. if a light is shone into a person's eye, the pupils constrict; if the light is accompanied by the ring of a bell, the pupils eventually will contract at the sound of the bell alone); in *operant conditioning* there is a new response to a stimulus that remains the same; the new response is learned because it is reinforced by a reward; *see also* behaviour modification

condom a very thin sheath of rubber, plastic or animal membrane that is rolled onto the erect penis before sexual intercourse to prevent sperm entering the vagina, so that fertilisation cannot occur; also provides some protection against infection by the human immunodeficiency virus and other sexually transmitted infections; *see* diagram on next page top

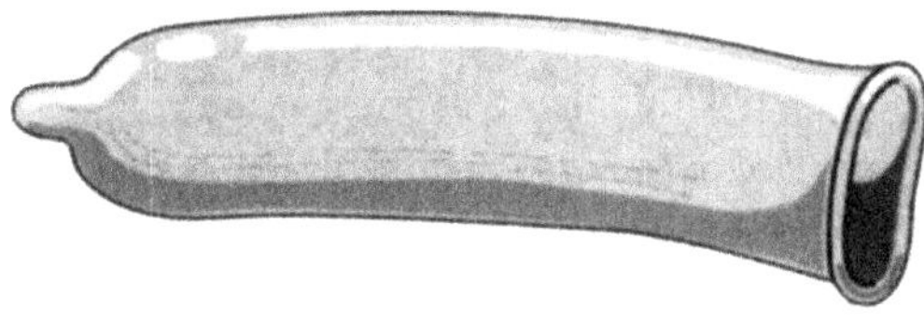

Condom

conduction of heat, the transfer of heat from one object to another through physical contact (e.g. heat is transferred from the body to clothing or to objects a person is holding or touching; about 3% of body heat is lost by conduction)

conduction myofibre muscle fibre in the heart specialised for conducting a nerve impulse to the heart muscle; part of the conduction system of the heart; stimulates contraction of the ventricles; also called *Purkinje fibre*; *see also* diagram (p. 200)

conductive deafness DEAFNESS due to a fault in the conduction of sound vibrations in the ear

conductivity the ability to carry an impulse resulting from a stimulus from one region of a cell to another; highly developed in both nerve and muscle fibres

condyle a projection from a bone that fits into another bone, thus forming a joint (e.g. on either side of the foramen magnum of the skull there is an occipital condyle that articulates with the first of the neck vertebrae)

condyloid joint a SYNOVIAL JOINT in which one surface of bone (the CONDYLE) is slightly convex, so that it fits into a slightly concave depression in another bone; this type of joint occurs between the radius and the carpal bones, between the metacarpal bones and the phalanges of the fingers, and between the metatarsal bones and the phalanges of the toes; allows movement in two directions (e.g. up and down and side to side)

cone one of the light receptors in the RETINA of the eye

congenital existing at birth

congenital disorder an alternative name for BIRTH DEFECT

congestion the abnormal accumulation of fluid, usually blood, but also bile or mucus, within the vessels of an organ

congestive heart failure a state in which the heart is unable to pump enough blood to supply the oxygen demands of the body

conjoined twins identical twins whose bodies are joined together at birth; occurs when the zygote that divides to form identical twins fails to completely separate; estimated to occur in about one in 200 000 births; formerly called Siamese twins

conjunctiva a delicate membrane covering the CORNEA of the eye and the inside of the eyelid; *see also* diagram of eye (p. 94)

conjunctivitis inflammation of the conjunctiva of the eye

connective tissue a binding and supporting tissue that has a lot of non-cellular material between the cells (e.g. ADIPOSE TISSUE, AREOLAR TISSUE, BONE, CARTILAGE, LIGAMENT, TENDON); BLOOD is sometimes classified as a connective tissue

connector neuron a nerve cell in the brain or spinal cord that carries messages between other nerve cells; links sensory and motor neurons; also called an *association neuron*, *interneuron* or *internuncial neuron*

consanguineous marriage the union of two close relatives, usually cousins

constipation a condition caused by infrequent defecation; faeces become hard and dry due to reduced movement of the intestines

consumers organisms that feed on other organisms, including primary consumers (herbivores and plant parasites) and secondary consumers (carnivores and parasites that feed on herbivores and plant parasites)

consumption the former name for TUBERCULOSIS

contagious disease a disease passed on by direct contact; sometimes used loosely to include any disease that can be passed from one person to another

continental drift an alternative name for PLATE TECTONICS

continuous variation the gradual change in a characteristic within a population (e.g. human skin colour, which varies gradually and continuously from very light skin colours to very dark skin colours); *see also* polygenic inheritance

contra- a prefix meaning opposite or against (e.g. contraception—prevention of conception)

contraceptive any mechanical device or chemical agent used to prevent conception (e.g. condom, intrauterine device, birth control pill)

contraceptive pill commonly referred to as 'the pill'; an oral CONTRACEPTIVE used to prevent PREGNANCY; usually contains OESTROGEN and PROGESTERONE that prevent OVULATION; also referred to as *birth control pill*

contractibility the ability to shorten; refers to the ability of muscle tissue to shorten or develop tension upon the application of a stimulus; also called *contractility*

contractility an alternative name for CONTRACTIBILITY

contraction the shortening of the fibres of muscle tissue

control a procedure carried out to give a comparison in an experiment (e.g. where the effect of a new drug is being tested, one group of subjects is given the drug while another group, the control group, is treated in an identical way but is not given the drug)

control group *see* control

controlled experiment an experiment in which only one VARIABLE is tested at a time and in which the experimenter controls the conditions; there are normally two identical set-ups, the only difference between them being the VARIABLE that is being tested; comparison between the control and the experimental set-ups enables the experimenter to determine whether the variable tested has influenced the result of the experiment

controlled variable a VARIABLE that is controlled in an experiment

contused wound an injury in which the skin is not broken; bleeding is internal, causing a bruise; also known as a *contusion*

convection the transfer of heat that occurs through the circulation of the heated particles of a liquid or gas; cool air in contact with the body is warmed, carried away by convection currents and replaced by more cool air; about 15% of body heat loss is due to convection

convergence 1. the inward turning of the eyeballs when looking at a close object; the closer the object under observation, the greater the convergence **2.** CONVERGENT EVOLUTION

convergent evolution the evolution of similar characteristics in two or more unrelated groups of organisms; often found in organisms living in similar environments (e.g. the development of flight in birds and in certain insects); also called *convergence*, *parallel evolution* or *parallelism*

convoluted rolled, coiled or twisted; *see* distal convoluted tubule, proximal convoluted tubule

convolutions 1. rounded ridges on the surface of the cerebrum of the brain; also called *gyri* (singular *gyrus*); *see* diagram (p. 260) **2.** the coiled parts of a kidney tubule; *see* nephron

convulsion a violent, involuntary contraction of an entire muscle group; may occur when nerves to muscles are stimulated by fever, hysteria, poisons, or withdrawal of certain drugs

coprolite fossilised faeces; much can be determined from coprolites of human ancestors (e.g. the type of food that was consumed)

copulation an alternative name for SEXUAL INTERCOURSE

cord blood stem cells *see* stem cells

core body temperature the temperature deep inside the body; about 37°C; also referred to as *core temperature*

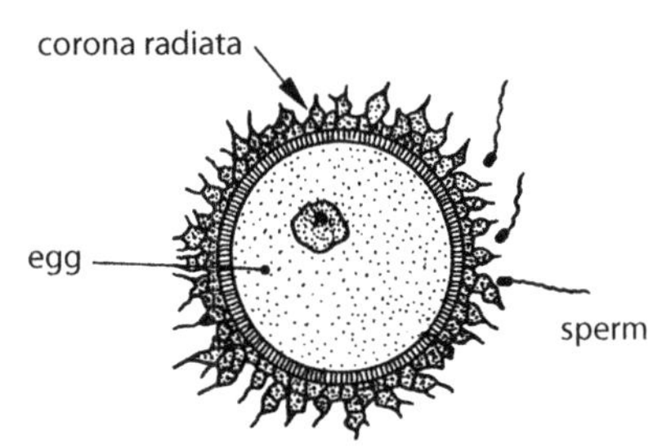

Corona radiata

core temperature the temperature deep inside the body; about 37°C

cornea the transparent, fibrous coat at the front of the eye; *see also* diagram of eye (p. 94)

cornified layer a layer of the EPIDERMIS of the skin

corona radiata the layer of follicle cells immediately surrounding an egg (ovum) while in the ovary; at ovulation the corona radiata leaves the follicle along with the egg; *see* diagram, top

coronary artery the artery that supplies blood to the muscles of the heart; *see also* diagram of major arteries (p. 20)

coronary heart disease any condition that reduces the flow of blood to the heart muscle; may be caused by build-up of deposits in a coronary artery or by a blood clot in a coronary artery; also known as *coronary artery disease*

coronary occlusion an obstruction of blood flow in one of the arteries that supply blood to the heart muscle; may cause a heart attack

coronary thrombosis an alternative name for HEART ATTACK

corpus a body; the major portion of an organ

corpus albicans a fibrous mass of scar tissue left on the ovary after the CORPUS LUTEUM degenerates

corpus callosum one of the tracts in the white matter of the cerebrum; a large bundle of transverse fibres connecting the CEREBRAL HEMISPHERES

corpus luteum the temporary endocrine gland that forms in the ovary after release of an egg; secretes oestrogens and progesterone; if the egg is fertilised, the corpus luteum remains active in the ovary, continuing to produce hormones that maintain pregnancy; if the egg is not fertilised, the corpus luteum degenerates and becomes the CORPUS ALBICANS; may be called the *yellow body*

corpuscle a small body; used to denote a number of structures; e.g. red corpuscle (RED BLOOD CELL), white corpuscle (white blood cell or LEUCOCYTE), RENAL CORPUSCLE, MEISSNER'S CORPUSCLE, PACINIAN CORPUSCLE

corpuscles of touch an alternative name for MEISSNER'S CORPUSCLES

cortex the outer layer of an organ (e.g. adrenal cortex, cerebral cortex, cortex of the kidney (renal cortex))

corticoid a substance whose function and properties are similar to those of CORTICOSTEROIDS

corticosteroid the steroid hormones released by the adrenal cortex, including the glucocorticoids (such as cortisol) and others (such as aldosterone)

corticosterone one of a group of hormones called GLUCOCORTICOIDS

cortisol one of a group of hormones called GLUCOCORTICOIDS

cortisone one of a group of hormones called GLUCOCORTICOIDS

costal relating to the ribs

costal cartilage the cartilage that is attached to the ribs; *see* diagram below

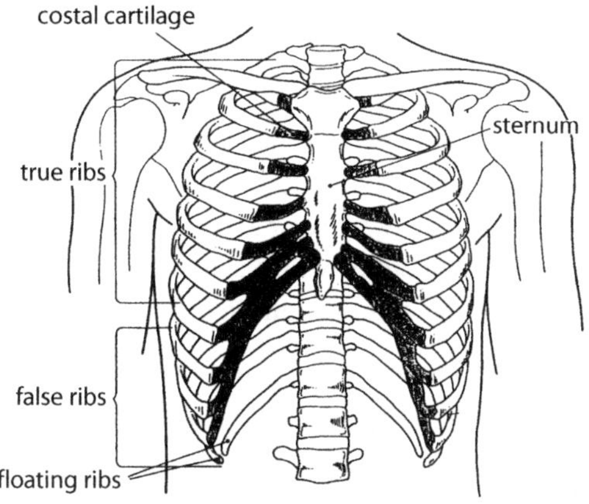

Costal cartilage and ribs

coughing a sudden expulsion of air from the lungs; the rush of air clears the airway of mucus and irritating particles

countercurrent exchange the transfer of heat from a fluid flowing in one direction to fluid flowing in the opposite direction in an adjacent vessel; occurs in the blood vessels of the limbs, where heat is conducted from warmer blood in vessels supplying the hands and feet to the cooler blood that is returning to the heart in nearby veins

countercurrent flow the mechanism in which two different fluids flow in opposite directions; frequently substances, and heat, are exchanged at points of contact along a concentration gradient

courtship behaviour the interactions between members of the opposite sex that take place before mating

covalent bond the form of molecular bonding characterised by the sharing of a pair of electrons between atoms

cover slip a small piece of very thin glass that is placed over a specimen on a microscope slide; also called a *cover slide* or *cover glass*

Cowper's gland an alternative name for BULBO-URETHRAL GLAND

cramp an involuntary, sustained contraction of a muscle, usually painful; may occur during exercise or at rest; precise cause is unknown but may be due to low levels of sodium and chloride ions in the blood

cranial relating to the skull (e.g. cranial capacity), or to the brain (e.g. cranial nerve)

cranial capacity the volume of the part of the skull occupied by the brain; used as an indication of brain size, particularly with fossil skulls

cranial nerve one of the 12 pairs of nerves that arise from the brain and the brain stem; each is designated by a Roman numeral and a name; *see* table of cranial nerves, opposite

cranium the part of the skull that encloses the brain and protects the organs of sight, hearing and balance; also known as the *braincase*

creatine phosphate a high-energy molecule found in muscle; breaks down into creatine and phosphate with the release of energy that is used to convert ADP to ATP; also called *phosphocreatine*

creatinine a nitrogenous waste formed from creatine in muscle and excreted by the kidney

creation the forming of the universe and the production of life by God; special creation is the belief that each species (or genus) was individually created by God

creation science a strategy used by opponents of evolution to try to give scientific credibility to the belief of special creation; those who believe in creation science are known as creationists

creationism the belief that all matter, including the various forms of life, the world and the universe, were created by God, usually in the way described in the Bible

cremation the shrivelling of a red blood corpuscle as a result of the withdrawal of water (e.g. if it is placed in a solution with a high concentration of dissolved substances)

Cretaceous period the youngest of three periods of geological time in the Mesozoic era; 145–65 million years ago; the supercontinent, Gondwana, of which the Australian plate was a part, began to break up during the Cretaceous period; *see* geological time scale (p. 111)

cretinism a condition resulting from undersecretion of thyroxine (the hormone from the thyroid gland) during development, characterised by slow growth, retarded mental development and retarded sexual development; can be due to lack of iodine in the diet, since thyroxine contains iodine

cri du chat syndrome a rare genetic disorder caused by a missing part of chromosome number 5; infants with the syndrome make a sound like a mewing kitten, hence the name, which is French for 'cry of the cat'

cricoid cartilage one of the cartilages of the LARYNX

crista (plural *cristae*) **1.** any crest or ridged structure **2.** a small elevation in the ampulla

Table 3 *A summary of the cranial nerves*

Nerve		*Function*
I	olfactory	sensory—smell (olfaction)
II	optic	sensory—vision
III	oculomotor	motor—movement of eyelid and eyeball; accommodation of lens; constriction of pupil sensory—amount of stretch in eye muscles
IV	trochlear	motor—movement of eyeball sensory—amount of stretch in eye muscles
V	trigeminal	motor—chewing sensory—touch, pain and temperature from cornea, skin of nose, forehead, scalp, nasal cavity, palate, teeth, cheek, lips
VI	abducens	motor—movement of eyeball sensory—amount of stretch in eye muscles
VII	facial	motor—muscles of facial expression; secretion of saliva and tears sensory—taste from two-thirds of tongue; amount of stretch in facial muscles
VIII	vestibulocochlear	sensory—hearing, equilibrium (balance)
IX	glossopharyngeal	motor—swallowing, saliva secretion sensory—taste from rear one-third of tongue; regulation of blood pressure; amount of stretch in muscles involved in swallowing
X	vagus	motor—muscles of pharynx and larynx; muscles of the viscera sensory—skin of external ear; taste from rear of tongue; sensations from viscera
XI	accessory	motor—swallowing; head movements sensory—amount of stretch in muscles involved in swallowing
XII	hypoglossal	motor—tongue movement and swallowing sensory—amount of stretch in muscles involved in tongue movements and swallowing

of each semicircular canal that serves as a receptor for balance and equilibrium **3.** a fold of the inner membrane of a MITOCHONDRION

critical period the specific developmental stage during which responses are easily learned (imprinting) and after which may not be learned; in humans, there are critical periods for learning certain social behaviours

Crohn's disease chronic inflammation of a part of the digestive tract, usually the lower part of the small intestine

Cro-Magnon people the first anatomically modern humans found in Europe; noted for the manufacture of advanced tools, particularly blades; lived during the upper Palaeolithic period (40 000–10 000 BP)

cross **1.** any mating between two organisms that produces offspring **2.** a hybrid offspring produced from the mating of two unlike parents

crossing over a process that may occur during meiosis; like (homologous) chromosomes, each consisting of a pair of chromatids, may become entangled and pieces of chromatid are exchanged; causes a change in the sequence of alleles along a chromosome; by permitting an exchange of alleles among chromatids, crossing over is one factor that results in genetic variation; the point where crossing over occurs is called a CHIASMA (plural *chiasmata*); *see also* gene linkage, recombination, diagram below

cross-sectional studies studies of human characteristics, frequently growth, that measure samples of people of different age-groups at a single point in time (e.g. studying the average height of girls by measuring the heights of all 6-year-old females, all 7-year-old females, all 8-year-old females and so on at a particular school on the one day); *see also* longitudinal studies

crown the portion of a tooth that projects above the gum; *see also* diagram of tooth (p. 267)

crown–rump length the length of a FOETUS from the top of the head to the bottom of the buttocks; used to estimate the age of the foetus

cruciate ligament a cross-shaped LIGAMENT; the best known one occurs in the knee joint

cryoprecipitate a frozen blood product that contains blood-clotting factors; used to treat bleeding disorders

cryosurgery a procedure in which abnormal TISSUE is frozen and destroyed, often using a probe containing liquid nitrogen

cryptorchidism a defect during human development in which the testes fail to descend into the scrotum

CT scan a method of TOMOGRAPHY

cuboidal epithelium epithelium with cube-shaped cells; *see* epithelial tissue

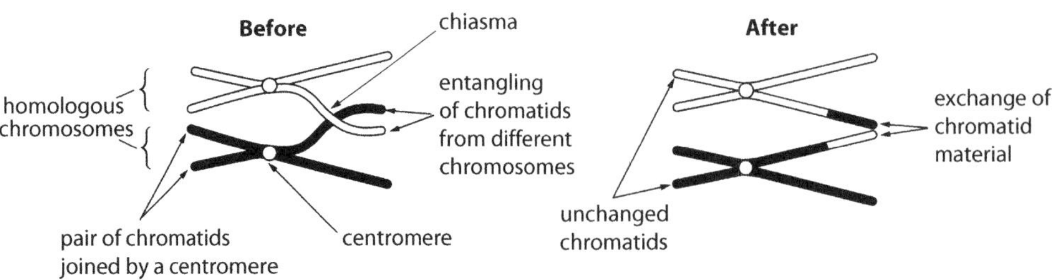

Crossing over

cultural anthropology a branch of ANTHROPOLOGY that studies variations in culture among human populations

cultural diffusion the spread of any aspect of culture from one population to another

cultural period a period during history or prehistory that can be identified by a particular type of cultural development such as a particular form of stone tool or a particular type of art

culture **1.** learned behaviour and knowledge, shared by members of a society, that is passed on from generation to generation; human culture is often distinguished by language and divisible into material culture (objects and physical conditions) and non-material culture (behaviour, belief and understanding) **2.** a growth of micro-organisms in, or on, a specially prepared medium

culture medium a substance on or in which micro-organisms or tissues can be grown; contains all the nutrients necessary for normal growth; may be solid (jelly) or liquid; solid media usually contain agar; often referred to as a *medium*

culture plate an alternative name for a PETRI DISH

cupula a dome-like structure, especially the mass of gelatinous material that covers the hair cells in the ampullae of the inner ear; when the head moves, endolymph in the semicircular canals of the ear flows and pushes on the cupula so that the sensory hairs are bent; the stimulated hairs send impulses to the brain so that equilibrium can be maintained

Cushing's syndrome a condition caused by oversecretion of glucocorticoids (especially cortisol and cortisone) from the adrenal cortex; fat is redistributed so that sufferers have spindly legs, a very round 'moon' face, an enlarged abdomen, flushed facial skin and poor wound healing

cusp a raised area on the chewing surface of a tooth

cutaneous relating to the skin

cuticle the thickened layer of dead skin at the base of a fingernail or toenail

cyanobacteria PROKARYOTES that contain a blue pigment in addition to the green pigment chlorophyll; chlorophyll enables the process of PHOTOSYNTHESIS to occur

cyanosis a slightly bluish, or dark purple, discolouration of the skin and mucous membranes that is due to an oxygen deficiency; seen in blue babies, heart attack victims, cases of carbon monoxide poisoning, extreme cold and other conditions

cyclic adenosine monophosphate (cAMP) a compound through which many hormones (such as adrenalin, antidiuretic hormone and thyroid stimulating hormone) exert their effect; cAMP is formed from ATP when the hormone attaches to its receptor at the cell membrane of the target cell; cAMP then brings about the response indicated by the hormone

cyclo-oxygenase (COX) a protein that acts as an enzyme; speeds up the production of PROSTAGLANDINS, hormone-like substances that regulate INFLAMMATION

cyclosis the circulation of CYTOPLASM inside cells; also referred to as *cytoplasmic streaming*; one reason for such streaming is for movement of the cell (e.g. white blood cells)

cyst a sac with a distinct wall containing fluid or some other material; some cause medical problems; others are quite harmless; some arise as outgrowths of the skin due to a blocked duct (e.g. a sebaceous cyst); others are internal growths (e.g. OVARIAN CYST)

cysteine one of the 20 amino acids that are common in proteins; not essential in the human diet; *see also* list of amino acids (p. 11)

cystic fibrosis (CF) a disorder controlled by a RECESSIVE allele carried on an AUTOSOME; mucus-secreting glands, particularly in the lungs and pancreas, become fibrous and produce abnormally thick mucus; results in chest infections, a lack of digestive enzymes and increased salt loss; incurable, but can be detected during foetal development

cystitis inflammation of the urinary bladder

cyt-, cyto-, -cyte prefix or suffix meaning cell (e.g. cytology—the study of cells; leucocyte—a white blood cell)

cytochrome an iron-containing protein that is important in cellular respiration; one form is cytochrome c, a UBIQUITOUS PROTEIN that is central to the process of RESPIRATION in MITOCHONDRIA; cytochrome alternates between a reduced form and an oxidised form and hence is important in the electron transport system of cellular respiration

cytokinesis division of the cytoplasm of a cell into two parts; occurs after nuclear division (mitosis or meiosis)

cytology the branch of biology that deals with the study of cells

cytoplasm the contents of a cell, excluding the nucleus; between the cell membrane and the nucleus; consists of the cytosol, organelles and inclusions; *see also* diagram of cell structure (p. 44)

cytoplasmic streaming an alternative term for CYCLOSIS

cytosine one of the 4 bases that form part of the DNA and RNA molecule; *see* deoxyribonucleic acid, ribonucleic acid

cytoskeleton a complex network of microtubules and microfilaments in the cytoplasm of a cell that is thought to give the cell its shape and assist cell movement and intracellular transport

cytosol the soluble part of cell cytoplasm; does not include any particles

cytotoxic T cell alternative name for KILLER T CELL

D

Darwinius masillae a 47 million-year-old fossil primate first described and classified in 2009, nicknamed Ida, but scientifically named *Darwinius masillae* in honour of Charles Darwin's 200th birthday year. It is believed to be the most complete and best preserved primate fossil ever uncovered; originally discovered by an amateur fossil hunter in 1983 at Messel pit, a world-renowned fossil site near Darmstadt in Germany but not described and classified until 2009. It is considered to be an early example of the primate lineage that diversified into monkeys, apes and hominins; this prehistoric primate is a young female with an opposable thumb and nails on the fingers and toes; her teeth are similar to those of monkeys and her forward facing eyes would have enabled stereoscopic vision; the talus (or heel) bone has the same shape as that of modern humans

data observations and measurements (e.g. data are the observations or measurements made during an experiment)

dating determining the age of a fossil or artefact; *absolute dating* or *chronometric dating* is the determination of the actual age of material by use of a dating method such as radiocarbon or potassium-argon dating; *relative dating* determines whether one object is older or younger than another; many methods of absolute dating depend on measuring the level of radioactive elements in the sample to be dated—since each radioactive element decays at a known rate, the age of the sample can be determined; such methods are known as *radioactive dating* (or *radiometric dating*); *see also* methods of dating by name, principle of superposition

daughter cells the new cells that are produced when a cell divides

dead air volume an alternative name for DEAD SPACE

dead space that part of the respiratory system where GAS EXCHANGE does not occur; *see* lung volumes

deafness the inability to hear, or significant hearing loss; may be *nerve deafness*, which is due to defects in the sound receptors or in the nerve pathways involved in hearing; or may be *conductive deafness*, due to a fault in the conduction of sound vibrations through the outer or middle ear

deamination the removal of the amino group (NH_2) from an organic compound, especially from an amino acid molecule; the remainder of the amino acid molecule can then enter the KREBS CYCLE; occurs in the liver, where *ammonia* (NH_3) is formed before being converted to urea for excretion; *see also* transamination

death the permanent end of all life functions in an organism

death rate of a population, the number of deaths in a year for every 1000 people in the population at the middle of that year

deciduous teeth TEETH that are shed and replaced by permanent teeth

decomposers organisms, such as some bacteria and fungi, that convert dead organic matter into inorganic materials; important in the cycling of matter because the conversion of organic matter to inorganic matter makes it available for use by plants and so the cycle continues

decussation the crossing of nerve fibres from one side of the central nervous system to the other; the two sets of fibres form an X shape

deep ecology an ecological point of view in which humans are seen as no more or less important than any other species; that is, all species have a right to exist; the opposite to an anthropocentric view of environmental issues; *see* anthropocentrism

deep vein thrombosis (DVT) formation of a blood clot in one of the deep veins, often in the leg; causes pain and swelling; may be fatal if the clot dislodges and travels to the lungs; may result from inactivity and lower air pressure, such as when on a long flight

defecation the elimination of solid wastes (FAECES) from the RECTUM; also spelt *defaecation*

deglutition the act of swallowing

deficiency disease a DISEASE resulting from a dietary deficiency of a specific NUTRIENT, especially a VITAMIN or MINERAL; may result from insufficient intake, digestion, absorption or utilisation of the nutrient; *see also* Table 7 on page 170 (minerals) and Table 9 on page 289 (vitamins)

dehydration an excessive loss of water and accompanying salts from the body; results when the body loses more fluid than it takes in

deleterious allele an allele that greatly reduces the chances of survival, or chance of reproduction, of any organism that inherits the allele (e.g. the allele for thalassaemia; *see* anaemia)

deletion a type of CHROMOSOMAL MUTATION where part of a chromosome is lost

delta cells cells in the islets of the PANCREAS that secrete growth hormone inhibiting factor

deme a strictly local population composed of individuals who live together in a common habitat and who usually choose their mates from among their own members; genetically isolated from other members of the species because there is little mating outside the group (e.g. certain religious groups in the United States such as the Dunkers and Hutterites)

dementia deterioration of intellectual and other mental processes; the progressive loss of the ability to think clearly; can be caused by a brain, tumour, disease of the blood vessels that supply the brain, or nutritional deficiencies, but the most common cause is Alzheimer's disease

demographic transition the change in population structure that occurs as a country becomes industrialised; characterised by decline in death rate followed by decline in birth rate; as countries become industrialised, the time delay between decrease in death rate and decrease in birth rate results in a period of rapid population growth; when the country is fully industrialised, birth and death rates are approximately the same and population stabilises; recent research suggests that there may not necessarily be a single path of population development as assumed by the idea of demographic transition

demography the study of the size, composition and distribution of populations

denaturation changing the three-dimensional shape of a molecule so that its physical and biological properties are changed; may be caused by excessive heat, strong acids or alkalis or organic solvents; usually refers to proteins and is usually not reversible (e.g. heating of egg white, which is the protein albumen, produces solid egg white)

dendrite an extension from the body of a nerve cell that carries nerve impulses into the cell body; dendrites are highly branched and there are usually several main dendrites to each nerve cell; also called *dendrons*; *see also* axon, diagram of nerve cell (p. 187)

dendrochronology determining the age of wood by counting the annual growth rings in the timber; variation in width of the annual growth rings also indicates changes in climate; by correlation with living trees and by progressive correlation with ancient timbers, the ages of some wooden remains can be precisely dated as far back as 9000 years; also called *tree ring dating*

dendron an alternative name for a DENDRITE

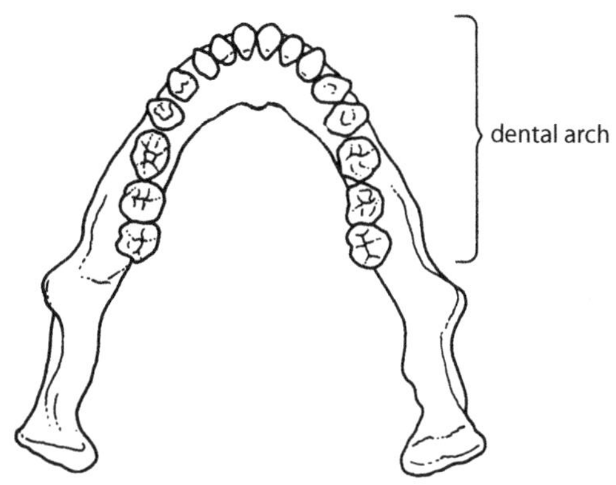

Dental arch

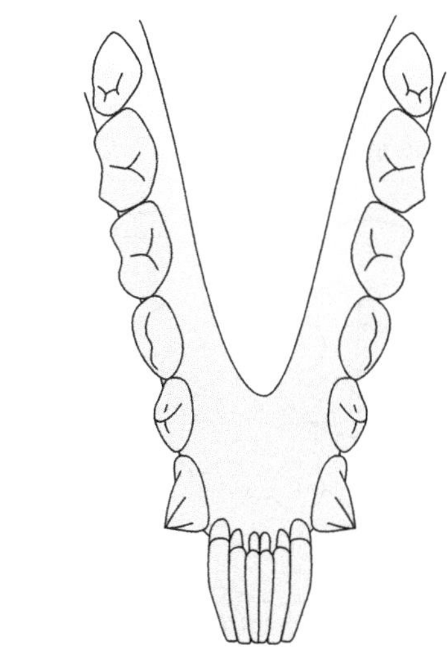

Dental comb of a lemur

dense bone an alternative name for compact bone; *see* bone

dense connective tissue a connective tissue with a very large number of fibres, such as ligament, tendon or fibrous membranes

dental arcade an alternative name for DENTAL ARCH

dental arch the shape of the pattern made by the teeth as they are set in the jaw (e.g. apes have a U-shaped dental arch, humans have an arch that is parabolic in shape); also called *dental arcade*; *see* diagram previous page

dental caries tooth decay; *see* caries

dental comb a structure formed when the front teeth of the lower jaw project forward almost horizontally; found in PROSIMIANS; used for GROOMING the fur

dental formula a shorthand way of listing the number and type of TEETH present, usually in half of each jaw; upper jaw numbers are shown over the lower jaw numbers; the human dental formula for the permanent teeth is:

$$\frac{2.\ 1.\ 2.\ 3.}{2.\ 1.\ 2.\ 3.}$$

meaning 2 incisors, 1 canine, 2 premolars and 3 molars in each half of the upper jaw and in each half of the lower jaw

dental plaque a mass of bacterial cells, carbohydrate and other debris that sticks to teeth and may lead to tooth decay

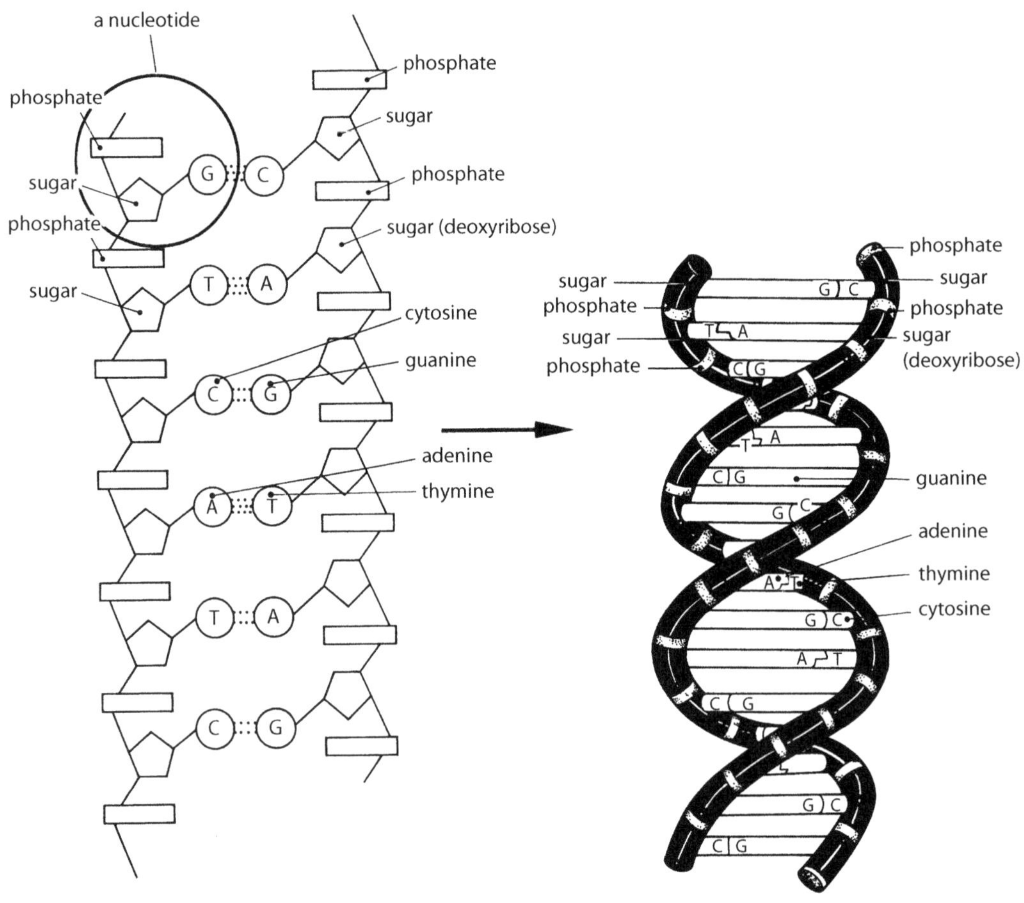

Structure of DNA

dentine the bone-like tissue that makes up most of a tooth and is under the enamel of the tooth; *see also* diagram of tooth (p. 267)

dentition the type, number and arrangement of TEETH in the mouth

deoxygenated containing a reduced amount of oxygen (e.g. deoxygenated blood is blood that has just been through the capillaries of the body and contains a relatively low proportion of oxygen, because blood is deoxygenated during its passage through the body tissues); *see also* oxygenated

deoxygenated blood blood that has passed through the tissues and thus carries a reduced amount of oxygen

deoxyribonuclease an enzyme that breaks down DNA; *see* nuclease

deoxyribonucleic acid (DNA) the chemical compound that carries the inherited information in a cell; found mostly in the nucleus; the compound of which genes and chromosomes are made; formed from subunits called NUCLEOTIDES; each nucleotide consists of a sugar (deoxyribose), a phosphate group and one of four bases (containing nitrogen); the four bases in DNA molecules are *adenine*, *thymine*, *guanine* and *cytosine*; a DNA molecule consists of a double strand of nucleotides joined by bonds between the nucleotide bases and twisted into the form of a DOUBLE HELIX; bonds between the bases will only form between adenine and thymine and between cytosine and guanine; this is important for making exact copies of a DNA molecule, a process known as REPLICATION; the order of the bases in the DNA molecule is the *genetic code*—the information required for the manufacture of specific proteins in a cell; *see also* protein synthesis, DNA diagram, previous page

deoxyribose a 5-carbon (pentose) sugar that is part of the DNA molecule; *see also* deoxyribonucleic acid

dependence a term used in reference to drug use, indicating inability to function normally without a drug; *see* drug dependence

dependent variable a VARIABLE that changes in response to changes in the independent variable

depolarisation movement of sodium ions into a nerve cell so that the difference in electrical charge (potential difference) between the inside and outside of the cell membrane (the membrane potential) is reduced; such a cell is then said to be *depolarised*; when depolarisation has occurred, an ACTION POTENTIAL (or nerve impulse) has been produced; *see also* polarised membrane, diagram below

depolarised describes a situation where there is a reduced difference in electrical charge between the inside and outside of the membrane of a nerve cell; *see* depolarisation

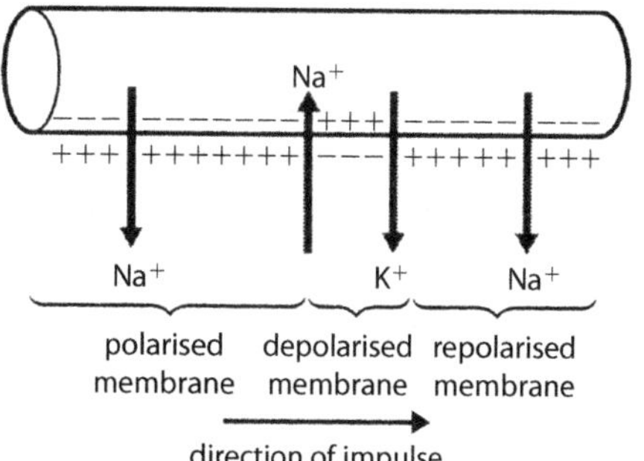

Depolarisation

Depo-Provera a hormonal contraceptive; contains the synthetic hormone progestin and must be injected every three months

depressed fracture a break of a bone in which the injured area of bone is forced inwards; *see* fracture

depression a mental illness in which a person has feelings of profound sadness or despair

derived trait a characteristic that has changed from its ancestral state (e.g. the large brain possessed by modern humans is a derived trait relative to the ancestral stock of apes and humans); also called a *derived character*

dermis the part of the skin below the EPIDERMIS; thicker than the epidermis and composed of connective tissue containing fibres; thickest on the palms and soles, very thin on the eyelids and penis; contains blood vessels, nerves, hair follicles and glands; *see also* diagram of skin structure (p. 250)

descending tract a bundle of motor fibres in the central nervous system; *see* tract

designer drug a drug synthesised in illegal laboratories that has similar properties to drugs that are derived from natural sources

detox abbreviation for DETOXIFICATION

detoxification removal of harmful substances from the body, such as when withdrawing from alcohol or other drugs

developed country a country with an industrial economy; often referred to as an industrialised country; the term is not favoured by cultural anthropologists; *see* development

developing country a country in the early stages of developing an industrial economy; the term is not favoured by cultural anthropologists; *see* development; may also be referred to as an *underdeveloped country* or *Third World country*

development **1.** the series of events that lead to the formation of an adult organism from a fertilised egg **2.** may be used more broadly to refer to all the age-related changes that occur to a person between fertilisation and death **3.** of a country, the change from an agricultural to an industrial economy; use of the term is declining because it implies that cultural development is unidirectional and suggests the progress and improvement of society

Devonian period one of 7 periods of geological time in the Palaeozoic era, between the Silurian and Carboniferous periods; 416–359 million years ago; *see also* geological time scale (p. 111)

diabetes a common name for DIABETES MELLITUS

diabetes insipidus a disorder in which the affected individual passes very large quantities of urine and gradually becomes dehydrated; caused by undersecretion of antidiuretic hormone (ADH)

diabetes mellitus a group of diseases, all of which result in an abnormally high level of glucose in the blood and excretion of glucose in the urine; may also be called *sugar diabetes*; *type 1 diabetes*, or *insulin-dependent diabetes*, develops rapidly, usually before the age of 20, and may also be called *juvenile-onset diabetes*; it is caused by a decline in the insulin-producing cells of the pancreas and is treated by injection of insulin at regular intervals; *type 2 diabetes*,

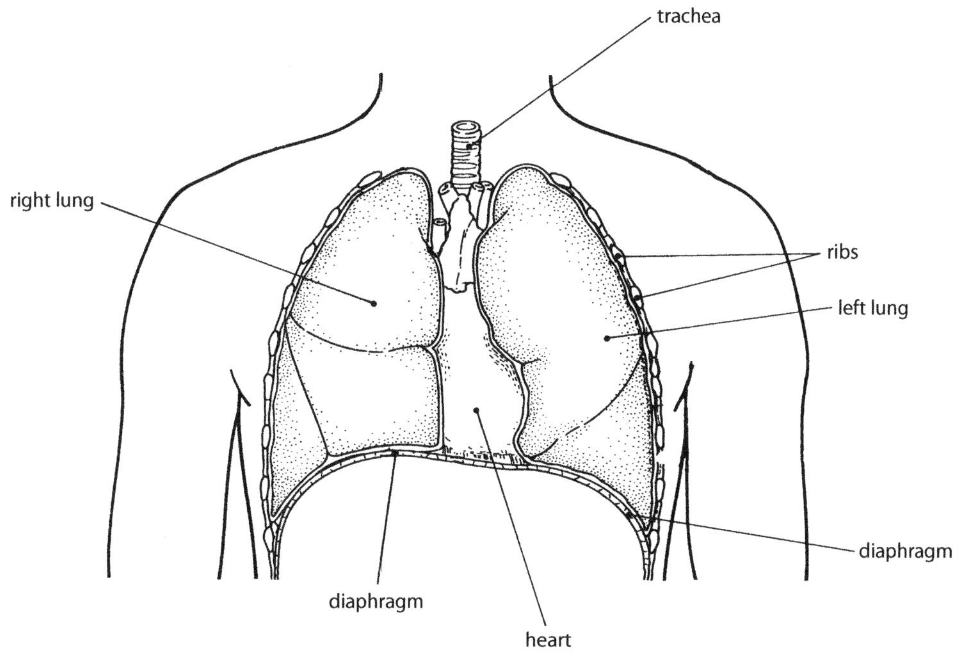

The human diaphragm

or *adult-onset diabetes*, is much more common and usually occurs in people over 40 who are overweight; symptoms are often mild and the high blood glucose level may be controlled by diet

diacylglycerol an alternative name for *diglyceride*; *see* fat

diagnosis the process of deciding the nature of a disease from its signs and symptoms, using observation, laboratory tests, X-rays or other means; also, the result of that process, i.e. the decision about the nature of the disease

diakinesis the final stage of prophase I of MEIOSIS; during diakinesis the chromosomes are fully contracted, the nuclear membrane breaks down, the nucleolus disappears and a spindle forms

dialysis a medical procedure that is used to remove wastes and excess water from the body when the kidneys have ceased to function

diaphragm **1.** the dome-shaped skeletal muscle between the chest and abdominal cavities that contracts to increase the volume of the chest cavity (and lungs) when breathing in; *see* diagram opposite **2.** a thin rubber cap that is fitted across the top of the vagina before sexual intercourse to prevent fertilisation **3.** a device on a light microscope for controlling the amount of light passing through the specimen

diaphysis the shaft of a LONG BONE; *see also* epiphysis and diagram (p. 90)

diarrhoea frequent passing of liquid faeces; irritation of the alimentary canal causes increased movement of the intestines so that there is not enough time for water absorption in the large intestine

diastema a gap in a row of teeth; usually refers to a gap next to the canine teeth that occurs in primates with canine teeth that are much longer than the other teeth; allows space for the canine on the opposing jaw; humans have no diastema because the canines are approximately equal in length to the other teeth

diastole the period of relaxation of the heart; *see* cardiac cycle

diastolic blood pressure the pressure of blood on the arterial walls while the ventricles are relaxed; *see* blood pressure

diencephalon the part of the brain consisting mainly of the thalamus and hypothalamus; *see also* diagram of brain (p. 35)

diet the food normally consumed; a *balanced diet* is one that contains all essential nutrients in appropriate amounts and the correct amount of energy for body size and level of activity; in everyday use, a diet usually means a reduction in energy intake in an effort to lose weight

dietary supplement a product that contains substances such as VITAMINS, MINERALS and PROTEINS to boost the usual intake of these substances; also known as a *food supplement* or *nutritional supplement*

differentially permeable membrane a membrane that permits the passage of certain substances (usually small molecules) but restricts the passage of others (large molecules); cell membranes are differentially permeable; also called *semipermeable*, *selectively permeable* or *partially permeable membrane*

differentiation of cells, the process by which new cells develop special characteristics to suit particular functions; the formation of the various cell types that make up a plant or animal

diffusion the movement of particles of a liquid or a gas so they are distributed evenly throughout the available space; usually taken to mean the net movement of ions or molecules from a higher to a lower concentration until they are evenly distributed (also called *net diffusion*); diffusion is one way that substances pass across a cell membrane; in some cases carrier molecules in the membrane transport substances across the membrane, a process called *facilitated diffusion*

diffusion gradient an alternative name for CONCENTRATION GRADIENT

dig a common name for EXCAVATION

digestion the mechanical and chemical breakdown of food resulting in small molecules that can be absorbed into the cells of the body

digestive juice a liquid secreted for the mechanical or chemical breakdown of food (e.g. pancreatic juice, gastric juice, intestinal juice, bile)

digestive system the group of organs involved in the digestion of food; the alimentary canal and its associated organs, such as the pancreas, liver and salivary glands; *see* diagram opposite

digestive tract an alternative name for the ALIMENTARY CANAL

digit a finger or toe

diglyceride a FAT with molecules containing two fatty acids

dihybrid an offspring produced by parents each of whom is pure-breeding for two pairs of contrasting characteristics; the dihybrid has two different alleles for one gene

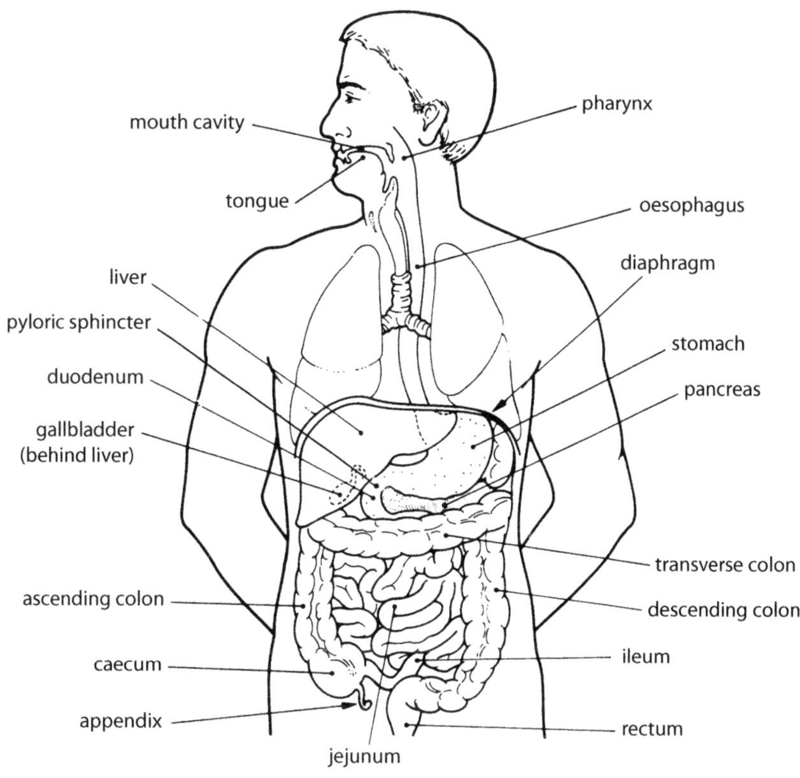

The digestive system

and two different alleles for a second gene (e.g. the genotype RrAa); *see also* mono-hybrid

dihybrid cross a mating between individuals in which two pairs of contrasting characteristics are being considered (e.g. ability to roll the tongue versus non-rolling *and* albinism versus normal pigment production); *see also* monohybrid cross

dilation becoming wider (e.g. dilation of skin blood vessels to promote heat loss; dilation of the CERVIX of the uterus during childbirth)

dimorphism the occurrence of two forms; *see* sexual dimorphism

dipeptidase an enzyme secreted into the small intestine; breaks down dipeptides into individual amino acids

dipeptide two amino acids joined together (by a PEPTIDE BOND); *see also* peptide

diphtheria an infectious disease caused by toxins that are produced by the bacterium *Corynebacterium diphtheriae*; epithelial cells in the throat die and mucous membranes become enlarged and leathery, eventually obstructing air passages and causing death; once common in children but now rare in Australia since a high proportion of children are immunised against the disease

diploid having two of each type of chromosome; in humans all body cells are diploid, but sperm and eggs are HAPLOID; the diploid chromosome number in humans is 46; diploid cells or organisms are represented by the symbol *2n*

diplotene the 4th stage of prophase I of MEIOSIS; during diplotene the like chromosomes begin to separate a little, except at chiasmata, where crossing over of chromatids has occurred

directional selection refers to NATURAL SELECTION; occurs when one extreme in a continuous characteristic is selected rather than the other extreme; that is, the selection is in one direction (e.g. in the event of a long-term decrease in environmental temperature, those members of a species of mammal that have a lower surface-to-volume ratio would be selected because they would lose heat less rapidly)

disaccharide two simple sugar molecules chemically combined; *see* carbohydrate

disc a common name for the cartilage between the vertebrae, correctly called an INTERVERTEBRAL DISC

disease any variation from the normal state of health; may be due to infection, injury, inherited abnormality or other causes

dislocation when a bone is displaced from a joint, often accompanied by damage to surrounding ligaments, tendons and the joint capsule

dissociation **1.** the separation of substances into ions when dissolved in water; also known as *ionisation* **2.** the break-up of a chemical combination into its component parts, such as when oxyhaemoglobin dissociates into oxygen and haemoglobin

distal further from the point of origin, as in 'distal convoluted tubule'; *see also* proximal

distal convoluted tubule the second set of convolutions of the kidney tubule (furthest from the glomerular capsule); receives the forming urine after it has passed through the loop of Henle; as with other parts of the tubule, reabsorbs some substances into the blood and secretes others into the forming urine; passes the urine into a collecting duct; *see also* diagram of kidney nephron (p. 183)

distribution the occurrence of a species, race or population; *geographical distribution* is the part of the earth occupied by a group; *temporal distribution* is the period of time over which a species, race or population existed on the earth (i.e. distribution in time)

diuresis increased or excessive output of urine from the kidneys

diuretic a substance that increases the volume of urine produced by the kidneys; inhibits the secretion of antidiuretic hormone

diurnal active during the day (e.g. chimpanzees are diurnal, feeding and moving during the day and sleeping at night); *see also* nocturnal

diurnal rhythm a 24 hour rhythm that is based on a regular pattern of light and dark; *see also* circadian rhythm

diverticulitis inflammation of diverticulae, which are pouch-like outgrowths from the wall of the colon; diverticulae occur due to weaknesses in the muscular wall of the colon and may become inflamed because material in them is not pushed along with the rest of the matter in the colon

division of labour certain activities in a society being performed by members of particular status such as sex or age (e.g. in many hunting and gathering societies there was a division of labour—the males hunted and the females collected plant material)

dizygotic twins TWINS that develop from two separate zygotes

DNA the abbreviation for DEOXYRIBONUCLEIC ACID

DNA fingerprinting a technique increasingly used in investigation of crimes that uses a person's DNA; it is very useful for identification because it is unique to that particular individual; also, DNA is a robust molecule whose structure can survive drying out, and only trace amounts of material are required for analysis, so that old bloodstains and dried or decomposed tissue may still contain DNA; the amount of variation in the DNA allows banding patterns of DNA fragments to be identified for an individual as a 'DNA fingerprint'; the technique can also be used to establish family relationships, as some of the pattern of bands of a child will appear in the DNA of either the mother or the father

DNA ligase an enzyme that reconnects broken pieces of DNA; used by cells for DNA repair and by molecular biologists for copying genes; originally called *DNA-joining enzyme*

DNA polymerases a group of enzymes that add nucleotides to the separated halves of a DNA molecule during replication; some forms of DNA polymerase have the ability to 'edit' the DNA sequence, checking for and removing any incorrect base together with its attached sugar and phosphate

DNA probe a fragment of DNA that is used to detect a complementary sequence of bases in a given sample of DNA

DNA profiling an alternative name for DNA FINGERPRINTING

DNA sequencing determining the precise order of NUCLEOTIDES in a sample of DNA

dolichocephalic having a long, narrow head; *see* cephalic index

dominance in genetics, where one form (allele) of a gene masks the effect of an alternative form when both are present (as they are in the heterozygote); the characteristic the individual has is determined by the *dominant allele* and is known as the *dominant characteristic*; the dominant allele is not masked by other alleles; e.g. the ability to roll the tongue is a dominant characteristic controlled by a dominant allele (symbolised by *R*); non-rolling is controlled by the recessive allele (*r*); a heterozygous individual (*Rr*) will have the dominant characteristic, that is, will be able to roll the tongue; *see also* recessive

dominance hierarchy the social order within a group of animals whereby the more dominant individuals get the best food and the best mates; also known as the *pecking order*, since such hierarchies are common in bird groups

dominant characteristic one of a pair of contrasting characteristics, controlled by an allele that is not masked by other alleles

donor insemination (DI) use of donated semen for insemination; *see* artificial insemination

dopamine a NEUROTRANSMITTER that occurs in nerve cells in the midbrain; usually has an inhibiting effect; is involved with nerve impulses that bring about large movements of skeletal muscles; also thought to be involved in some emotional responses; Parkinson's disease is caused by a decrease in production of dopamine by certain brain cells

dorsal located on, or near, the back of an animal; towards the backbone (e.g. dorsal surface, which in most animals is also the upper surface but in humans, because we stand erect, is the back surface)

dorsal root one of the two ROOTS that link a spinal nerve to the spinal cord; *see also* ventral root; diagram (p. 255)

dorsal root ganglion a group of nerve cell bodies located in the dorsal ROOT of a spinal nerve

double circulation a circulatory system in which blood is pumped to the lungs, then returns to the heart to be pumped to the rest of the body; *see also* pulmonary circulation, systemic circulation, diagram below

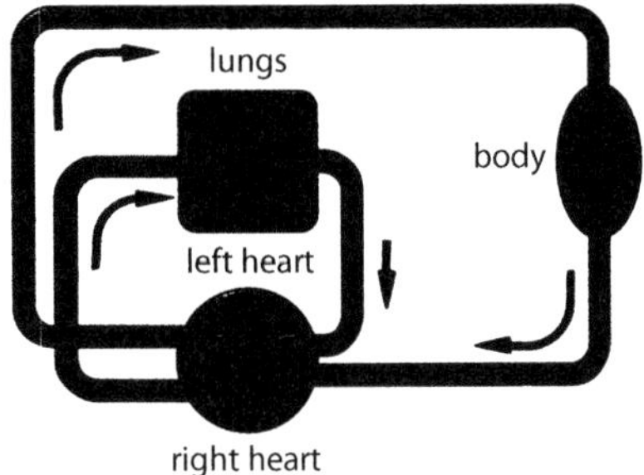

Double circulation

double helix the three-dimensional structure of the DNA molecule; it consists of two polynucleotide chains joined by hydrogen bonds between the nucleotide bases; the whole molecule is twisted into a double helix with a complete turn every 10th base; *see also* diagram showing structure of DNA (p. 72)

double-blind experiment a research procedure in which neither the subject nor the experimenter knows which subjects are in the EXPERIMENTAL GROUP and which are in the CONTROL GROUP; a person not involved with the subjects assigns them to groups so that any conclusions drawn by the experimenter will be free of any bias that he or she may have; widely used in psychological research and in testing medicinal drugs; also known as *double-blind procedure*

doubling time **1.** for human populations, the number of years for a population to double in size (if the current rate of population growth is sustained) **2.** in cultures of micro-organisms, the time for the number of cells in the culture to double

Down's syndrome an alternative name for TRISOMY-21

droplet infection transmission of disease by tiny droplets exhaled from the lungs in breathing, coughing or sneezing; when the moisture evaporates, the bacteria or viruses are suspended in the air and could be inhaled by another person; diseases spread by droplet infection include tuberculosis, diphtheria, whooping cough, measles, mumps, colds and influenza

drug a chemical substance used with the intention of changing the physical or mental functioning of the body; may be medicinal, for preventing, curing or alleviating the effects of disease (e.g. antibiotics, pain-killers), or may be taken because of the effect it has on the nervous system (e.g. alcohol, marijuana)

drug delivery the way in which a particular medicinal drug is introduced into a person's body

drug dependence a state in which a person is so reliant on a drug that continued use is necessary to avoid severe physical or psychological disturbance; *physical dependence* is when the person suffers acute physical discomfort if the drug is not used regularly; *psychological dependence* is when a person believes the drug is essential to normal well-being

dryopithecines the general term used to refer to the very large group of apes that lived in a relatively stable environment 23–12 million years ago (some authorities suggest 17–14 million years ago); covered a large geographical area including Africa, Asia and Europe; often referred to as *Miocene apes*; includes the genus *Proconsul*

Dryopithecus a genus of fossil apes that lived in Africa 23–17 million years ago; more monkey-like than ape-like in appearance, with limbs of roughly the same length; probably ate fruit and lived in trees; more frequently referred to as *Proconsul*

Duchenne muscular dystrophy an inherited wasting disease of the leg muscles and later the arms, shoulders and chest; *see* muscular dystrophy

ductless gland an alternative name for ENDOCRINE GLAND

ductus arteriosus the blood vessel that enables blood in the pulmonary artery of a foetus to bypass the lungs and flow directly into the aorta; necessary because the lungs of a foetus do not function until birth, at which time the ductus arteriosus closes; *see also* diagram of foetal circulation (p. 212)

ductus deferens an alternative name for VAS DEFERENS

ductus venosus the blood vessel in a foetus that enables most of the blood to bypass the liver and flow into the inferior vena cava; *see also* diagram of foetal circulation (p. 212)

duodenal ulcer an ULCER in the first part of the small intestine

duodenum the first part of the small intestine, about 25 cm long; *see also* diagram of digestive system (p. 77)

duplication (mutation) a type of CHROMOSOMAL MUTATION that results in part of a chromosome being duplicated

dura mater one of the membranes covering the central nervous system; *see* meninges

dwarfism a condition in which a person does not grow to normal height; results from the under-secretion of growth hormone from the anterior lobe of the pituitary gland; bones and other tissues fail to grow at the normal rate; also referred to as *pituitary dwarfism*

dynamic equilibrium a state reached when the rates of forward and reverse reactions are equal; the system is dynamic because the individual components react continuously; it is in equilibrium as no net change occurs; a stable, balanced or unchanging system results; frequently referred to as a STEADY STATE

dysentery a severe disease of the last part of the small intestine (the ileum) and the colon; the mucosa becomes ulcerated and inflamed, resulting in abdominal cramps, fever and severe diarrhoea; faeces often contain blood; caused by certain amoebic or bacterial infections

dysmenorrhoea painful menstruation caused by forceful contractions of the uterus; often accompanied by headache, fatigue, nausea, vomiting and diarrhoea

dyspepsia the medical term for INDIGESTION

dysphoria a feeling of depression or distress; the opposite of EUPHORIA

dysplasia abnormal changes in the size, shape or organisation of adult cells; may precede cancer

E

ear the receptor for sound and also for balance and equilibrium; divided into three regions—the outer, middle and inner ear; *see* diagram below; *see also* parts of the ear by name

eardrum the common name for TYMPANIC MEMBRANE

ECG the abbreviation for ELECTROCARDIOGRAM; *see also* electrocardiograph

eclampsia a condition that occurs if TOXAEMIA of pregnancy is untreated

ecstasy an illegal hallucinogenic drug sold on the street as small white or

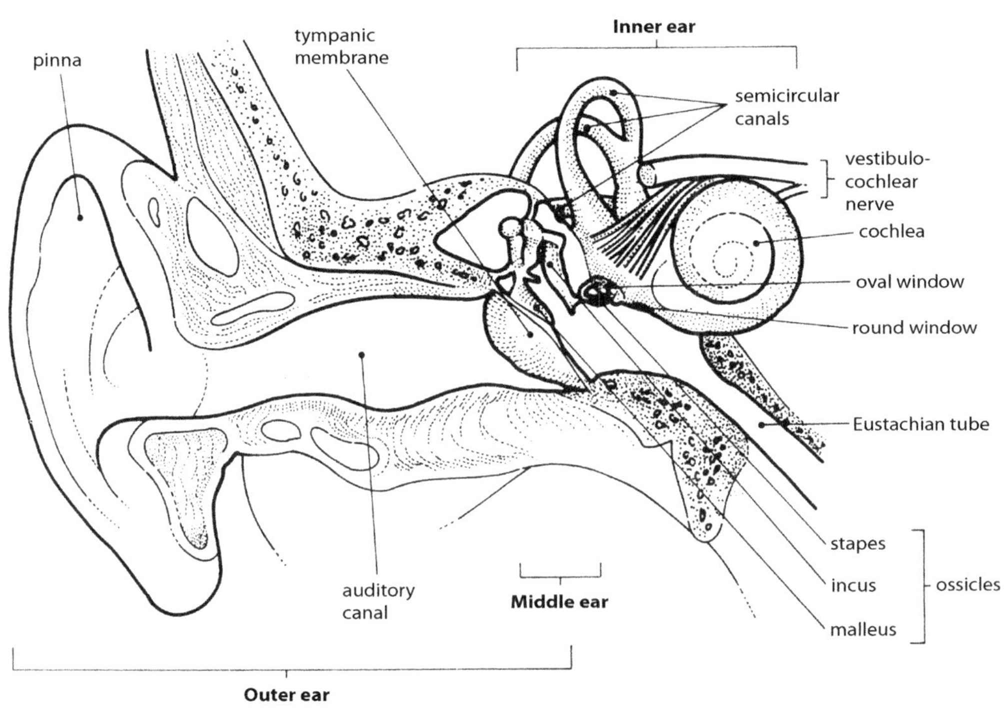

The ear

yellow tablets; also known as MDMA or MethyleneDioxyMethAmphetamine; has a chemical structure similar to both hallucinogens and amphetamines

ectoderm one of the PRIMARY GERM LAYERS

ectomorph a human body type; *see* somatotype

ectoparasite an external PARASITE

ectopic pregnancy a pregnancy in which the embryo implants at a site other than in the lining of the uterus; most such pregnancies occur in a uterine tube but may occur in the ovary, abdomen or cervix

eczema a skin disorder in which the skin becomes thickened, scaly and itchy

edema an alternative spelling for OEDEMA

Ediacaran the earliest of the 7 periods of geological time in the Palaeozoic era; from 620 to 542 million years ago; named after the Ediacara Hills of South Australia; *see* geological time scale (p. 111)

EEG the abbreviation for ELECTROEN-CEPHALOGRAM

effector a muscle or gland that carries out a response to a stimulus

effector neuron an alternative name for MOTOR NEURON

efferent a term referring to conducting or carrying out of an organ, used in reference to nerves and blood vessels (e.g. an efferent nerve carries nerve impulses *out of* an organ); *see also* afferent

efferent arteriole the blood vessel that leaves the glomerulus of the kidney; it has a smaller diameter than the AFFERENT ARTERIOLE and thus helps to raise blood pressure in the glomerulus; *see also* diagram of kidney nephron (p. 183)

efferent division the part of the peripheral nervous system that carries impulses *out of* the brain and spinal cord; *see* nervous system

ejaculation the process of expelling semen from the penis; muscular contractions of the wall of the urethra propel the semen from the penis

elastic cartilage CARTILAGE containing elastic fibres

elasticity the ability of MUSCLE FIBRES to return to their original length after being stretched

elastin a PROTEIN that makes up elastic fibres; occurs in the skin and in tissues that form the walls of blood vessels

electrocardiogram (ECG) a record of the electrical changes that occur during the beating and relaxation of the heart; *see* diagram below

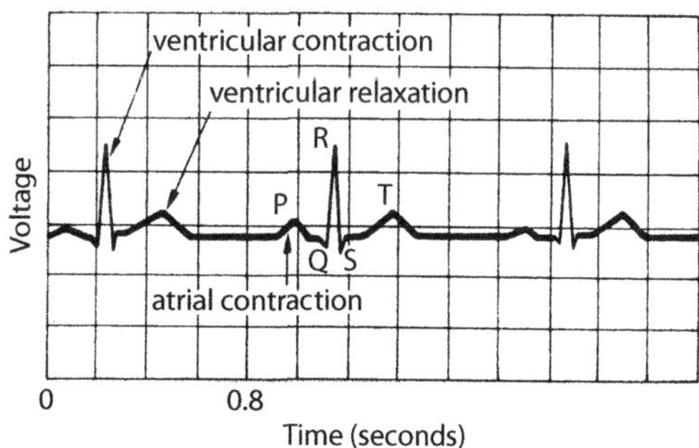

Electrocardiogram

electrocardiograph (ECG) an instrument used to obtain an ELECTROCARDIOGRAM; electrodes are attached to the skin of the arms, legs and chest to record the small changes in voltage that occur as the heart beats; used to help diagnose heart disease

electrochemical change the change in electrical voltage that is brought about by changes in the concentration of IONS inside and outside the cell membrane of the NEURON

electroencephalogram (EEG) a record of the electrical activity in the brain; electrical potentials known as *brain waves* occur in the cerebral cortex and can be detected by sensors at the surface of the skull; sensors are placed on the head to detect the electrical waves, which are then amplified by a machine called an electroencephalograph; the machine produces a trace of the waves on paper or on a screen; the trace is an electroencephalogram; particularly useful in the diagnosis of epilepsy

electroencephalograph (EEG) a machine that records electrical activity in the brain; also known as *brain electrical activity mapping* (*BEAM*); *see* electroencephalogram

electrolyte a compound that has at least one IONIC BOND so that when it dissolves in fluid it dissociates into positive and negative ions; acids, bases and salts are electrolytes; most electrolytes are inorganic but a few, such as some proteins, are organic; many electrolytes are essential minerals; they also control osmosis of water between body compartments and they help maintain the pH required for cell activities

electromagnetic receptor an alternative name for PHOTORECEPTOR

electron a negatively charged particle that is in motion around the nucleus of an atom

electron micrograph *see* electron microscope

electron microscope a microscope that uses a beam of electrons instead of light rays and electromagnets instead of glass lenses; images are of much greater magnification and resolution than those of a light microscope, thus providing more detail; the *transmission electron microscope* passes electrons through the specimen and focuses them on a fluorescent screen or photographic plate; the *scanning electron microscope* produces a beam of electrons that systematically scans the object examined, producing three-dimensional views of the surface of the object; the photograph produced from an electron microscope is called an *electron micrograph*; *see also* resolution

electron spin resonance a signal produced by excited electrons used in absolute dating of fossils and artefacts; natural radiation in soil excites electrons in the crystalline structure of buried objects; the longer an object remains buried, the greater is the accumulation of excited electrons; signal strength is a measure of the time since the electrons were raised to their excited state; *see also* thermoluminescence

electron transfer the exchange of electrons between molecules in a chemical reaction; *reduction* occurs when a molecule loses or donates an electron; *oxidation* occurs when a molecule accepts an electron

electron transport system a series of chemical reactions occurring in the mitochondria of a cell; energy from carrier molecules is transferred to ATP for storage; oxygen is required for the process and water is produced; also called the *electron transport chain*; *see also* Krebs cycle

electrophoresis a process in which large charged molecules in a solution are made to migrate by passing an electric current through the solution; used in particular to separate proteins—the charged particles move through a supporting medium (such as filter paper or a gel) at differing rates depending on the size, shape and charge of the molecule

element a substance made up of atoms that are all of the same type and cannot be chemically broken down to simpler substances (e.g. carbon, hydrogen, oxygen, nitrogen, phosphorus, sulphur, calcium)

elimination the expulsion of undigested food remains from the anus; also called *defecation*; should not be confused with EXCRETION

ellipsoid joint an alternative name for a CONDYLOID JOINT

embolism an obstruction or closure of a blood vessel by debris or foreign material that has been transported by the blood; *see also* embolus

embolus foreign material or debris that is transported by the blood (e.g. a blood clot, a bubble of air, fat from broken bones, a mass of bacteria); can obstruct the flow of blood in a blood vessel; *see also* embolism

embryo the early stages of development of an organism; in humans the embryo is the stage up to 8 weeks after fertilisation, after which it is called a FOETUS; in other animals the embryo represents all the stages of development until birth or hatching; *see also* pre-embryo

embryoblast alternative name for INNER CELL MASS

embryology the study of the early development of an organism; in humans, from fertilisation to the end of the 8th week of pregnancy

embryonic disc the group of cells in the very early embryo that develops into the PRIMARY GERM LAYERS; *see* diagram below

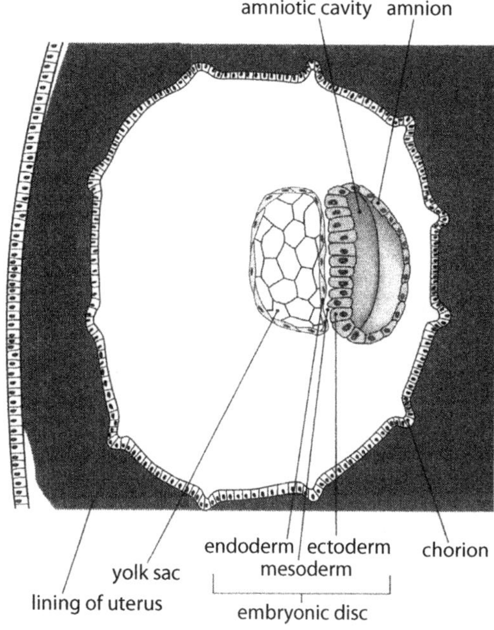

Embryonic disc

embryonic membranes an alternative name for FOETAL MEMBRANES

embryonic period the first 8 weeks of pregnancy, during which time the developing individual is referred to as an *embryo*

embryonic stem cells *see* stem cells

emetic a substance given to induce vomiting, often after consumption of a poison; a well known emetic is syrup of ipecac (or ipecacuanha)

emigration the movement of individuals out of a population (e.g. the emigration of people from South-East Asia to Australia to found the Australian Aboriginal people)

emphysema a lung disorder in which the walls of the air sacs lose their elasticity and begin to break down, thus reducing the surface available for exchange of gases; caused primarily by smoking

emulsification the process of forming an emulsion—a suspension of very fine droplets of one liquid in another liquid; bile is important in the emulsification of fats in the small intestine

enamel a hard, white material covering the crown of a tooth; *see also* diagram of tooth (p. 267)

encephalitis inflammation of the brain; most frequently caused by the invasion of the nervous tissue by viruses, but may be due to other agents such as bacteria and fungi; may result in paralysis, coma and, in extreme cases, death

endemic a disease that is always present in a given country; new cases occur at a relatively constant but low rate over time (e.g. cholera is endemic to South-East Asia)

endergonic reaction a chemical reaction that requires energy to begin the reaction (e.g. the building of proteins from amino acids); *see also* activation energy, exergonic reaction

endo- a prefix meaning within (e.g. endoparasite—an internal parasite)

endocast an impression of the inside of the brain case made of rock or some other solid material; may show the size, shape and some of the surface details of the brain; some fossils are natural endocasts but endocasts can also be made from fossilised or unfossilised skulls

endocrine disrupters synthetic chemicals that the body mistakes for products of endocrine glands; can interfere with growth, development, intelligence, behaviour and reproduction; examples are PCBs and DDT; also called *hormone disrupters*

endocrine gland a gland that secretes into the extracellular fluid surrounding its cells, so that the secretion may then pass into the capillaries to be transported by the blood (e.g. thyroid gland, pituitary gland, testes, ovaries, adrenal gland); also called *ductless gland*; *see* table of hormones (p. 128) for names of endocrine glands; *see also* exocrine gland

endocrine system the body system involved in chemical communication between cells, made up of endocrine glands; *see* table of hormones (p. 128) for names of endocrine glands

endocrinology a field of study concerned with the structure and functions of the endocrine glands, and the diagnosis and treatment of disorders of the endocrine system

endocytosis the process by which a cell takes in materials by enfolding and enclosing them to form a membrane-bound vesicle; includes *phagocytosis* (the engulfing of solid particles, particularly the ingestion of micro-organisms, other foreign matter and cell debris) and *pinocytosis* (taking in liquid droplets); *see* diagram opposite

endoderm one of the PRIMARY GERM LAYERS

endogamy **1.** marriage or mating between persons who are closely related (e.g. first cousins); such a marriage would be described as *endogamous* **2.** marriage or mating between persons belonging to the same social group as required by law or

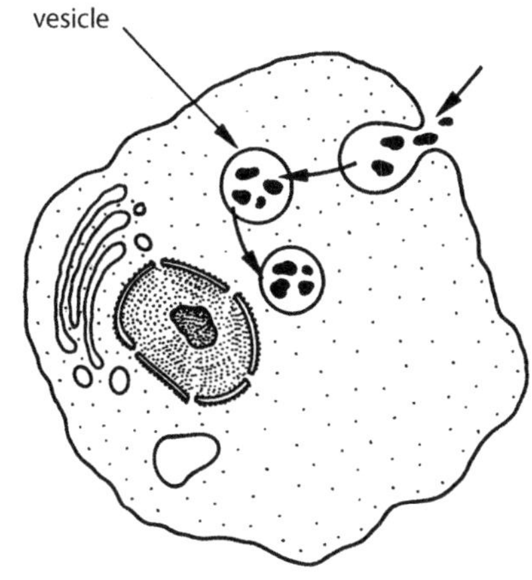

Endocytosis

custom (e.g. in certain societies persons are expected to marry within their own social group); *see also* exogamy

endogenous produced or originating from within; e.g. endogenous insulin is produced by the pancreas (rather than being injected); the opposite of EXOGENOUS

endogenous retrovirus (ERV) a RETROVIRUS that has become part of an organism's GENOME and exists in every cell of the body; passed on from generation to generation

endolymph fluid filling the membranous LABYRINTH of the inner ear

endometriosis a condition in which tissue from the lining of the uterus (the endometrium) grows outside the uterus, often on the ovaries or uterine tubes; may be caused by some tissue migrating up the oviducts during menstruation and growing on organs in the abdominal cavity

endometrium soft mucous membrane lining the uterus; the endometrium thickens during the menstrual cycle and if fertilisation occurs, the embryo implants itself in the lining; if fertilisation does not occur, the endometrium breaks down and is lost at menstruation

endomorph a human body type; *see* somatotype

endoparasite an internal PARASITE

endopeptidase a digestive enzyme that breaks down a PEPTIDE BOND within a protein chain (e.g. PEPSIN and trypsin)

endoplasmic reticulum (ER) a network of membranes forming channels through the cytoplasm of a cell; used for storage, support, synthesis and transport within the cell; *rough* (or *granular*) *endoplasmic reticulum* has ribosomes attached to it; *smooth* (or *agranular*) *endoplasmic reticulum* has no ribosomes; *see also* diagram of cell structure (p. 44)

endorphins a group of polypeptides that seem to be produced in the anterior lobe of the pituitary gland but also occur in parts of the brain; part of a larger group of chemical messengers in the brain called *neuropeptides*; have a similar chemical structure to morphine, the pain-killer derived from opium; thought to be the body's natural pain-killers by binding to the same receptors in the brain as morphine; have also been linked with learning, memory, control of body temperature and hormone regulation; *see also* enkephalins

endoscope a tube, with lenses and illumination, that is used to look inside hollow organs such as the rectum or the bladder

endoscopy use of an endoscope to examine body cavities

endosteum a membrane that lines the marrow cavity of LONG BONE and the trabeculae of spongy bone

endosymbiotic hypothesis the idea that the organelles of eukaryotic cells once lived as independent units before being taken in by larger cells; the organelles were probably taken in by feeding but not digested, and gave the cell a survival advantage, leading to their gradual incorporation into the cell structure through natural selection; *see also* eukaryote

energy the capacity to do work; forms of energy include light, heat, kinetic (movement) energy and electrical energy; the various forms of energy can be converted into other forms

energy value of food, a measure of the quantity of energy contained in a gram of food; carbohydrates and proteins have an energy value of about 17 kJ/g, fats about 37 kJ/g

enkephalins a group of polypeptides that occur in the central nervous system; part of a larger group of chemical messengers in the brain called *neuropeptides*; discovered in 1975, the enkephalins have a similar structure to morphine, the pain-killer derived from opium; thought to be the body's natural pain-killers by binding to the same receptors in the brain as morphine; also spelt *encephalins*; *see also* endorphins

enteritis inflammation of the intestines; *see* gastroenteritis

enterokinase an enzyme secreted by the mucosa of the small intestine; *see* kinase

environment all the elements in the surroundings that may impact on the development, action or survival of a CELL, ORGAN, or ORGANISM

environmental resistance all the environmental factors that prevent a species from reproducing at its maximum rate; the physical and environmental factors that limit population growth in a particular area

enzyme an organic substance (usually a protein) that increases the speed of a chemical change without being altered or destroyed in the change; an organic catalyst; each chemical reaction in a cell is controlled by its own particular enzyme; part of the surface of the enzyme molecule is an active site that binds to one of the reacting molecules (the substrate); after the reaction the enzyme is released intact

enzyme amplification a series of chemical reactions in which the product of one step is an ENZYME that produces an even greater number of product molecules at the next step, which in turn results in an even *greater* amount of the product; frequently triggered by a HORMONE

enzyme–substrate complex a temporary compound made up of an enzyme molecule and a reacting molecule; as a result of the combination a change occurs in the reacting molecule, the substrate; also called a *substrate–enzyme complex*; *see also* enzyme and diagram (p. 5)

Eocene the 2nd epoch of geological time in the Cainozoic era, between the Palaeocene and Oligocene epochs; 56–34 million years ago; the first true primates, early prosimians, appeared during this epoch; *see also* geological time scale (p. 111)

eon the longest division of geological time, containing two or more ERAS

epicanthic fold the fold of skin covering all or part of the upper eyelid; a characteristic of Asian peoples; thought to be an adaptation for protection of the eye against cold winds or perhaps against the high levels of ultraviolet radiation at high altitudes; also called *Mongolian eye fold*; *see* diagram opposite

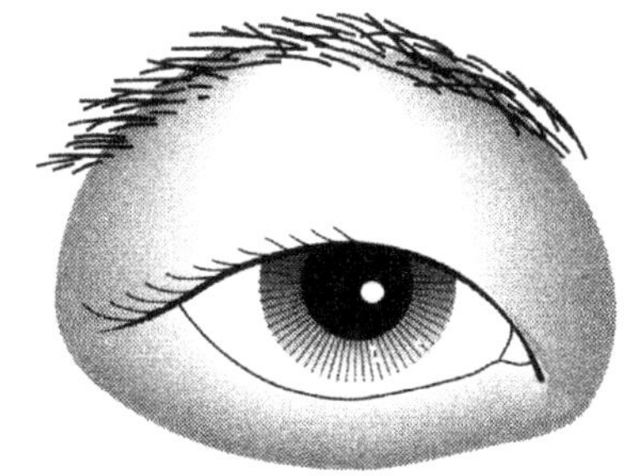

Epicanthic fold

epidemic an increase in the number of cases of a disease; numbers affected increase to a much higher level than normal, resulting in rapid spread of the disease through a population

epidemiologic transition a change in disease patterns that has occurred in many industrialised countries; infectious diseases are replaced by non-infectious diseases as the leading causes of death

epidemiology the field of medical science concerned with the occurrence, distribution and cause of disease in human populations

epidermis the outermost layer of cells of an organism; in humans, the outermost layer of the skin; composed of stratified epithelium; made up of 4 layers or strata; the deepest layer is the *stratum germinativum* (or *Malpighian layer*), where mitosis occurs to produce new skin cells and where the pigment melanin develops in the cells; this layer is made up of the *stratum basale*, where most cell division occurs, and above it the *stratum spinosum*, where cells become more flattened in shape; above the stratum germinativum is the *stratum granulosum*, a layer of flattened cells in which granules develop, the nucleus disintegrates and the cells die; the next layer, the *stratum lucidum*, is prominent only in areas of thick skin; at the outside is the *stratum corneum* (or *cornified layer*), consisting of dead cells in which the cytoplasm has been replaced by keratin; the outermost cells of this layer are continually worn away, being replaced by cells from deeper layers; see also diagram of skin structure (p. 250)

epididymis a highly folded tubule that fits against the rear surface of each testis and in which the sperm undergo maturation; the tubes of the epididymis are joined to the testis at one end and to the vas deferens (sperm duct) at the other end; *see also* diagram of male reproductive system (p. 235)

epigenetic factors inheritable factors that affect the development or function of an ORGANISM which do not involve its DNA sequence; they alter the expression of a GENE, without changing its structure

epiglottis a flap of tissue at the back of the throat; covers the opening of the windpipe when swallowing so that food goes down the oesophagus and not down the windpipe; *see also* diagram of respiratory system (p. 236)

epinephrine an alternative name for ADRENALINE

epiphyseal line *see* epiphyseal plate

epiphyseal plate a thin region of cartilage occurring across a LONG BONE between the head and the shaft; allows the bone to increase in length until early adulthood, when it is replaced by bone to form the *epiphyseal line*

epiphysis (plural *epiphyses*) the expanded end of a LONG BONE; *see also* diaphysis, diagram next page

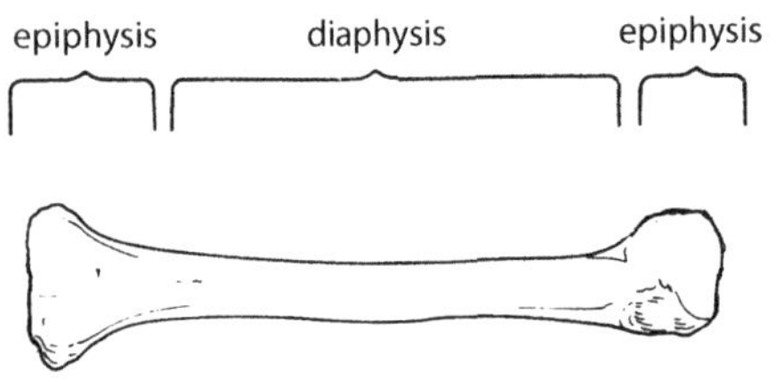

Structure of long bone

episiotomy a cut made to widen the opening of the vagina during childbirth; performed to prevent tearing of tissues

epistasis when a gene at one position on a chromosome has a controlling effect over a gene (or genes) at another position

epithelial tissue tissue that forms the outer part of the skin and lines hollow organs and ducts; a covering tissue; also called *epithelium* (plural *epithelia*); *simple epithelium* consists of a single layer of cells and occurs in places specialised for filtration or absorption or where there is little wear and tear; different types of simple epithelium are classified according to the shape of the cells; *squamous epithelium* consists of a layer of flattened cells; it lines the air sacs of the lungs, the insides of the heart and blood vessels and the insides of the mouth and vagina; *cuboidal epithelium* is made up of roughly cubical cells; it lines the tubules of the kidney, covers the surface of the ovary and forms part of the retina of the eye; *columnar epithelium* is made up of elongated or column-shaped cells; it lines the inside of the alimentary canal from the stomach to the anus, the inside of the gall bladder and the ducts of many glands; *glandular epithelium* is specialised for secretion of substances; it is found in glands such as the salivary glands, sweat glands, pituitary gland and thyroid gland; *stratified epithelium* consists of several layers of cells; it protects underlying tissues from wear and tear and from the external environment; the surface layer of cells is continually worn away (e.g. the outer part of the *skin*); *pseudostratified epithelium* consists of cells that are all attached to the basement membrane, but some are shorter than others and do not reach the free surface; found in large ducts of some glands and in regions of the male urethra

epithelium an alternative name for EPITHELIAL TISSUE

Epivir an antiretroviral drug used to reduce levels of HIV in the blood and to preserve immune system functionality; when used with ZIDOVUDINE it is believed to delay the onset of resistance to zidovudine or even to restore sensitivity to that drug; first became available in Australia in 1996

epoch a subdivision of the geological time scale; each of the periods in the Cainozoic era (the most recent era) is divided into epochs; *see* geological time scale (p. 111)

Epstein–Barr virus the virus that causes glandular fever and has links with certain cancers including Hodgkin's disease, Burkitt's lymphoma and nasopharyngeal carcinoma

equator (of a cell) the centre of the cell

era a major subdivision of geological time, usually divided into two or more PERIODS; there are 6 eras in the earth's geological history: Cainozoic, Mesozoic, Palaeozoic, Proterozoic, Archaeozoic (Archaean) and Hadean (Azoic); the last three listed eras may be referred to as EONS in some time scales; *see* geological time scale (p. 111)

erection the hardening and lengthening of the penis during sexual arousal; brought about by spaces in the penis becoming filled (engorged) with blood

erythro- a prefix meaning red (e.g. erythrocyte—red blood cell)

erythroblastosis a blood disease affecting unborn or newborn infants; may occur when an Rh^- mother has an Rh^+ baby; if the mother has been sensitised by a previous exposure to the Rh antigen, she may produce antibodies against the baby's Rh^+ red blood cells; the baby may be born with JAUNDICE and may require a blood transfusion; full name *erythroblastosis foetalis* or *haemolytic disease of the newborn*

erythrocyte an alternative name for RED BLOOD CELL

erythropoiesis production of red blood cells (by STEM CELLS) in the bone marrow

erythropoietin (EPO) a hormone, produced in the blood plasma, that stimulates the red bone marrow to produce more red blood cells; an enzyme called *renal erythropoietic factor* is released by the kidneys when levels of oxygen are lower than normal, and this stimulates production of erythropoietin; a synthetic version of erythropoietin has been illegally injected by athletes to enhance performance by increasing the oxygen-carrying capacity of their blood

essential amino acid one of the 10 AMINO ACIDS that must be included in the human diet; *see* table of amino acids (p. 11)

essential fatty acid a FATTY ACID that the body cannot make but is essential for normal body functioning and so must be consumed in the diet (e.g. linoleic acid)

essential nutrient any substance consumed in food that is necessary for normal growth and maintenance of the body

Essure a small spiral device that is used to provide permanent contraception by blocking the uterine tubes

ethical behaviour behaviour that follows ethical principles or values

ethics a set of values or moral principles

ethology the study of animal (and human) behaviour

eugenics the deliberate change of the hereditary characteristics of a population, usually by social controls such as preventing persons with certain disabilities from having children

eukaryote any organism whose cells possess a membrane-bound nucleus containing the genetic material; all plants and animals (except blue-green algae and bacteria) are eukaryotes; eukaryotic cells contain many highly organised structures called organelles, which carry out processes within the cell; also spelled *eucaryote*; *see also* prokaryote

euphoria a feeling of elation, great pleasure and happiness

euphoriant any substance that produces EUPHORIA; many chemical substances that are misused are euphoriants (e.g. heroin, ecstasy)

Eustachian tube a tube connecting the middle ear with the upper part of the throat; allows air pressure to equalise on each side of the eardrum such as when ascending or descending a high hill; swallowing or yawning opens the tube to allow air into, or out of, the middle ear; also known as the *auditory tube*; *see also* diagram of ear (p. 82)

euthanasia the painless killing of an individual who is suffering from an incurable illness that is causing great pain or the loss of many normal functions; often referred to as *mercy killing*

eutherian mammal a placental mammal; a mammal belonging to the subclass Eutheria, characterised by the young developing in the uterus and being nourished through a placenta

evaporation the change of phase from a liquid to a gas, which requires energy (e.g. when perspiration on the skin changes to water vapour—evaporates—it has a cooling effect because the heat energy needed to bring about the change of phase is absorbed from the skin)

'Eve' the hypothetical ancestor of all humans; the hypothesis of a single female ancestor is based on the fact that small amounts of DNA exist in the mitochondria of cells, and that this mitochondrial DNA (mtDNA) is only inherited through the mother; eggs transmit mtDNA but sperm do not, thus there is no shuffling of genes in the mtDNA during reproduction; the 'Eve' hypothesis relies on this fact to trace ancestry through females back to some original source; researchers in the field have suggested that the mtDNA of all modern females stems from a single female who lived in Africa between 290 000 and 140 000 years ago

evolution gradual change in the characteristics of species over time; also defined as change in gene frequencies in a population over time; there is natural variation in the characteristics of a population, and the individuals with more favourable variations tend to have more descendants than those with unfavourable variations, so that over time the favourable variations tend to become characteristic of the whole species; evolution results in the species becoming better adapted to its environment; culture also evolves and anthropologists study both the biological and cultural evolution of the human species, although cultural evolution is not necessarily directed towards the survival of the species; *see also* adaptation

evolutionary forces the mechanisms that can cause changes in allele frequencies from one generation to the next; the four evolutionary forces are mutation, natural selection, genetic drift and gene flow

evolutionary tree a diagram or map showing the evolutionary history of organisms; the evolutionary tree of the human species shows the relationships between species believed to be the ancestors of modern humans, including those that have become extinct

ex- a prefix meaning outside, out of or from (e.g. exocrine glands secrete to the outside—to the body surface or a duct)

ex vivo outside the body; refers to experiments or observation made on tissues outside the body; an alternative term to *in vitro*; the opposite of *in vivo*

excavation in archaeology or anthropology, **1.** the site of an investigation by digging (e.g. a hole in the floor of a cave that has been carefully dug to look for fossils or artefacts); commonly called a *dig* **2.** the act of excavating; the removal of earth to expose underlying fossil or cultural material

excretion the removal from the body of the wastes of metabolism (e.g. carbon dioxide is excreted by the lungs and nitrogen compounds like urea are excreted by the kidneys); note that excreted materials are *produced* by the organism, so that removal of indigestible remains in defecation is not considered to be excretion

excretory system usually refers to the organs involved in the filtration and removal of waste materials from the blood—the kidneys, ureters, bladder and urethra; these organs are more correctly called the *urinary system* because the lungs and skin are also involved in excretion

exergonic reaction a chemical reaction that releases energy (e.g. the breakdown of glucose molecules in cellular respiration); where heat energy is released, may also be called an *exothermic reaction*

exocrine gland a gland that secretes into a duct or directly onto a surface (e.g. sweat gland, salivary gland, pancreas); *see also* endocrine gland

exocytosis movement of large molecules out of a cell in membrane-bound vesicles that fuse with the membrane, releasing their contents to the outside

exogamy **1.** marriage or mating between persons who are not related; such a marriage is said to be *exogamous* **2.** marriage or mating between persons who do not belong to the same social group as required by law or custom (e.g. in certain societies persons are expected to marry outside their village or tribe); *see also* endogamy

exogenous produced or originating outside an organism; e.g. a hormone supplement is an exogenous form of the hormone; the opposite of ENDOGENOUS

exon a segment of DNA in a gene that is transcribed into messenger RNA, which in turn determines the sequence of amino acids in a protein; between the exons in a gene are DNA segments called *introns*; introns are also transcribed into messenger RNA but are then removed so that the exons join up to form the functional messenger RNA molecule; *see also* protein synthesis

exopeptidase a digestive enzyme that breaks off an amino acid from the end of a PEPTIDE chain; produced by the pancreas

exophthalmia abnormal protrusion of the eyeball so that the eyelids will not cover it; may be caused by a number of diseases

exothermic reaction a chemical reaction that releases heat; *see* exergonic reaction

experiment a procedure carried out to test a hypothesis

experimental group the subjects in an experiment who are given some special treatment in relation to the independent variable (e.g. in an experiment to test the effectiveness of a new drug, the group of subjects who are given the drug); *see also* control

experimental variable an alternative name for *independent variable*; *see* variable

expiration breathing out; pushing air out of the lungs; also called *exhalation*; *see also* inspiration

expiratory reserve volume the volume of air that can be forced out of the lungs after a normal exhalation; *see* lung volumes

expired air resuscitation (EAR) mouth-to-mouth (or mouth-to-nose) RESUSCITATION

exponential growth growth of a population at an ever-increasing rate

extended family **1.** a social unit composed of a group of sisters and their offspring (the matrilineal form) or a group of brothers and their offspring (patrilineal form) **2.** in more common usage, a family in which the grandparents, aunts, uncles and cousins all live together in the same home, or close by in neighbouring houses; *see also* nuclear family

extensibility the ability of tissue to be stretched when pulled (e.g. as in muscular tissue)

extension straightening; movement that increases the angle between articulating bones; *see also* diagram showing movements at joints (p. 1)

extensor a muscle that extends or straightens a limb (e.g. the triceps muscle in the upper arm is an extensor that contracts to straighten the arm at the elbow)

external auditory meatus an alternative name for AUDITORY CANAL

external cardiac compression attempting to restore heartbeat by pushing rhythmically on a patient's breastbone; *see also* resuscitation

external ear an alternative name for the OUTER EAR

external fertilisation an alternative name for IN-VITRO FERTILISATION

extinct describes groups of organisms that used to exist but now have no living representatives (e.g. all members of the genus *Australopithecus* are now extinct)

extracellular fluid fluid outside the body cells, found mainly in two places—in the small spaces between cells (where it is called intercellular fluid) and the liquid part of the blood (the plasma); *see* diagram below

eye the structure that contains nerve endings that are stimulated by light; responsible for sight; *see* diagram below

eyepiece an alternative name for the OCULAR of a microscope

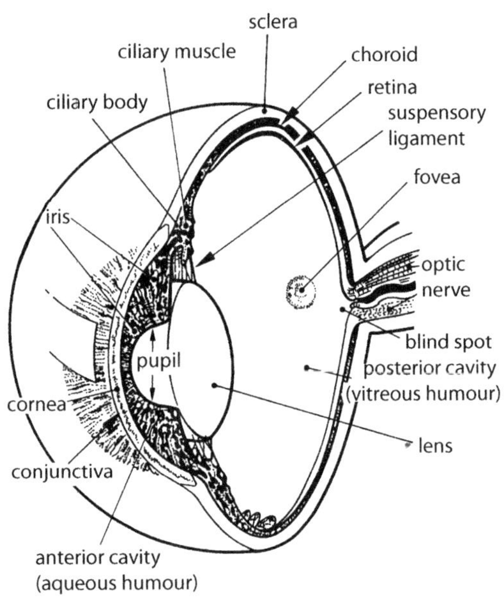

The eye

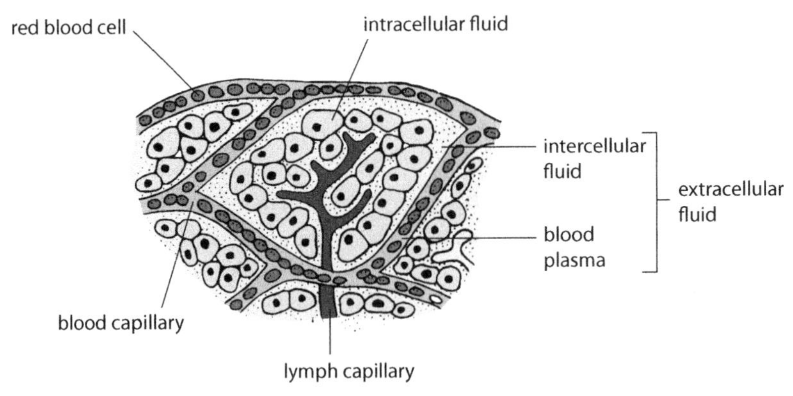

Extracellular fluid

F

facilitated diffusion a form of carrier-mediated transport in which the speed of diffusion of molecules across a cell membrane is increased by the use of carrier proteins to transport the substance across the membrane

facioscapulohumeral muscular dystrophy *see* muscular dystrophy

factor VIII a blood-clotting protein; a lack of this factor results in HAEMOPHILIA A (classic haemophilia)

facultative reabsorption active reabsorption of water from the kidney tubule; the quantity reabsorbed is controlled by ANTIDIURETIC HORMONE; *see also* obligatory reabsorption

FAD the abbreviation for FLAVIN ADENINE DINUCLEOTIDE

faeces semisolid material from the alimentary canal that is passed out through the anus; contain undigested food, water, salts, cells that have worn away from the inside of the alimentary canal, bacteria, mucus and compounds excreted from the liver (the bile pigments)

faint a temporary loss of consciousness due to reduced blood supply (and therefore oxygen supply) to the brain

fair test a test or experiment in which all of the VARIABLES that may influence the result are controlled (kept constant) except the variable being tested

fallen arches an alternative name for FLAT FEET

Fallopian tube an alternative name for UTERINE TUBE

false ribs RIBS not directly attached to the breastbone

family **1.** one of the levels of biological classification; a family is a subdivision of an order and usually contains a number of genera; most family names end in -idae (e.g. humans belong to the family Hominidae, gibbons to the family Hylobatidae)
2. parents, their children and relations; *see* nuclear family, extended family

family planning the processes involved in regulating the number of children a couple may have and the time span between children; in some countries there are strong social pressures (or even laws) that determine the desirable number of children for a couple (e.g. in China the 'One Child Campaign', adopted in 1979, encouraged couples to have only one child); in Australia, the Family Planning Association provides information to couples on the methods available to prevent fertilisation

family studies studies undertaken to assess the influence of heredity on certain characteristics; family members are examined over a number of generations to see how much they resemble each other with respect to a particular characteristic

fantasy a drug sold as a powder that can be mixed into drinks; when ingested, it causes an altered state of consciousness, and possibly hallucinations, blurred vision, heart palpitations and breathing difficulty; the chemical name is gamma hydroxybutyric acid, also known as *GHB*, *GBH*, *ketamine* or *special K*

fast-twitch fibres MUSCLE FIBRES that contract quickly and produce more power than other muscle fibres; suitable for short bursts of activity

fat a compound formed from one molecule of glycerol and one, two or three FATTY ACID molecules; the fat stored in the body has three fatty acid molecules and is known as *triglyceride* (or *triacylglycerol*); it is stored in adipose tissue and is used as a long-term energy reserve for the body; it also provides heat insulation and protective cushioning for some organs; a *diglyceride* (or *diacylglycerol*) is a fat consisting of two fatty acids chemically combined with glycerol; a *monoglyceride* (or *monoacylglycerol*) is a fat consisting of a glycerol molecule and a single fatty acid molecule; a *saturated fat* has a high proportion of saturated fatty acid molecules; it is usually solid at room temperature and is usually from animal sources; foods such as pork, beef, lamb, milk, milk products and eggs contain saturated fat; high consumption of saturated fat has been linked with increased risk of cardiovascular disease; *unsaturated fat* is made up of a high proportion of unsaturated fatty acids; *monounsaturated fats* contain a high proportion of monounsaturated fatty acids; *polyunsaturated fats* contain a high proportion of polyunsaturated fatty acids; unsaturated fats are usually liquid at room temperature and usually come from plants (e.g. maize, sunflower, soybean oils); consumption of unsaturated fat rather than saturated fat tends to lower blood cholesterol

fat-soluble vitamin a vitamin that is soluble in fats but not in water and so can only be ingested in foods containing fat, and are absorbed in the small intestine along with dietary fats; can be stored in cells (unlike water-soluble vitamins, which cannot be stored); vitamins A, D, E and K are fat-soluble

fatigue tiredness

fatty acid a type of organic acid found in FATS; each molecule of the fat stored in the body contains three fatty acid molecules; if a fatty acid molecule contains the maximum number of hydrogen atoms, it is said to be *saturated*; *unsaturated fatty acid* molecules are not completely saturated with hydrogen atoms (i.e. the molecule contains one or more double bonds between carbon atoms); *monounsaturated fatty acids* could contain one more hydrogen atom; *polyunsaturated fatty acid* molecules could hold more than one extra hydrogen atom

faunal correlation a method of relative dating in which fossil sites can be assigned an approximate age, based on the similarity of fossil remains to those that have been found at other sites and for which the age is known

fecundity potential reproductive capacity (applies especially to females); may also be defined as the number of people capable of having children

feedback one factor in a system being influenced by another factor (or factors) in the system that is responding to stimulus from the first factor; *see* feedback system

feedback system a situation in which the response to a stimulus changes the original stimulus; feedback may be negative or positive; in *negative feedback* the response to a stimulus brings about a change that is opposite to, or reduces the effect of, the original stimulus; negative feedback occurs in many cases where the products of a process act as regulators of the process (e.g. if blood glucose level begins to fall, the liver releases more glucose into the blood, and blood glucose then rises—a change opposite to the original stimulus; if body temperature begins to rise, the body responds by sweating—this tends to lower body temperature, thus changing the original stimulus); *positive feedback* reinforces the original stimulus; it does not lead to the maintenance of stable conditions and so is unusual in body functioning (e.g. during childbirth, the pressure of the baby's head at the opening to the mother's uterus causes contraction of the uterine muscles—the greater the pressure, the stronger the contractions, which in turn cause greater pressure and so on)

Fehling's test a chemical test used to detect the presence of reducing sugar, particularly glucose, in food samples, body fluids or other solutions; Fehling's A (containing copper sulphate) and Fehling's B (containing sodium hydroxide and sodium potassium tartrate) are mixed with the test solution and boiled; if the solution contains reducing sugar, copper I oxide is formed and the colour changes from blue to brick-red; *see also* Benedict's test

Femidom a female condom; a polyurethane sheath that lines the vagina and is effective as a birth control device as well as giving some protection against sexually transmitted diseases

femoral artery the main artery in the thigh; carries blood to the lower leg; *see also* diagram of major arteries (p. 20)

femur the thigh bone; the longest and strongest bone of the body; connected at its upper end to the pelvis by a ball-and-socket joint and at its lower end to the tibia by a hinge joint at the knee; *see also* diagram of skeleton (p. 249)

fenestra ovalis an alternative name for the OVAL WINDOW between the middle and inner ear

fenestra rotunda an alternative name for the ROUND WINDOW between the middle and inner ear; *see* oval window

fertile able to initiate, sustain or support REPRODUCTION; (of an egg or sperm cell) capable of developing into a new individual

fertilisation the fusion of a sperm and an egg to produce a zygote that will develop into a new organism; involves two processes, the formation of the fertilisation membrane and the fusion of the egg and sperm nuclei (more correctly called *pronuclei*) to form the zygote; in humans also known as *conception*

fertilisation membrane the membrane formed when the plasma membranes of the egg and the sperm fuse during the process of fertilisation; the formation of the fertilisation membrane acts as a barrier to penetration by any further sperm

fertility the ability to produce offspring; in population studies, fertility is the number of live children produced by a population

fertility rate of a population, the number of live births per 1000 females of childbearing age in the population; for statistical purposes childbearing age is regarded as between 15 and 44

fever a rise in body temperature above the normal level of 37°C; frequently occurs following infection by micro-organisms

fibre **1.** a type of carbohydrate in the diet; soluble fibre reduces blood cholesterol; insoluble fibre promotes movements of the alimentary canal **2.** long narrow cells (such as muscle cells), or long narrow extensions of cells (such as those of nerve cells) **3.** one of the microtubules that make up the spindle during mitosis and meiosis **4.** some proteins (such as collagen or elastin) that are found in connective tissue

fibrillation unco-ordinated contractions of individual muscle fibres to the extent that the muscle cannot contract smoothly; usually used to describe such contractions in heart muscle, which result in ineffective heartbeat; may be confined to the atria (*atrial fibrillation*) or the ventricles (*ventricular fibrillation*); can sometimes be stopped by passing an electric current through the heart using a device called a *cardiac defibrillator*

fibrin fibrous insoluble protein that forms when blood clots; *see* blood clotting

fibrinogen a soluble protein that makes up about 7% of the proteins in the blood plasma; is converted to fibrin when blood clots; *see* blood clotting

fibroblasts large cells that form the fibres that make up part of connective tissues

fibrocartilage CARTILAGE with bundles of fibres

fibrositis a condition resulting from the inflammation of fibrous tissue, characterised by pain, stiffness or soreness, especially of the fibrous sheath around the muscles; fibrositis of muscles in the lumbar region (the small of the back) is called lumbago

fibrous capsule the external layer of an ARTICULAR CAPSULE; *see* diagram of structure of synovial joint (p. 264)

fibrous joint an immovable JOINT

fibrous protein PROTEIN with molecules in the form of long strands or flat sheets

fibula the smaller, more slender bone of the lower leg; lies next to the tibia and slightly behind it; *see also* diagram of skeleton (p. 249)

field diameter the diameter of the FIELD OF VIEW of a microscope; as the magnification increases, the field diameter decreases; if the field diameter is known, the size of objects viewed with the microscope can be estimated

field of view the area that can be seen when looking through a microscope

fight-or-flight response a response preparing the body for increased activity, brought about by stimulation of the sympathetic division of the autonomic nervous system and by increased secretion of the hormone adrenaline; the response involves making extra glucose and oxygen available to the skeletal muscles and the brain (the organs involved in reacting to danger); some of the responses are increased heart rate, dilation of blood vessels in skeletal muscle and brain, constriction of blood vessels in the alimentary canal, release of glucose from the liver, and increased breathing rate; also called *alarm reaction*

filtrate **1.** the fluid remaining after filtration has taken place **2.** in the kidney, the fluid that passes into the glomerular capsule from the glomerulus; in a healthy person that would be a sample of all the materials present in the blood except the cells, platelets and most proteins; *see also* diagram of kidney nephron (p. 183)

filtration **1.** passing liquid through a filter or a membrane that acts like a filter **2.** in the kidney, passage of liquid from the glomerulus into the glomerular capsule; the blood pressure forces water and dissolved substances through the capillary walls into the capsule; also known as *glomerular filtration*

filtration slits spaces between cells covering the capillaries of the glomerulus of a kidney NEPHRON; *see* podocytes

fimbriae finger-like projections that form a fringe around the open end of each uterine tube; movements of the fimbriae and their cilia are thought to cause currents of fluid from the abdominal cavity to enter the uterine tube and thus at ovulation carry the egg from the ovary into the tube; *see also* diagram of female reproductive system (p. 235)

first-class protein PROTEIN containing all essential amino acids

'First Family' fossil remains found at Hadar, in Ethiopia, consisting of an entire group of individuals who may have died all at once in a catastrophe of some kind, possibly a flash flood; the group consisted of at least 4 children and 9 adult males and females; members of this 'first family' have been classified as *Australopithecus afarensis* and provide some evidence of the variation that existed between members of this species; many researchers now believe that the fossils of these individuals were together because of animal scavenging; dated to around 3 million years ago

first filial generation (F1) a term used in Mendelian genetics to refer to the first generation produced by a mating of two individuals; *see also* second filial generation

first meiotic division the first of two cell divisions that take place during meiosis; during the first division the number of chromosomes is halved

fission track dating an absolute dating method used to date the glass of volcanic deposits; when uranium-238 decays, fission tracks that can be observed under a microscope are produced in the glass; the tracks in the sample are counted and the sample placed in a nuclear reactor to make the remainder of the uranium-238 decay; the new tracks can then be counted and, since the rate of decay of uranium-238 is known, the age of the deposit from which the glass came can be calculated; dates arrived at can be compared with those obtained from POTASSIUM–ARGON DATING

fissure a deep downfold in the cerebral cortex of the brain (shallow grooves between the folds are called *sulci*); *see also* diagram (p. 260) and diagram of brain (p. 35)

fitness the ability of an organism to survive and reproduce, passing its genes on to the next generation; individuals with characteristics favoured by NATURAL SELECTION have high fitness

fixator a muscle that contracts to immobilise a joint; *see* synergist

fixed joint an alternative name for an *immovable joint*; *see* joint

flagellum (plural *flagella*) a long hair-like structure that protrudes from a cell; used to propel the cell through a liquid; the only human cell that has a flagellum is the sperm; *see* diagram of sperm (p. 254)

flat feet weakening of the tendons and ligaments of the feet so that the height of the longitudinal arch is reduced; also called *fallen arches*; *see also* arches

flavin adenine dinucleotide (FAD) a co-enzyme involved in the Krebs cycle; during the Krebs cycle, which is part of aerobic respiration, the co-enzymes FAD and NAD are reduced to become $FADH_2$ and $NADH_2$; they then contain the energy from the breakdown of glucose; in the electron transport chain, the energy in the two co-enzymes is transferred to ATP

flexion bending; movement that decreases the angle between the articulating bones; *see also* diagram showing movements at joints (p. 1)

flexor a muscle that bends a limb at a joint (e.g. the biceps muscle in the upper arm is a flexor that contracts to bend the arm at the elbow)

floating ribs RIBS that are unattached at the front

fluid-mosaic model the generally accepted description of the cell membrane as consisting of two layers of lipid molecules in which proteins are embedded; the proteins are able to move about within the lipid bilayer

flukes parasitic worms that live inside the body; they have suckers to attach to their host

fluorine dating determining the relative age of buried bones by the amount of fluoride they have absorbed from water in the soil; the more fluoride that is present, the older is the specimen; only useful with bones found at the same location because the amount of fluorine in groundwater varies from place to place

foetal alcohol syndrome (FAS) abnormalities in a newborn baby due to excessive alcohol consumption by the mother during pregnancy; symptoms include retarded physical and mental growth and reduced head and brain size

foetal membranes the membranes outside a developing embryo that protect and nourish the embryo and later the foetus; consist of the yolk sac, amnion, chorion and allantois; also called the *embryonic membranes*; the outermost membrane, the *chorion*, encloses the embryo and all the other membranes; it produces the hormone human chorionic gonadotrophin (HCG) and eventually forms the main part of the placenta; the *amnion* or 'bag of waters' is a thin protective membrane inside the chorion and also surrounds the developing embryo; the space between the amnion and the developing baby is the *amniotic cavity*, which becomes filled with *amniotic fluid* secreted by the amnion; between the chorion and amnion is the *allantois*, a membrane that eventually becomes part of the umbilical cord; the 4th membrane is the *yolk sac*; in fish, reptiles and birds it contains nourishment for the embryo, but in mammals nutrients come from the placenta, and the yolk sac remains small, eventually becoming part of the umbilical cord; *see also* diagram of foetal membranes opposite top

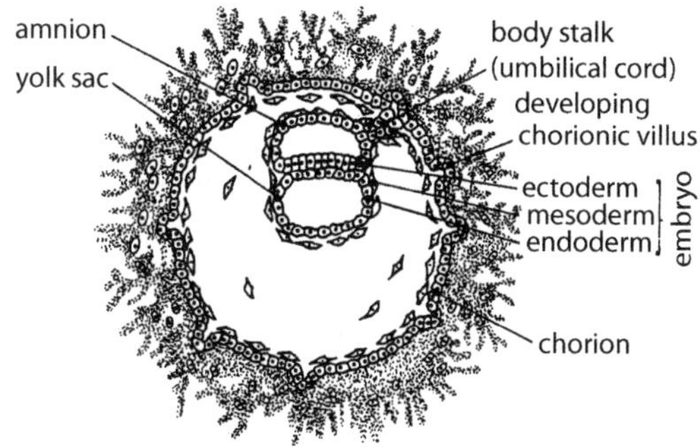

Foetal membranes

foetal monitoring regular checking of the foetal heartbeat during LABOUR

foetoscope 1. a type of STETHOSCOPE used for listening to the foetal heartbeat 2. an instrument used to view the foetus in the womb

foetoscopy a technique by which the foetus is examined through a small, telescope-like instrument called a *foetoscope*; *see also* endoscopy

foetus the term used to describe the developing individual after the 8th week of pregnancy, once the embryo has become recognisably human; also spelt *fetus*

folacin an alternative name for FOLIC ACID

folate an alternative name for FOLIC ACID

folic acid one of the B complex of vitamins; water soluble; made by bacteria in the gut but may be obtained from leafy green vegetables and liver; deficiency can lead to production of very large red blood cells (*macrocytic anaemia*); increased intake by females before and during pregnancy can help prevent neural tube disorders in a developing foetus; also called *folate* or *folacin*

follicle a small cavity, sac or gland; *see also* hair follicle, ovarian follicle, Graafian follicle

follicle-stimulating hormone (FSH) a hormone secreted by the anterior (front) lobe of the pituitary gland; stimulates development of eggs and secretion of oestrogens from the ovaries of females and initiates sperm production in males; secretion is controlled by a releasing factor from the hypothalamus; *see* table of hormones (p. 128)

fontanelles membrane-covered spaces between the bones of the skull of a foetus or young child, which allow the skull to be compressed during birth to pass through the birth canal and which permit rapid growth of the brain during infancy; *see also* diagram of fontanelles below

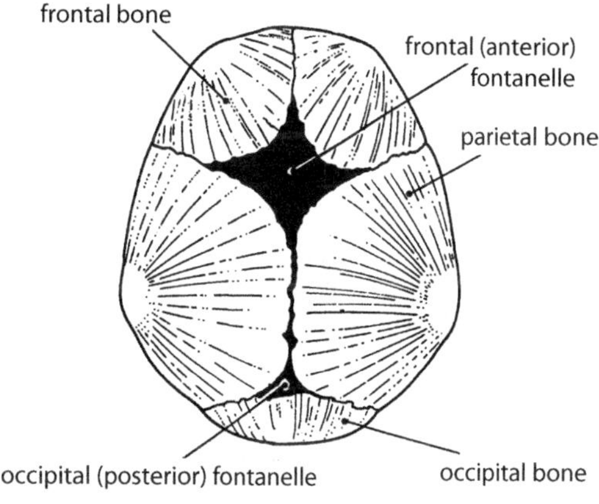

Fontanelles

food chain a sequence that shows how matter and energy are passed from organism to organism in a community, such as:

grass → grasshoppers → frogs → snakes → hawk; the arrows in the food chain represent the flow of energy and matter

food poisoning an illness caused by the consumption of food contaminated by bacteria or the products of bacteria; *Salmonella* are bacteria found in the alimentary canals of animals and are frequently the cause of food poisoning if good food-handling practices are not followed

food test a chemical test used to determine whether certain components are present in a food sample (e.g. Biuret test for protein, Benedict's test for reducing sugars)

food web the complex, interconnected feeding relationships among all organisms in a community; a number of interacting FOOD CHAINS

foramen (plural *foramina*) an opening (e.g. foramen ovale, foramen magnum)

foramen magnum the opening beneath the cranium through which the spinal cord passes as it descends from the brain; *see* diagram of foramen magnum, below

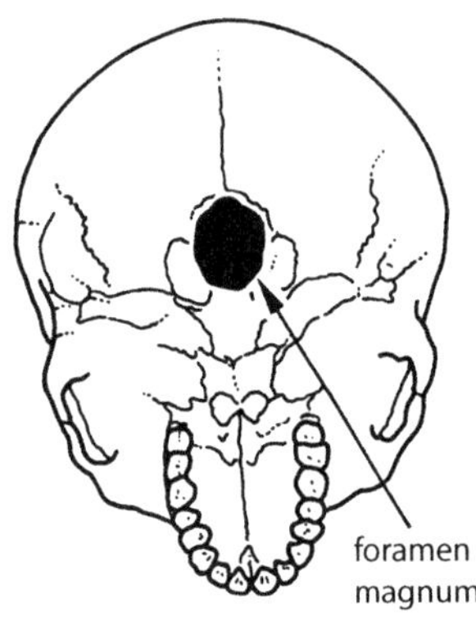

Foramen magnum

foramen ovale an opening between the atria of the foetal heart that allows blood to flow directly from the right atrium into the left atrium and therefore bypasses the lungs, which are not yet functioning; failure of the foramen ovale to close at birth can lead to a condition known as *blue baby* because blood flow through the lungs is reduced and the tissues do not receive enough oxygen; *see also* diagram of foetal circulation (p. 212)

forensic science any science that provides impartial scientific evidence for use in a court of law; more specifically, the methodical gathering and analysis of evidence discovered at a crime scene to establish facts that can be presented in court

foreskin an alternative name for PREPUCE

formed elements any cells or cell-like structures in the blood; red cells, white cells or platelets

fortified food food to which vitamins and/or minerals have been added in order to promote health and prevent disease

forward-facing eyes *see* binocular vision, and the diagram on page 30

fossil evidence of, or remains of, an organism that lived long ago, usually preserved in sedimentary rock; commonly thought of as skeletal material, but also includes such things as teeth, footprints, dung and pollen grains

fossil record the evidence provided by fossils of the development of life on earth; the record is incomplete because of the relatively small number of organisms fossilised and the even smaller number discovered by humans

fossilisation the process by which an ORGANISM that existed in some earlier age was preserved as a FOSSIL

founder effect a type of genetic drift occurring when a new population is formed by a small number of individuals; if the genetic make-up of the founders is not representative of the gene pool of the original population, the new population will have characteristics different from the original (e.g. Australian Aborigines have no B allele for the ABO blood group and thus groups B and AB do not occur in Aborigines—this may have been due to the founder effect of Australia being settled by a small founding population in which the B allele was absent); *see also* random genetic drift

fovea a small depression at the centre of the retina of the eye; *see* yellow spot

fracture any break or crack in a bone; *partial fracture*—the break does not extend right across the bone; *complete fracture*—the break is across an entire bone and the bone is broken into two pieces; *simple fracture* (or *closed fracture*)—the broken ends of the bone do not break through the skin; *compound fracture*—the broken ends of the bone protrude through the skin; *comminuted fracture*—the bone is splintered into many small pieces at the site of the break, and small fragments of bone are found between the two main parts; *depressed fracture*—an injured region of bone is forced inwards; frequently occurs when a person receives a heavy blow to the head; *greenstick fracture*—partial fracture in which one side of a bone is broken and the other bends; can only occur in children and is a frequent occurrence in young children; *impacted fracture*—the broken ends of the bone are forced into each other; frequently occurs when a person falls and lands on the end of a bone; *spiral fracture*—the bone is broken such that it is twisted apart at the break; *transverse fracture*—the break is at right angles to the long axis of the bone; *see* diagram next page

fraternal twins TWINS that develop from two separate zygotes; also called dizygotic twins

free nerve endings receptors not surrounded or protected by supporting structures; bare dendrites; common in the skin; thought to be mostly pain receptors, but those around hair follicles are sensitive to touch and some are heat and cold receptors; *see* diagram of skin structure (p. 250)

freely movable joint an alternative name for SYNOVIAL JOINT

frequency the number of times an event occurs in a given time

frequency distribution an arrangement of data to show how often each score or each category occurs; e.g. if the number of students wearing spectacles in each year at a school were counted it could give the following data: Year 8—15, Year 9—12, Year 10—4, Year 11—19, Year 12—9; the frequency distribution could be plotted as the histogram, below

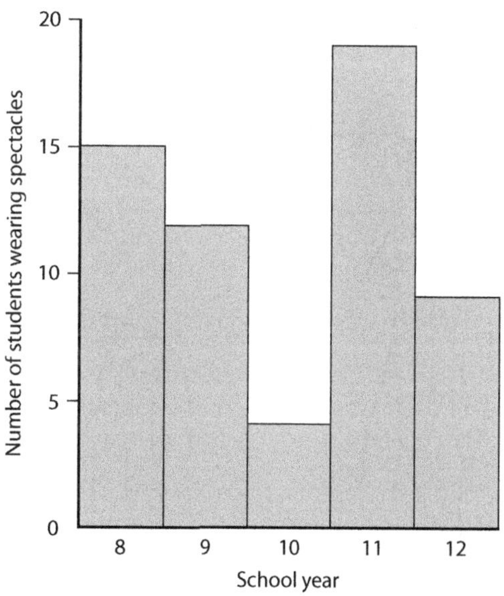

Frequency distribution

frequency table a table drawn up to show the distribution of the FREQUENCY of the occurrence of a particular factor according to some specified set of class intervals

friction ridges ridges of skin that form the fingerprints; also found on the palms, toes and soles of the feet; unique to primates; similar ridges are also found on the naked undersurface of the grasping tail of some of the New World monkeys; *see also* diagram, opposite top

frontal lobe one of the 4 lobes of each cerebral hemisphere of the brain; *see also* diagram opposite bottom

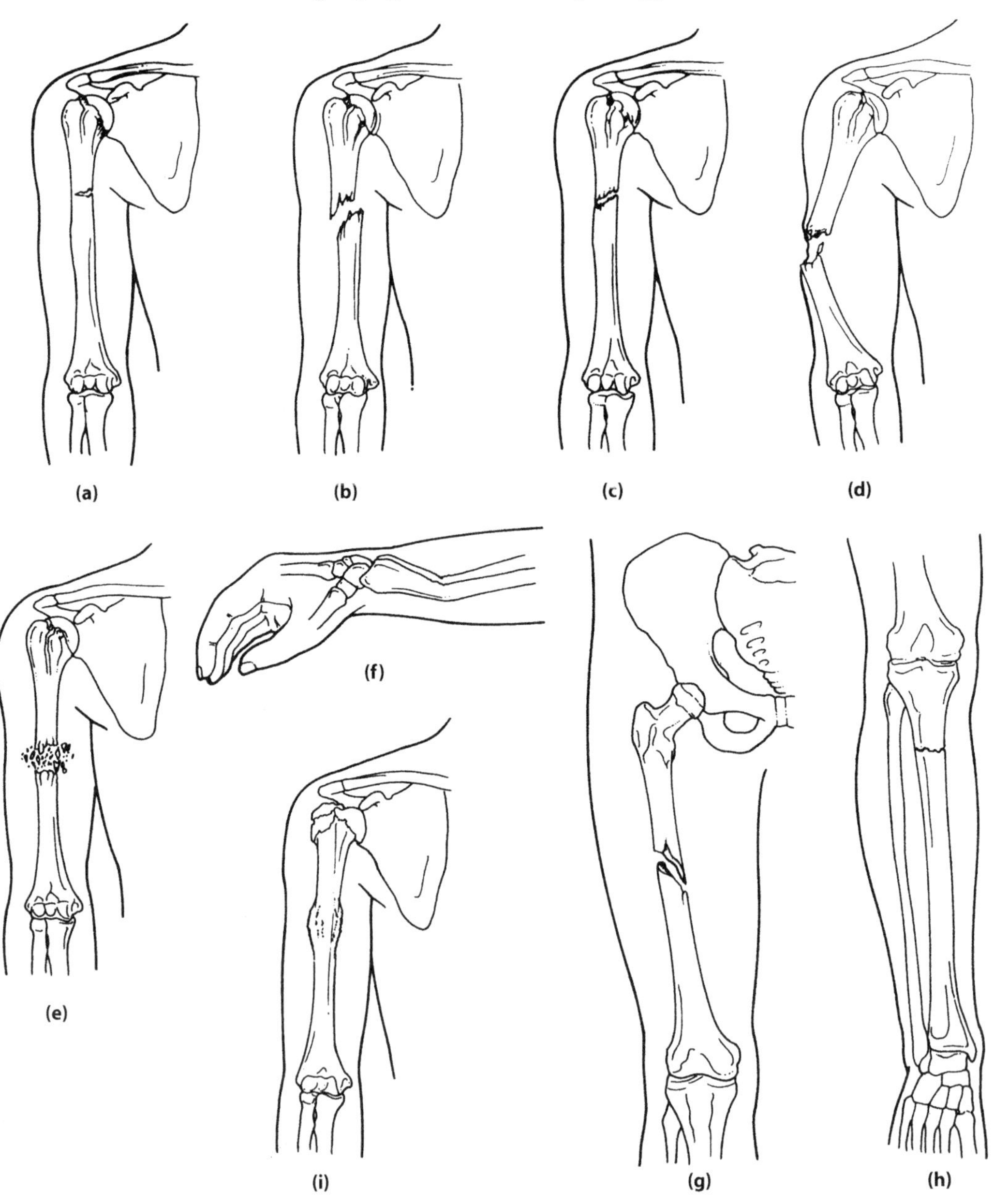

Bone fractures (a) partial; (b) complete; (c) simple; (d) compound; (e) comminuted; (f) greenstick; (g) spiral; (h) transverse; (i) impacted

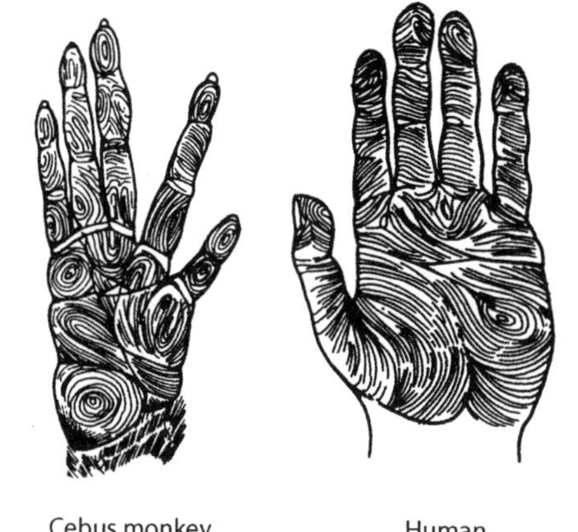

Friction ridges

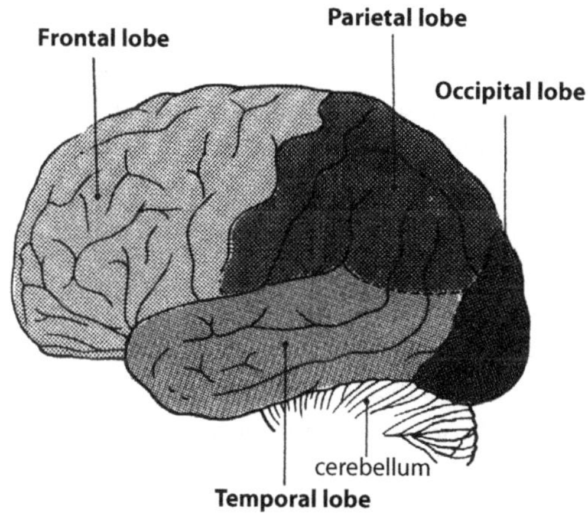

Lobes of the cerebrum

fructivore an alternative name for FRUGIVORE

fructose a simple sugar (monosaccharide); has the same chemical formula as glucose ($C_6H_{12}O_6$) but the atoms are arranged in a different way; found in fruit juices, honey and nectar; when combined with a glucose molecule it forms a molecule of sucrose (table sugar)

frugivore an animal that feeds on ripe fruit; also known as a *fructivore*; such animals are said to be *frugivorous* or *fructivorous*

frugivorous describes an animal that feeds on ripe fruit

FSH an abbreviation for FOLLICLE-STIMULATING HORMONE

functional food a food that is claimed to have health-giving properties beyond that of normal nutrition

fungus a plant-like organism that does not carry out photosynthesis, but instead consumes living or dead organic matter; some fungi are parasitic on humans (e.g. tinea, thrush and ringworm)

G

G1 phase the first stage of interphase in the CELL CYCLE

G2 phase the part of interphase in the CELL CYCLE following replication of DNA

galactose a simple sugar (monosaccharide); combines with glucose to form lactose, the sugar in milk; like glucose and fructose, has the chemical formula $C_6H_{12}O_6$

gall bladder a hollow bag that stores the bile secreted by the liver; located under the liver and opening into the small intestine through the bile duct; *see also* diagram of digestive system (p. 77)

gallstones stones formed in the gall bladder by the crystallisation of cholesterol or calcium carbonate; the stones grow in size and number and may eventually block the flow of bile from the gall bladder; also called *biliary calculi*

gamete a sperm or an egg cell; a cell that joins with another, in sexual reproduction, to form a new organism; differs from other cells of the body in that it has only one of each type of chromosome; has half the usual chromosome number—the haploid number (in humans, 23)

gamete intrafallopian transfer (GIFT) a form of IN-VITRO FERTILISATION

gametogenesis the formation and development of the gametes; involves meiosis (so that the gametes are haploid) and maturation; *spermatogenesis* is the process by which the testes produce spermatozoa (sperm); *oogenesis* is the formation and development of an egg (ovum) in the ovary—one of the four cells from meiotic division develops into a mature egg; the other three cells (the polar bodies) disintegrate; *see* diagrams, opposite left

gamma globulin a type of GLOBULIN

ganglion (plural *ganglia*) a group of nerve cell bodies that is outside the brain or the spinal cord (except for the basal ganglia, which are found in each cerebral hemisphere of the brain)

gas exchange the process in which oxygen is taken in and carbon dioxide given out; occurs in the lung's blood capillaries

gast-, gastro- prefixes meaning stomach (e.g. gastric juice is digestive juice in the stomach)

gastric glands glands in the internal lining of the stomach that secrete the components of gastric juice

gastric juice the digestive juice secreted by the glands of the stomach; contains mucus, hydrochloric acid and the enzyme gastric protease (pepsin)

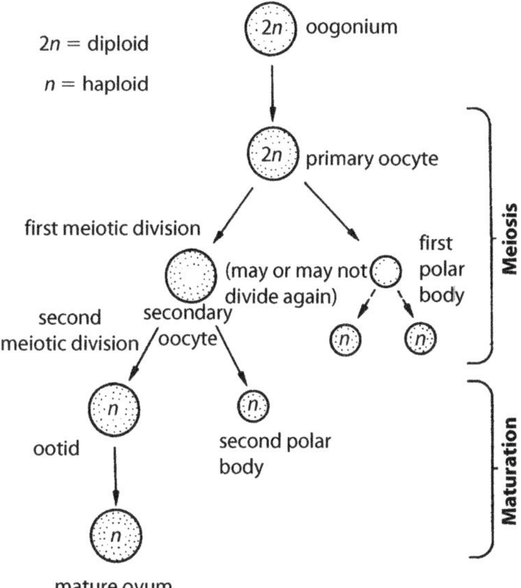

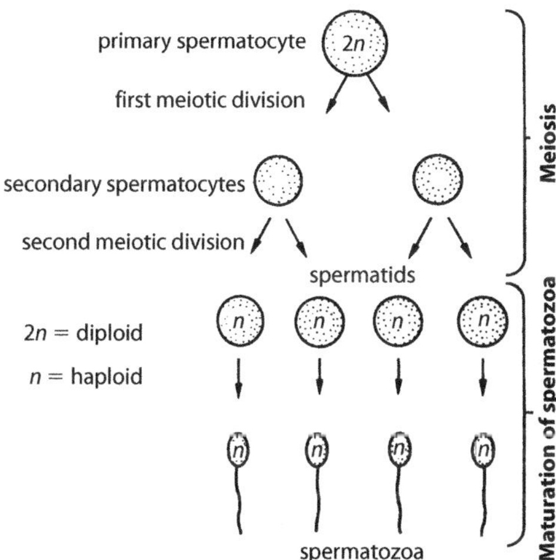

Gametogenesis (a) oogenesis; (b) spermatogenesis

gastric pits depressions in the internal lining of the stomach into which the gastric glands pass their secretions

gastric protease a protein-digesting enzyme that is part of the digestive juices in the stomach; breaks proteins into polypeptides; secreted in an inactive form, *pepsinogen*, which is converted to gastric protease by contact with hydrochloric acid; also called *pepsin*

gastric ulcer an ULCER in the lining of the stomach

gastrin a hormone secreted by the lining of the stomach and small intestine; stimulates secretion of gastric juice and increased movement of the alimentary canal

gastritis inflammation of the lining of the stomach; causes NAUSEA and discomfort after eating

gastrocnemius the muscle of the lower leg; with the underlying soleus muscle, makes up the calf of the leg; one end is attached to the femur; the other, through the Achilles tendon, is attached to the heel bone; contraction may cause bending of the leg at the knee or movement of the foot; *see* diagram, below

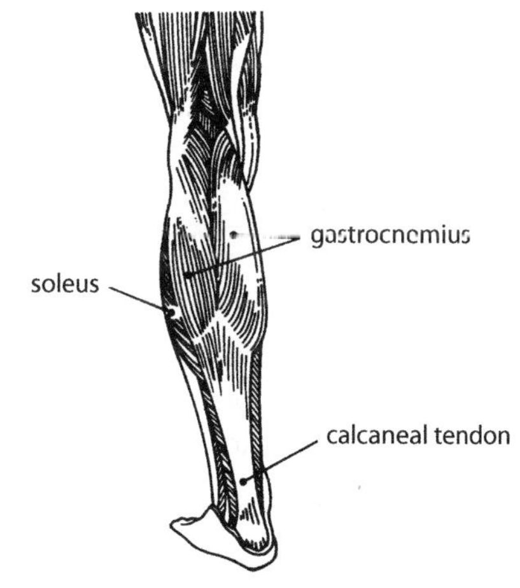

Gastrocnemius muscle

gastroenteritis inflammation of the mucosa of the stomach and intestines; symptoms are severe abdominal pain, nausea, vomiting, diarrhoea and fever; may be caused by irritants such as excessive alcohol, viral or bacterial infections, food allergies or overeating; if the inflammation is confined to the intestines, it is called *enteritis*

gastrointestinal tract an alternative name for ALIMENTARY CANAL

gated channel a pathway in the CELL MEMBRANE that can open or close under the influence of particular stimuli; gated channels control when substances are able to pass through the membrane

gene the factor that determines an inherited characteristic; located in the chromosomes; each gene is a specific sequence of DNA nucleotides that codes for the production of an enzyme or a protein; a *structural gene* controls the production of a specific enzyme or structural protein; a *regulatory gene* controls the production of a protein that, in turn, controls the activity of other genes; when cells divide, each new cell contains a full set of the original genes; when humans reproduce, a subset of each parent's genes are transmitted to their offspring via the sperm and the egg; *see also* allele

gene amplification the production of multiple copies of a gene or a piece of DNA; occurs naturally in certain cancers; can be performed in the laboratory using the POLYMERASE CHAIN REACTION

gene expression the production of a protein using the information encoded in a gene

gene flow a means of evolutionary change resulting from the movement of genes from one population to another through immigration, emigration and interbreeding; gene flow introduces new genes into a population; also acts to make populations more genetically similar to one another; sometimes called *gene migration*

gene linkage the tendency for certain genes to be inherited together because they are located in the same chromosome; such genes do not show INDEPENDENT ASSORTMENT; also called *genetic linkage* or *linkage*; *see also* crossing over, recombination

gene migration an alternative name for GENE FLOW

gene mutation an alteration to a single gene; *see* mutation

gene pool the sum of all the alleles carried by the members of a population; the total genetic variability present in a given population

gene probe an alternative name for GENETIC PROBE

gene regulation regulation of the stages of GENE EXPRESSION by controlling and determining which GENES are switched on and off, when, and for how long

gene therapy treating disease by replacing, manipulating or supplementing non-functional genes in cells and tissues; an experimental procedure used to treat hereditary diseases in particular

genetic code the sequence of bases in a DNA molecule; determines the proteins that a cell can make

genetic counselling counselling provided to couples who are at risk of having children affected by an inherited disease; the family tree of each partner is studied, enabling the risk to be determined

genetic disease a state of ill health resulting from changes to an individual's genetic material; usually refers to diseases that are inherited, but non-inherited forms result from MUTATION; may result from an absent or defective GENE or by changes to a CHROMOSOME; also called a *genetic disorder*, *hereditary disease* or *inherited disorder*

genetic drift an abbreviation for RANDOM GENETIC DRIFT

genetic engineering a branch of biotechnology concerned with genetically reprogramming cells so that their chemistry is altered to produce characteristics of medical or industrial significance; usually involves the deliberate transfer of genetic material from one organism to another (e.g. the hormone insulin is now obtained from bacteria that have been genetically engineered to produce the hormone); also referred to as *recombinant DNA technology*

genetic markers characteristics that can be used to study relationships between populations; can be used as a measure of BIOLOGICAL DISTANCE; blood groups are often used as genetic markers

genetic probe a fragment of DNA (or RNA) labelled with radioactive isotopes or a fluorescent marker that will bind to a specific sequence of BASES in another DNA (or RNA) molecule; used as a tag in order to identify or isolate a particular GENE

genetic reassortment an alternative name for INDEPENDENT ASSORTMENT

genetic recombination *see* recombination

genetic screening techniques for checking whether an individual has a particular genetic condition (e.g. although the symptoms of Huntington's disease do not usually appear until middle age, people who may be at risk through family history, can be tested to see whether they carry the dominant harmful allele)

geneticist a scientist who specialises in GENETICS

genetics the scientific study of INHERITANCE and the way that genes express themselves in an individual

genic mutation alternative name for *gene mutation*; *see* mutation

genital herpes a viral infection of the genital and anal regions; *see* herpes

genital warts one of the sexually transmitted diseases; results in warts in the genital area; transmitted during sexual intercourse and caused by the human papilloma virus

genitals the sex organs, especially the external organs such as the penis, testes and labia; also called *genitalia*

genome the complete set of genetic material in a cell; an organism's complete set of DNA; *see also* Human Genome Project

genotype the genes present in a cell or an individual; usually taken to refer to the specific characters under consideration; e.g. for tongue rolling there are two alleles (rolling, *R*, and non-rolling, *r*); since each individual has two alleles for each character, there are three possible genotypes—*RR* (homozygous dominant), *Rr* (heterozygous) and *rr* (homozygous recessive); *see also* phenotype

genus (plural *genera*) one of the levels of biological classification; the principal division of a FAMILY; consists of one or more similar species, e.g. the genus *Pan* includes two species, *Pan troglodytes* (the common chimpanzee) and *Pan paniscus* (the pygmy chimpanzee or bonobo); humans belong to the genus *Homo*

geographic range an alternative term for GEOGRAPHICAL DISTRIBUTION

geographical barriers features of the landscape that prevent a species from interbreeding; include oceans, mountain ranges, large lake systems, deserts and expansive ice sheets

geographical distribution the part of the earth in which a species, race or population occurs; *see also* distribution

geographical isolation the separation of one gene pool from another by geographical barriers such as oceans, mountain ranges, ice sheets or deserts; interbreeding between populations on either side of the barrier is prevented or restricted; in time may lead to development of subspecies or even separate species; has resulted in the development of human populations with differing characteristics

geological time scale a scale that divides up the period of time from the beginning of the earth to the present day; evidence from different types of rocks and from fossils is used to make up the divisions of time on the scale; major divisions of time are ERAS which are subdivided into PERIODS, which are further subdivided into EPOCHS; *see* geological time scale, opposite

geomagnetic dating a method of absolute dating that relies on the fact that the earth's magnetic field has changed direction at known times in the past; any metals in sediments take up the direction of the earth's field at the time they are deposited, and the magnetic field of the metals then remains fixed; the direction of the magnetic field in metals in sedimentary deposits can be used for dating or as a check against dates determined by other methods

geriatrics the field of medicine that specialises in the prevention and treatment of diseases in the aged

germ the common name for a micro-organism that causes disease; *see also* microbe

germ cell a HAPLOID CELL capable of fusing with another haploid cell from the opposite sex to form a zygote; a sperm or an egg cell; alternative name for a *gamete*

germ layer one of the layers of cells found in the early embryo; *see* primary germ layers

germ line a group of cells that gives rise to the sex cells of an organism; only germ-line cells have the potential to divide by meiosis, producing gametes

germinal mutation an alternative name for GERM-LINE MUTATION

germ-line mutation a change in the hereditary material of an egg or sperm cell that becomes incorporated into the DNA of every cell in the body of the offspring; passed on from parents to offspring; also called *germinal mutation* or *hereditary mutation*

gerontology the scientific study of ageing processes, especially in humans

gestation the period of time between fertilisation and birth; in humans, a period of 40 weeks; also known as PREGNANCY

gibbon the common name for the small, slender, long-armed, tree-dwelling anthropoid apes belonging to the genus *Hylobates*; found throughout South-East Asia; gibbons, and the closely related siamangs, are known as the LESSER APES

Table 5 *The geological time scale*

Eras	*Periods*	*Epochs*	*Millions of years (BP)*
Cainozoic (Cenozoic)	*Quaternary*	*Holocene (Recent)*	*0.011 (11 000 years)*
		Pleistocene	*1.8*
	Tertiary	*Pliocene*	*5.3*
		Miocene	*23*
		Oligocene	*34*
		Eocene	*56*
		Palaeocene	*65*
Mesozoic	*Cretaceous*		*145*
	Jurassic		*200*
	Triassic		*251*
	Permian		*299*
Palaeozoic	*Carboniferous*		*359*
	Devonian		*416*
	Silurian		*444*
	Ordovician		*488*
	Cambrian		*542*
	Ediacaran		*620*
Proterozoic			*2500*
Archaeozoic (Archaean)			*3800*
Hadean (Azoic)			*4550*

GIFT *see* gamete intrafallopian transfer

gigantism the abnormally large development in size or stature of the body, or parts of the body; frequently due to the oversecretion of growth hormone from the anterior lobe of the pituitary gland; also referred to as *giantism*

gingivitis inflammation of the gums

girdle the bones to which the arms and legs are attached; *see* pectoral girdle, pelvic girdle

glacial period an alternative name for ICE AGE

gland an organ or tissue that produces one or more SECRETIONS; exocrine glands secrete into a duct and endocrine glands secrete into the EXTRACELLULAR FLUID; *see also* Table of endocrine glands on page 128

glandular epithelium epithelium specialised for secretion; *see* epithelial tissue

glandular fever a contagious disease that affects the lymphoid tissues, resulting in abnormal lymphocytes; caused by the Epstein–Barr virus; occurs mainly in children and young adults; also called *infectious mononucleosis*

glans penis the slightly enlarged head of the penis

glaucoma a disease of the eye caused by an excessive accumulation of fluid inside the eyeball; if not treated, can cause blindness due to prolonged pressure on the nerve cells of the retina

glial cells an alternative name for NEUROGLIA

gliding joint a type of SYNOVIAL JOINT

globular protein PROTEIN with molecules of spherical shape

globulin a group of proteins that are insoluble in water but soluble in salt solution, and are coagulated by heat; present in many plant and animal tissues; make up 38% of the proteins in the blood plasma; occur in the antibodies in the blood; best known is gamma globulin, which can combine with the protein of the measles virus, the hepatitis virus and the tetanus bacterium

glomerular capsule the double-walled cup-like structure at the end of each kidney tubule; collects water and other substances filtered from the blood; also called *Bowman's capsule*; *see also* diagram of renal corpuscle top right and diagram of kidney nephron (p. 183)

glomerular filtration FILTRATION of blood in the kidneys

glomerulonephritis inflammation of the glomeruli of the kidney; also called *Bright's disease*

glomerulus the tightly coiled mass of capillaries that is surrounded by the expanded part of each kidney tubule; water and dissolved substances filter out of the blood from the capillaries of the glomerulus; *see also* diagram of renal corpuscle, below and diagram of kidney nephron (p. 183)

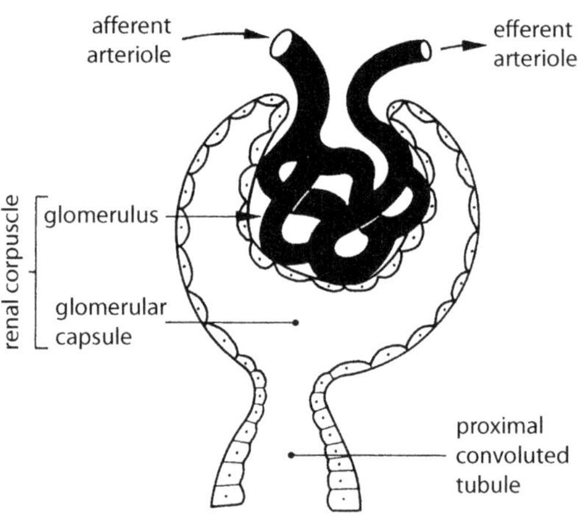

A renal corpuscle

glottis the opening between the VOCAL FOLDS; the entrance to the larynx; *see also* diagram below

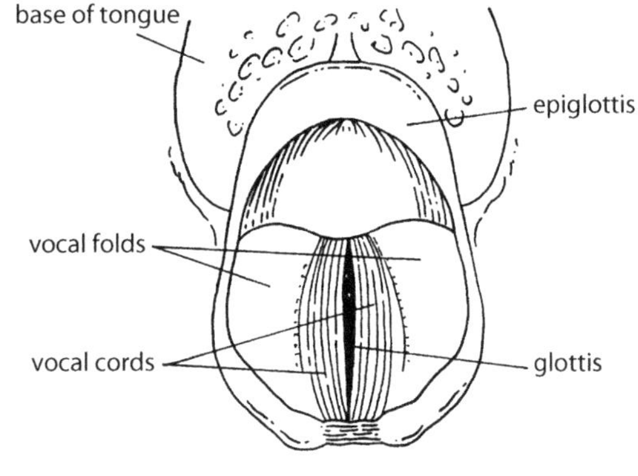

Glottis

glucagon a hormone secreted by the alpha cells of the islets of Langerhans of the pancreas; increases blood sugar level by promoting the conversion of glycogen to glucose in the liver; secretion of glucagon is controlled by blood sugar level through a feedback mechanism; *see also* insulin, table of hormones (p. 128)

glucocorticoids a group of hormones that are secreted by the adrenal cortex; three important glucocorticoids are *cortisone*, *cortisol* (*hydrocortisone*) and *corticosterone*; work with other hormones to promote normal metabolism by making sure adequate energy is available; also work to provide resistance to stress and act as ANTI-INFLAMMATORY agents

gluconeogenesis the process of producing new sugar (glucose) molecules from fats and amino acids; occurs in the liver and is stimulated by a number of hormones such as cortisol, thyroxine, adrenaline and glucagon

glucose $C_6H_{12}O_6$; the most common simple (monosaccharide) sugar; the main energy source for all cells

glutamic acid one of the 20 amino acids that are common in proteins; not essential in the human diet; also known as *glutamate*; *see also* list of amino acids (p. 11)

glutamine one of the 20 amino acids that are common in proteins; not essential in the human diet; *see also* list of amino acids (p. 11)

gluteal muscles the three pairs of muscles that make up the buttocks, joined to the pelvis at one end and to the femur at the other; hold the trunk erect on the legs and also move the leg at the hip; the largest pair is the *gluteus maximus*; *see also* diagram, right

gluten a protein that occurs in wheat, oats, barley and rye

gluten intolerance an alternative name for COELIAC DISEASE

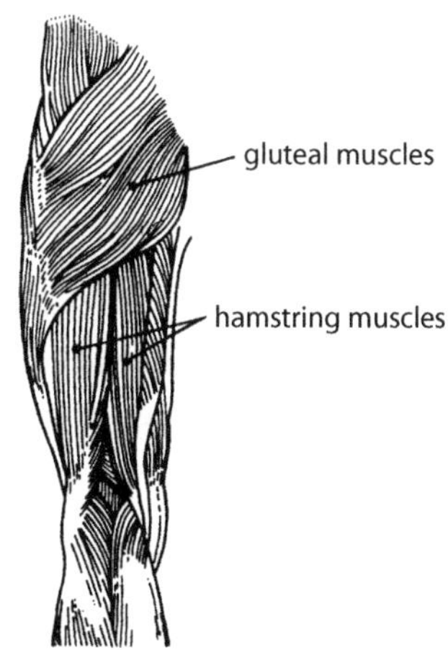

Gluteal muscles

gluteus maximus the largest of the GLUTEAL MUSCLES

glycaemic index (GI) a measure of the speed with which glucose is released from a carbohydrate-based food, and thus the speed with which blood sugar rises; glucose has a GI of 100; fast-releasing foods raise blood sugar quickly and are high on the index; foods low on the index give a slower, more sustained release of glucose

glycerol $C_3H_8O_3$; a component of a fat (or lipid) molecule; fat molecules consist of one glycerol molecule and up to three fatty acids; also known as *glycerine*

glycine one of the 20 amino acids that are common in proteins; not essential in the human diet; *see also* list of amino acids (p. 11)

glycogen a polysaccharide made up of thousands of glucose molecules bonded together in branching chains; functions as a store of glucose molecules in liver and muscle cells

glycogenesis the process whereby glucose molecules are chemically combined in long chains to form glycogen molecules; most glycogen is stored in the liver and in skeletal muscle fibres; glycogenesis is stimulated by the hormone insulin

glycogenolysis the process of converting glycogen back to glucose; glycogen stored in the liver is converted into glucose and released into the bloodstream; usually occurs between meals

glycolysis the first stage of cellular respiration, in which glucose molecules are converted to two pyruvic acid molecules; releases enough energy to form two molecules of ATP; *see also* anaerobic respiration

goblet cell a mucus-secreting cell found in the epithelial tissue of the intestines and the respiratory tract; has an enlarged opening shaped like a wine glass (goblet) near the surface of the tissue; also called a *chalice cell*; *see* diagram, below

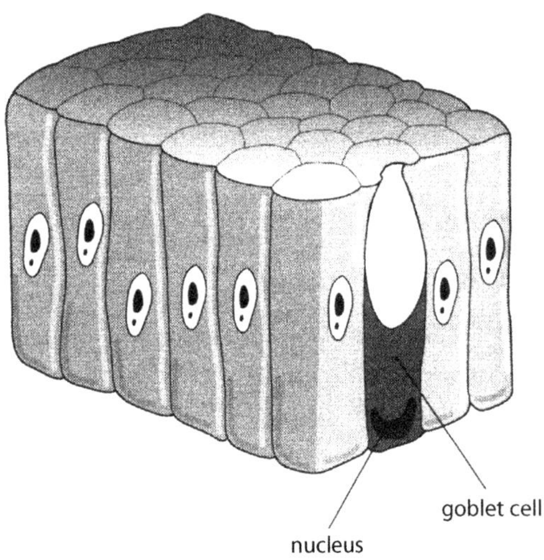

Goblet cell

goitre a swelling of the neck caused by an enlargement of the thyroid gland; usually caused by lack of iodine in the diet (thyroxine, the hormone secreted by the thyroid, contains iodine); *see also* cretinism

Golgi apparatus a structure in the cytoplasm of a cell consisting of a stack of flattened channels; packages materials for secretion from the cell; common in cells that produce secretions; also called *Golgi body* or *Golgi complex*; *see also* diagram of cell structure (p. 44)

gonads an alternative name for PRIMARY SEX ORGANS

gonadocorticoids sex hormones (androgens and oestrogens) secreted by the adrenal cortex; normally of such low concentration that they have no significant effect

gonadotropic hormones hormones secreted by the anterior (front) lobe of the pituitary, or the placenta, that affect the functioning of the gonads (sex organs) (e.g. follicle-stimulating hormone and luteinising hormone from the pituitary, and human chorionic gonadotropin from the placenta); also known as *gonadotropins* or *gonadotrophins*; *see also* table of hormones (p. 128)

gonadotropin *see* gonadotropic hormones

gonorrhoea one of the most common sexually transmitted diseases, transmitted by the bacterium *Neisseria gonorrhoeae* during sexual intercourse; infects the mucous membranes of the reproductive organs; often referred to as 'the clap'

gorilla the common name for the largest of the anthropoid apes, *Gorilla gorilla*; a herbivorous, ground-dwelling primate found in the tropical and mountain forests of western and central equatorial Africa

gossypol a chemical derived from cottonseed oil which was used as the basis for a male contraceptive pill because it inhibits an enzyme needed for sperm development; the drug was shelved when it was discovered that prolonged use could lead to shrunken testes, and that in some males fertility did not return after they stopped using the drug

gout an inherited genetic defect that results in an increased level of uric acid in the blood; this may lead to inflammation, swelling and intense pain in the joints; often referred to as *gouty arthritis*

Graafian follicle a fluid-filled spherical structure in the ovary containing an immature egg and its surrounding tissues that secrete oestrogens; forms deep inside the ovary and gradually moves to the outside as it enlarges and matures; eventually releases a mature egg by rupture of the wall; in humans the term Graafian follicle is normally used to describe a mature follicle; *see also* ovarian cycle, diagram of Graafian follicle below

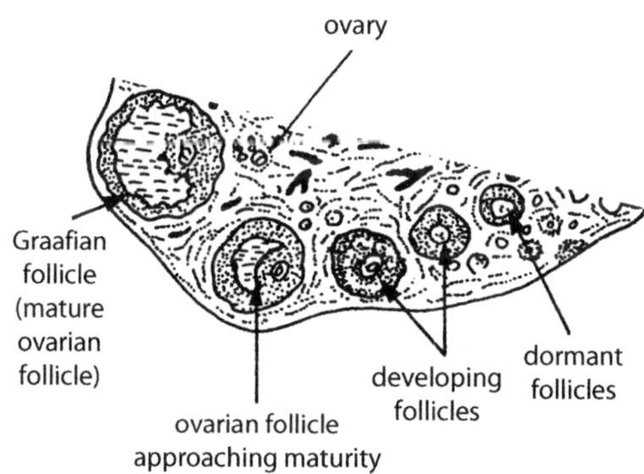

Graafian follicle

gracile slender in form; used to refer to the body build of primates, especially the more recent ancestors of humans such as the australopithecines (e.g. *Australopithecus afarensis* and *Australopithecus africanus* are gracile forms and contrast with the heavier, more robust forms *Australopithecus robustus* and *Australopithecus boisei*); *see also* robust

gradualism the idea that evolution occurs slowly by the accumulation of small changes over long periods of time; *see* punctuated equilibrium, for an opposing view of evolution

graft a piece of tissue taken from one part of the body to replace diseased or injured tissue removed from another part of the body (e.g. a skin graft can be used to cover an area where the skin has been burnt)

Gram's stain a stain containing crystal violet and iodine that is used to distinguish between two main groups of bacteria; bacterial cells such as *Staphylococcus* and *Streptococcus* absorb the stain, become deep purple in colour and are known as Gram positive; bacterial cells that do not stain, such as *Gonococcus*, are Gram negative

granular endoplasmic reticulum ENDOPLASMIC RETICULUM with ribosomes attached

granulocyte a type of white blood cell; *see* leucocyte

graph a diagram showing the relationship between two or more variables; a *line graph* is a graph in which the relationship between two variables is shown by a line joining points where values have been determined; a *histogram* is a graph in which one variable is represented by rectangles; the height of each rectangle represents the value of the variable; often used to show distributions of frequency; *see* illustration on next page for the types of graphs described

Graves' disease a medical condition in which an overactive THYROID GLAND secretes excessive amounts of thyroid hormones (HYPERTHYROIDISM); named after Robert Graves, the 19th century Irish doctor who first described the condition

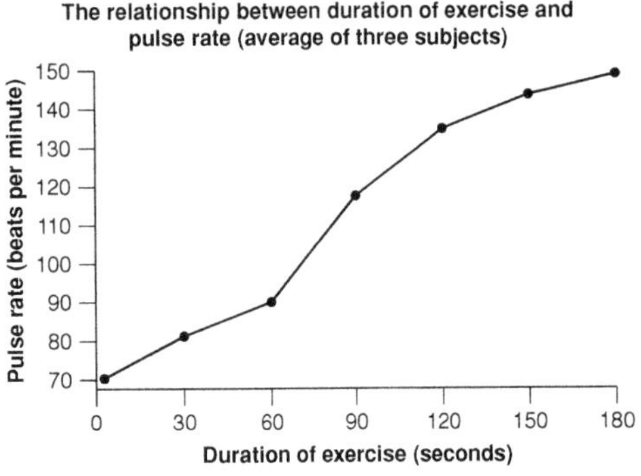

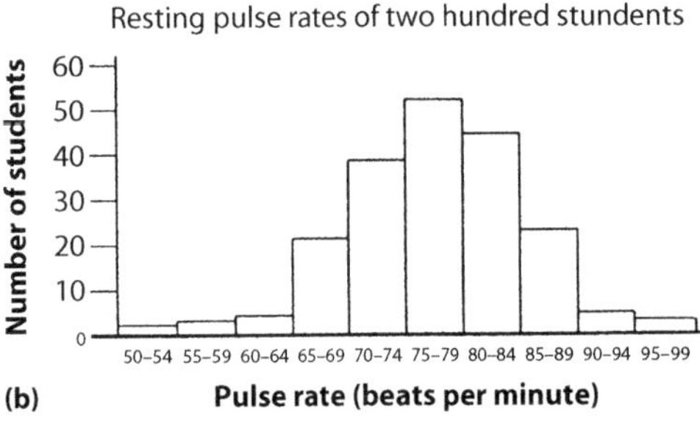

(a) Line graph (b) Histogram

great ape a gorilla, chimpanzee or orang-utan; *see also* ape

Great Rift Valley a valley in East Africa that is the junction between two major plates in the earth's crust; runs from the southernmost tip of Turkey, through Israel, along the Red Sea and into Africa, as far as the mouth of the Zambesi river; the Great Rift Valley is the most outstanding location in the world for fossils of early humans and human ancestors; evidence of over 4 million years of human evolution is buried in sediments that have undergone faulting and erosion so that fossils are exposed at the surface

greater vestibular glands an alternative name for BARTHOLIN'S GLANDS

greenstick fracture a broken bone in which one side of a bone is broken and the other bends; *see* fracture

grey matter the part of the brain and spinal cord made up of nerve cell bodies and unmyelinated fibres; in the spinal cord, the grey matter is in an H shape in the centre of the cord; in the brain, grey matter is at the surface; *see also* white matter

grooming the handling and cleaning of another individual's fur or hair to remove parasites and foreign matter; an important part of primate behaviour and social life; serves as a form of communication that soothes and provides reassurance

gross anatomy the branch of anatomy that deals with structures that can be studied without using a microscope; also referred to as *macroscopic anatomy*

growth an increase in the size of an organ or organism due to increasing number of cells or increased size of cells

growth hormone (GH) a hormone secreted by the anterior lobe of the pituitary; stimulates body cells to grow and multiply, especially the skeleton and skeletal muscles; also causes cells to change from breakdown of carbohydrates to breakdown of fats for energy; also known as *somatotropin* or *somatotropic hormone* (*STH*); *see* table of hormones (p. 128)

growth plate alternative name for EPIPHYSEAL PLATE

growth rate of a population, the death rate subtracted from the birth rate, expressed as a percentage; also referred to as the *rate of natural increase*

guanine one of the 4 bases that form part of the DNA and RNA molecule; *see* deoxyribonucleic acid, ribonucleic acid

gustation the sense of taste

gustatory cells taste-sensitive cells, grouped together as taste buds, which occur mostly on the tongue but also in the throat and on the soft palate; each gustatory cell has a *gustatory hair* that projects from the surface through an opening in the taste bud called a *taste pore*; the hairs detect the taste stimuli; *see* diagram, below

gustatory hair *see* gustatory cells

gut an alternative name for ALIMENTARY CANAL

gynaecologist a medical practitioner specialising in GYNAECOLOGY

gynaecology the diagnosis and treatment of disorders of the female reproductive system

gyri ridges on the surface of the cerebrum of the brain; also called CONVOLUTIONS

gyrus singular of GYRI

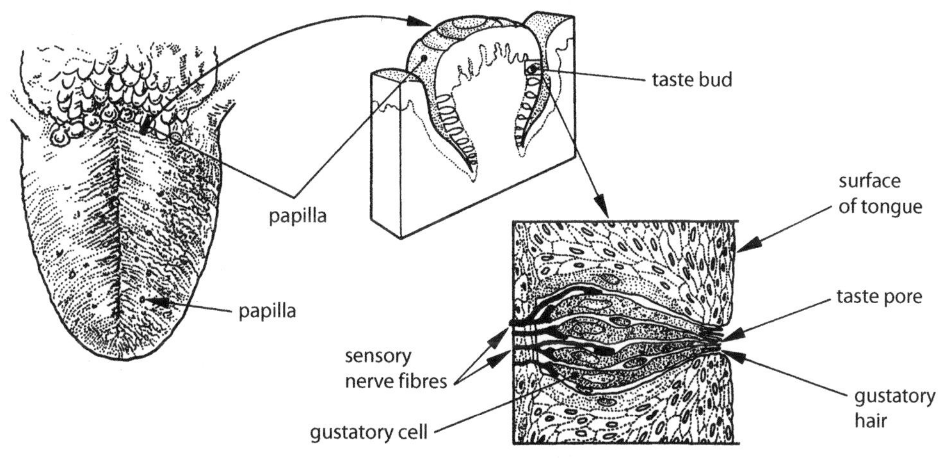

Gustatory cells

H

H zone part of the microscopic structure of skeletal muscle; *see* sarcomere

habilines a group of hominids known from fossils found in eastern and southern Africa, dated to between 2.4 and 1.6 million years ago; members of the species *Homo habilis* and *Homo rudolfensis*

Hadean one of the major subdivisions of geological time; the first era, from about 4550–3800 million years ago; molten material was solidifying to form rock during this era, and thus it is completely devoid of fossils; also known as the *Azoic era*; also referred to as an EON in some time scales; *see* geological time scale (p. 111)

haem-, haemo- prefixes meaning blood (e.g. haematocyte—a blood cell)

haematocrit the percentage of the blood volume that is made up of red blood cells; determined by centrifuging a blood sample and measuring the ratio of the packed red cells to the total sample volume; in males, normally 40–54% (average 46%); in females, 38–47% (average 41%); in anaemic persons can be as low as 15%

haematocyte the general name for a blood cell

haematopoiesis production of blood cells; occurs in red bone marrow; also called *haemopoiesis*

haemochromatosis an inherited disorder in which the body absorbs too much iron from food; the iron is stored in the liver, pancreas and other organs, causing liver damage, diabetes and other problems

haemodialysis an alternative term for DIALYSIS

haemoglobin the oxygen-carrying molecule in red blood cells; gives the blood its red colour; each red blood cell has up to 300 million molecules of haemoglobin; when combined with oxygen is bright red but becomes more bluish when oxygen is lost; the molecule consists of globin, a protein containing 4 polypeptide chains, each of which is combined with a haem group that contains an iron atom; each haem group can combine with an atom of oxygen; when combined with oxygen it is known as *oxyhaemoglobin*; haemoglobin is also capable of carrying some carbon dioxide in combination with the protein; it is then known as *carbaminohaemoglobin*; around 20% of the carbon dioxide carried by the blood is transported as carbaminohaemoglobin; *see also* diagram of chemical structure, opposite

haemoglobin A the normal form of HAEMOGLOBIN found in adult humans

haemoglobin S the most common type of abnormal HAEMOGLOBIN;

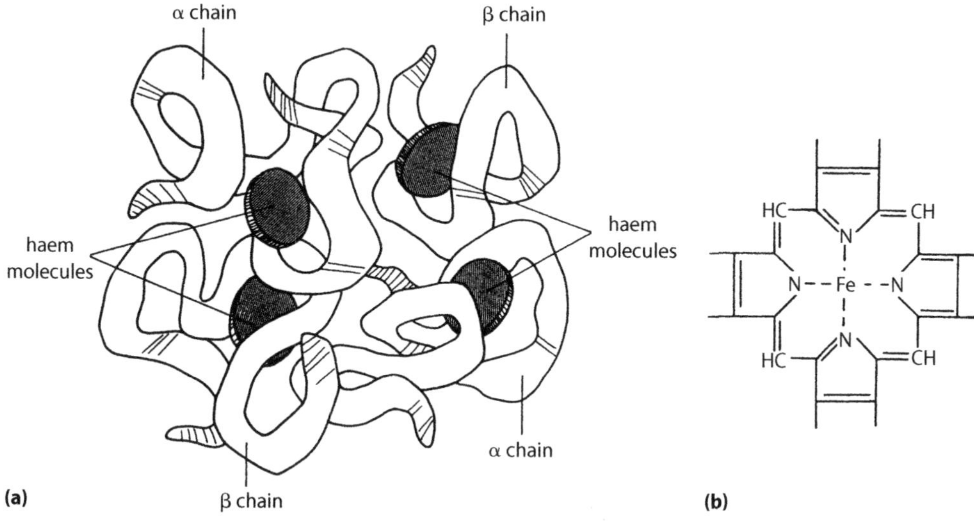

(a) A haemoglobin molecule; (b) a haem group

differs from normal adult haemoglobin (HAEMOGLOBIN A) by a single AMINO ACID (a VALINE replacing a GLUTAMINE); forms the basis of SICKLE-CELL ANAEMIA

haemolytic disease of the newborn an alternative name for *erythroblastosis foetalis*; *see* erythroblastosis

haemophilia refers to a number of blood disorders in which excessive bleeding occurs after only a minor cut or scratch, or internal bleeding may occur without any injury; due to insufficient production of certain chemical factors involved in blood clotting; symptoms of each disorder are the same but the clotting factor involved is different; treated by transfusions containing the deficient factor; haemophilia is an inherited disease caused by a recessive allele carried on the *X*-chromosome (it is *X*-linked)

haemophilia A HAEMOPHILIA caused by an inherited deficiency of FACTOR VIII; occurs almost exclusively in men; frequently referred to as *classic haemophilia*

haemopoiesis an alternative name for HAEMATOPOIESIS

haemorrhage bleeding; loss of blood from blood vessels

haemorrhoids dilated or abnormally enlarged blood vessels in the anal region; also called *piles*

hair an outgrowth of the EPIDERMIS; hairs may also be called *pili*; hair's main function is protection, and it is found on most parts of the body, excluding the lips, palms, soles, nipples, penis and labia minora, and the inner surface of the labia majora; each hair has a shaft projecting above the skin surface and a root that extends into the DERMIS; hairs consist of a column of cells in which the cytoplasm has been replaced by keratin; formed by cell division at the base of the HAIR FOLLICLE; when a hair is lost a new one is produced within the same follicle

hair follicle a layer of cells that forms a sac around the root of a hair; extends from the epidermis into the dermis of the skin; at the base of each follicle is a group of dermal cells, the papilla, which contains blood vessels to nourish the follicle cells as they repeatedly divide to form the hair; oil-secreting glands (sebaceous glands) open into the hair follicle; *see* diagram of skin (p. 250)

half-life the time required for one-half of any quantity of radioactive material to decay into stable non-radioactive material (e.g. carbon-14 has a half-life of 5730 years—it takes 5730 years for half of a given quantity of carbon-14 to decay to nitrogen; uranium-238 has a half-life of 4500 million years)

hallucinogen one of a group of drugs that affect mood, perception and thinking, e.g. mescaline (from the peyote cactus), psilocybin (from some mushrooms), and LSD and ecstasy (both made in laboratories); also called *psychedelic drug*

hammer one of the bones of the middle ear; *see* auditory ossicles

hamstrings a group of three muscles that lie at the back of the thigh, attached at one end to the pelvis and at the other to the tibia; contract to bend the leg at the knee; also used to describe the tendons at the back of the knee; see diagram (p. 113)

hand-axe a general term for a fist-sized lump of stone chipped into a shape resembling a flattened pear, the sharpened edges of which were formed by striking flakes from both sides; used extensively by early humans to cut up carcasses already skinned by sharp stone flakes; the earliest hand-axes appeared about the same time as *Homo erectus* and became the predominant tool in the kit of early stone age hunters

haploid the chromosome number found in mature sex cells—eggs or sperm—symbolised by *n*; made up of one chromosome from each homologous pair; in humans, such cells have 23 chromosomes; *see also* diploid

haploid cell a cell containing one of each type of chromosome, that is, half the number of chromosomes of a normal body cell; an egg or sperm cell; in humans, haploid cells have 23 chromosomes; haploid cells or organisms are represented by the symbol *n*

haplorrhine a primate without a moist nose (e.g. a tarsier, monkey, ape or human); in some classifications, a member of the suborder Haplorrhini; a suborder proposed by some taxonomists so that the tarsiers can be separated from animals like lorises and lemurs, because tarsiers are considered to be closer to anthropoids than to the other prosimians

Hardy–Weinberg equilibrium a mathematical model put forward by G. H. Hardy and W. Weinberg; demonstrates that gene frequencies will remain constant for generation after generation, in a large, randomly interbreeding population in which selection pressure, mutation pressure and gene flow are absent

Hashimoto's disease a medical condition in which inflammation results in an underactive THYROID GLAND and thus decreased secretion of thyroid hormones (HYPOTHYROIDISM); named after Hakaru Hashimoto, the Japanese doctor who first described the symptoms of the disease in 1912; also known as *chronic lymphocytic thyroiditis*

hashish a drug produced from the resin and pollen of the Indian hemp plant *Cannabis sativa*; usually smoked; contains more of the active ingredient than

MARIJUANA; hashish oil is a concentrated liquid extracted from the plant

Haversian canal a circular channel running through the centre of a Haversian system of compact bone; contains connective tissue, blood vessels and nerves; the Haversian canals branch throughout the bone and join up with the marrow in the centre of the bone and the fibrous sheath around the outside of the bone (periosteum); also called the *central canal*; *see also* diagram of microscopic structure of bone (p. 34)

Haversian system the basic unit of structure of compact bone; consists of a Haversian canal surrounded by concentric layers of hard matrix and bone cells; also called an *osteon*; *see also* diagram of microscopic structure of bone (p. 34)

heart a hollow, muscular organ that pumps blood; *see* diagram below

heart attack damage to part of the heart due to a decrease in blood supply (and therefore oxygen supply) to that part of the heart muscle; usually due to a blood clot becoming lodged in a branch of the artery that supplies the heart muscle (coronary artery); also called a *coronary thrombosis* or *myocardial infarction*; *see also* thrombosis

heart rate the number of times the heart beats per minute

heart sounds the sound of the heartbeat; usually heard through a stethoscope; caused by the turbulence of blood flow as the valves close; the first sound (*lubb*) is a long, booming sound caused by the closing of the atrioventricular valves; the second sound (*dupp*) is a short, sharp sound and is caused by the closing of the semilunar valves

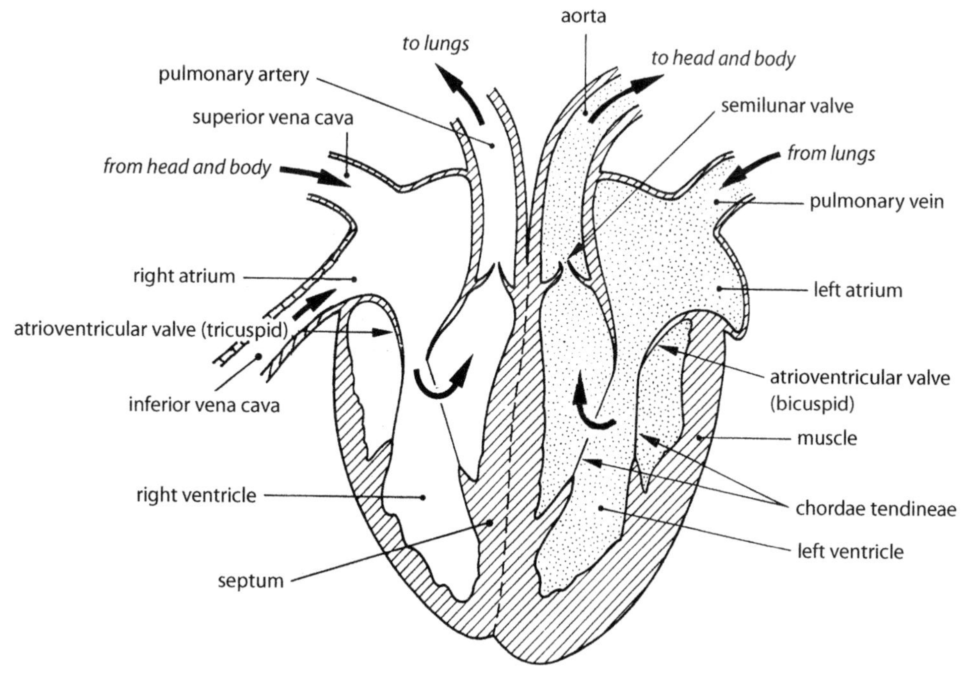

The heart

heat cramps muscle spasms that result from the loss of substances dissolved in sweat, especially ELECTROLYTES; one of the effects of exposure to excessive heat

heat exhaustion the collapse of a person after exposure to heat, during which the body's heat-regulating mechanism continues to function normally; characterised by cool, clammy skin and profuse perspiration; fluid and salt loss results in muscle cramps, dizziness, vomiting and fainting; also called *heat prostation*; not as serious as heatstroke

heat prostation an alternative name for HEAT EXHAUSTION

heatstroke the failure of a person's temperature-regulating mechanisms when exposed to excessive heat; characterised by reduced perspiration; a serious condition, because the body cannot easily lose heat and body temperature rises rapidly; also called *sunstroke*

Heimlich manoeuvre an alternative name for ABDOMINAL THRUST MANOEUVRE

Helicobacter pylori a bacterium that infects the stomach and DUODENUM; causes GASTRITIS and ulcers of the stomach and duodenum

helper T cell a type of T CELL

hemiplegia paralysis of both the upper and lower limbs and the trunk of one side of the body; *see also* paraplegia, quadriplegia

hemizygous having only one allele of a gene instead of the normal two; because males have only one *X*-chromosome, they are hemizygous for all the genes carried in that chromosome; *see also* homozygous, heterozygous

hepar-, hepato- prefixes referring to the liver, e.g. hepatitis is an inflammation of the liver

heparin an anticoagulant; helps to prevent BLOOD CLOTTING by inhibiting the conversion of prothrombin to thrombin; produced naturally by mast cells, particularly in the lungs and liver; also given artificially to prevent clotting during procedures like open-heart surgery

hepatic artery the artery that carries blood to the stomach and liver; *see also* diagram of major arteries (p. 20)

hepatic portal vein the vein that carries blood from the stomach and intestines to the liver; *see also* diagram of major veins (p. 284)

hepatitis inflammation of the liver; may be caused by viruses, drugs or chemicals; there are a number of viral types; *hepatitis A* is caused by the hepatitis A virus and is spread by contamination of food or utensils with faeces; it is a less serious disease than that caused by the hepatitis B virus; *hepatitis B* is spread by transfer of contaminated blood, especially via syringes, or by any other contaminated body fluid such as semen, saliva or tears; the chronic liver inflammation can last for many years or even for life; *hepatitis C* is another viral infection that results in severe impairment of normal liver functioning; a blood test for hepatitis C was developed in 1989 along with a treatment program that was able to control the infection; *hepatitis E* is spread through water contaminated with faeces; it causes anorexia, nausea and vomiting; a *hepatitis F* has been identified and it is spread through contaminated faeces but little else is known about it; *hepatitis G* is caused by a virus that is spread by blood-to-blood contact; the virus does not cause any apparent symptoms

hepatocytes liver cells; the basic building blocks of the liver; perform functions such as secreting bile, removing proteins from the blood, regulating blood volume, producing blood-clotting agents and metabolising carbohydrates, proteins and fats

herbivorous describes an organism that feeds on plants (e.g. a sheep or cow)

herd immunity a type of immunity that occurs because such a high proportion of people in a population are immunised, that those who are not immune are protected

hereditary describes characteristics capable of being transmitted genetically from parents to offspring (e.g. eye colour is a hereditary characteristic)

hereditary disease an alternative name for GENETIC DISEASE

hereditary spastic paraplegia (HSP) a degenerative disorder of the SPINAL CORD; results in progressive weakness and stiffness of the legs

heredity the transmission of inherited characteristics from parents to offspring

hernia the protrusion of an organ through a weak area in a membrane or cavity wall, often the abdominal cavity; also called a *rupture*; a *hiatus hernia*, or *hiatal hernia*, is where the lower oesophagus, stomach or small intestine protrudes into the chest cavity in the area where the oesophagus passes through the diaphragm

heroin a drug derived from OPIUM

herpes an infection with herpes virus that causes small blisters in mucous membranes or the skin; can remain latent for long periods, becoming active at irregular intervals; *Herpes simplex* type 1 produces cold sores, especially around the mouth; *Herpes simplex* type 2 causes *genital herpes*, which is transmitted during sexual contact—it produces persistent and painful blisters of the genital and anal regions

Herpes simplex *see* herpes

Herpes zoster *see* chicken pox

heter-, hetero- prefixes meaning different, e.g. heterozygous—having different alleles for a characteristic

heterozygous the condition in which an individual possesses different alleles for a given characteristic (e.g. an individual may have one allele for ability to roll the tongue, *R*, and one allele for non-rolling, *r*, and would thus have the genotype *Rr* for that characteristic; such an individual is known as a *heterozygote*); *see also* homozygous, hemizygous

hexose sugar a sugar with molecules containing 6 carbon atoms (e.g. galactose, glucose, fructose); hexoses are joined to make complex carbohydrates; *see also* pentose sugar

hiatus hernia a HERNIA of the diaphragm; also called *hiatal hernia*

hiccup a rapid, reflex inspiration that is suddenly stopped by the GLOTTIS closing; the reflex involves the phrenic and vagus nerves and the medulla of the brain; has no known physiological function; also known as a *singultus*; also spelt *hiccough*

high energy bond a chemical bond containing much more energy than is usual in a bond (e.g. in an ATP molecule, the bonds between the AMP and the 2nd and 3rd phosphate groups)

high power the highest magnification on a light microscope

hilus an area, usually a depression, where nerves and blood vessels enter and leave an organ (e.g. the *renal hilus*, where the renal artery and renal vein join the kidney); also called *hilum*

hinge joint a type of SYNOVIAL JOINT

hip girdle an alternative name for PELVIC GIRDLE

histamine a substance that is present in many tissues, especially certain connective tissue cells, blood platelets and certain white blood cells; is released in response to injury to cells that contain it; causes vasodilation and increased permeability of blood vessels in the injured area—a part of the inflammatory response; *see also* inflammation

histidine one of the 20 amino acids that are common in proteins; essential in the human diet; *see also* list of amino acids (p. 11)

histogram a GRAPH in which one variable is represented by rectangles; *see* diagram (p. 111)

histology the study of tissues; in particular, the study of the microanatomy of cells, tissues and organs

histones proteins found in the nucleus of a cell; DNA wraps around the histones to form chromosomes

HIV/AIDS HUMAN IMMUNODEFICIENCY VIRUS and the symptoms that result from the infection—ACQUIRED IMMUNE DEFICIENCY SYNDROME

hobbit common name for *Homo floresiensis*

Hodgkin's disease a cancer that usually begins in the lymph nodes; can be managed in the early stages by carefully applied X-rays, and in later stages by drug therapy; cause is not known, but it is associated with the Epstein–Barr virus (the virus that causes glandular fever), which has links with the cancer Burkitt's lymphoma

holandric gene a gene that is located on the *Y*-chromosome and which therefore occurs only in males

Holocene the later of the two epochs in the Quaternary period of geological time; spanning the past 11 000 years; also called the *Recent* epoch; *see* geological time scale (p. 111)

hom-, homeo-, homo- prefixes meaning the same as, similar or like (e.g. homeothermic—maintaining the body at the same temperature)

home base a camp site where hunters brought back food for sharing with other members of their group

home range an area through which an animal or group of animals moves and feeds; *see* territory

homeostasis the maintenance of a relatively constant internal environment despite fluctuations in the external environment; all the organ systems of the body contribute to homeostasis; also spelt *homoeostasis; see also* steady state

homeothermic able to maintain a body temperature independent of the external environmental temperature; mammals and birds are homeothermic; also spelt *homoeothermic* or *homothermic*; also known as *warm blooded*

hominid a primate belonging to the family Hominidae; includes orang-utans, gorillas, chimpanzees and humans

Hominidae the family to which all present-day and extinct ancestors of orang-utans, chimpanzees, gorillas and humans belong

hominin a primate belonging to the tribe Hominini; includes modern humans and extinct ancestors of humans

Homininae a subfamily that includes chimpanzees and humans

hominine a primate belonging to the subfamily Homininae; includes chimpanzees and humans

Hominini a tribe that includes modern humans and our extinct ancestors

hominoid human-like in appearance; a member of the superfamily Hominoidea, which includes modern humans and the anthropoid apes

Hominoidea one of the superfamilies of anthropoids; includes apes and humans; possess a shoulder structure adapted for climbing and hanging, lack a tail, are generally larger than monkeys and have the largest brain size to body size ratio among primates; a member of this superfamily is called a *hominoid*; *see also* classification of primates (p. 219)

Homo a genus of hominid with a number of recognised species, *Homo habilis*, *Homo erectus* and *Homo sapiens*; the genus dates from about 2.5 million years ago; the major features characterising *Homo* are large brain size, reduction in size of face and teeth, and an increased use of tools and other cultural adaptations

Homo antecessor a species of *Homo* named in 1997 from a fossil skull discovered in a cave in Spain; claimed to be the ancestor of modern humans and of the Neanderthals because the face was similar to modern *Homo sapiens* but the brow and jaw had Neanderthal characteristics; highly controversial

Homo erectus a species of the genus *Homo* that lived between 1.6 million and 200 000 years BP; first appeared in Africa and later spread to Asia and possibly into Europe; possessed a larger brain than *Homo habilis* but not as large as *Homo sapiens*; the first hominids to use fire; 'Peking man' and 'Java man' are examples of *Homo erectus*

Homo ergaster a species of *Homo* represented by fossils found in Eastern Africa and dated from 1.8 to 1.4 million years ago; characterised by a smaller jaw, a more projecting nose and more modern proportions of the arms and legs than earlier forms of *Homo*; classified by many as *Homo erectus*, but proponents of *Homo ergaster* claim that the African fossils vary enough from those found in Asia to warrant a different name

Homo floresiensis a possible species in the genus *Homo* found on the Indonesian island of Flores; characterised by a small brain and small body; lived until relatively recent times

Homo habilis the oldest known species in the genus *Homo*; dated to 2–1.5 million years BP; to date found only in Africa; in overall appearance, this species is similar to the australopithecines but has a larger cranial capacity; considered to be adept at tool making

Homo heidelbergensis a name now being used by some authorities to describe a type of *Homo* that migrated into Europe about 500 000 years ago; *Homo heidelbergensis* had more modern features than *Homo erectus* and later evolved into Neanderthals; other scientists believe that all

Homo fossils found outside Africa are either *H. erectus* or *H. sapiens*

Homo modjokertensis a fossil found at Modjokerto in Java in 1936; now considered to be an example of *Homo erectus*

Homo neanderthalensis a regional population that existed in Europe during the last of the ice ages; associated with Mousterian culture and the Middle Palaeolithic period of 30 000 to 125 000 years ago; recent evidence suggests that the Neanderthals may date back as far as 230 000 years ago; once considered to be a variant of archaic *Homo sapiens* and classified as *H. sapiens neanderthalensis*, but DNA evidence indicates they were a separate species—*H. neanderthalensis*; some authorities include people of the time from West Asia as Neanderthals and refer to those in Europe as 'classic' Neanderthals; 'classic' Neanderthal features include a stocky build, a large head and brain and a projecting nose; the relationship between Neanderthals, and later *Homo sapiens* populations in Europe is still disputed; *see also Homo antecessor*

Homo rudolfensis a species of *Homo* represented by fossils found in Eastern Africa and dated from 2.5 to 1.9 million years ago; characterised by a long, broad face with flatter brow ridges and a larger, rounder braincase than *Homo habilis*

Homo sapiens modern humans; the species to which all present populations of humans belong; all hominids within the last 200 000 years are considered to belong to this species; usually classified as two groups, archaic *H. sapiens* and anatomically modern *H. sapiens*; archaic *H. sapiens* were an early variant of *H. sapiens* dating from approximately 200 000 to 32 000 years BP; archaic forms had roughly the same brain size as modern humans, but the skull was a different shape; skulls had a sloping forehead, the cranium was not as high and brow ridges were more pronounced; anatomically modern *Homo sapiens* is the modern form of the human species; dating to at least the last 100 000 years, and possibly even earlier; distinguished from the archaic form by smaller faces with less prominent brow ridges, high foreheads, a well developed bony chin, and longer but less robust limbs

homoeostasis an alternative spelling for HOMEOSTASIS

homoeothermic an alternative spelling for HOMEOTHERMIC

homologous chromosomes the pairs of CHROMOSOMES containing genes that control the same characteristics although the alleles of those genes may differ; the body cells of each individual are diploid and have homologous pairs of chromosomes, because one member of the pair comes from the male parent and one from the female; the gametes, being haploid, have only one member of each homologous pair of chromosomes

homologous organs organs with similar structures, but not necessarily similar functions; such organs derive from a common ancestry; *see also* diagram of homologous bones opposite

homozygous the condition in which an individual has the same alleles for a given characteristic at a given point in a chromosome (e.g. an individual may have two alleles for ability to roll the tongue, *RR*, or two alleles for non-rolling, *rr*; such an individual is known as a *homozygote* for that characteristic); the homozygous condition is also known as *pure breeding*

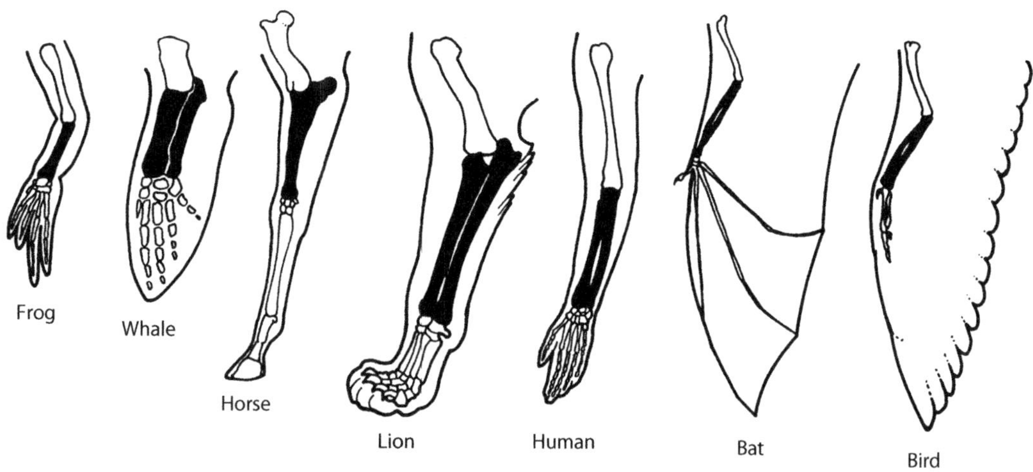

Homologous bones in the forelimbs of vertebrates

because two organisms homozygous for the same characteristic will always produce offspring with that characteristic; *see also* heterozygous, hemizygous

homunculus a distorted human figure that represents the size of parts of the brain devoted to different parts of the body; *see* diagram next page

hookworm a parasitic worm with hooked mouthparts that attaches itself to the inner lining of the intestine

hormonal contraception using reproductive hormones to control the production of eggs or sperm and thus CONCEPTION; most commonly used to refer to the CONTRACEPTIVE PILL or *birth control pill*

hormone a chemical secreted by an endocrine gland, and often carried in the blood, that affects the functioning of a cell or organ; chemical messengers; *local* (or *tissue*) *hormones* act on cells around the area where they are produced (e.g. the prostaglandins); *circulating hormones* are carried by the blood to act on distant targets (e.g. adrenaline, insulin and oestrogen); hormones may also be classified as exocrine, paracrine or autocrine; *exocrine hormones* are usually secreted into the blood, *paracrine hormones* usually act over a short distance and travel to the site of action in the extracellular fluid, *autocrine hormones* bring about a response by interacting with receptors on the surface of the cell that secretes them; *see also* individual hormones by name, table of hormones on pages 128–130

hormone clearance the breakdown of HORMONE molecules once they have produced the required effect; most breakdown occurs in the LIVER and KIDNEYS; degraded hormones are excreted in the BILE or URINE

hormone disrupters substances that interfere with the normal functioning of the endocrine system

hormone replacement therapy (HRT) low doses of oestrogen prescribed for women to relieve the symptoms of menopause

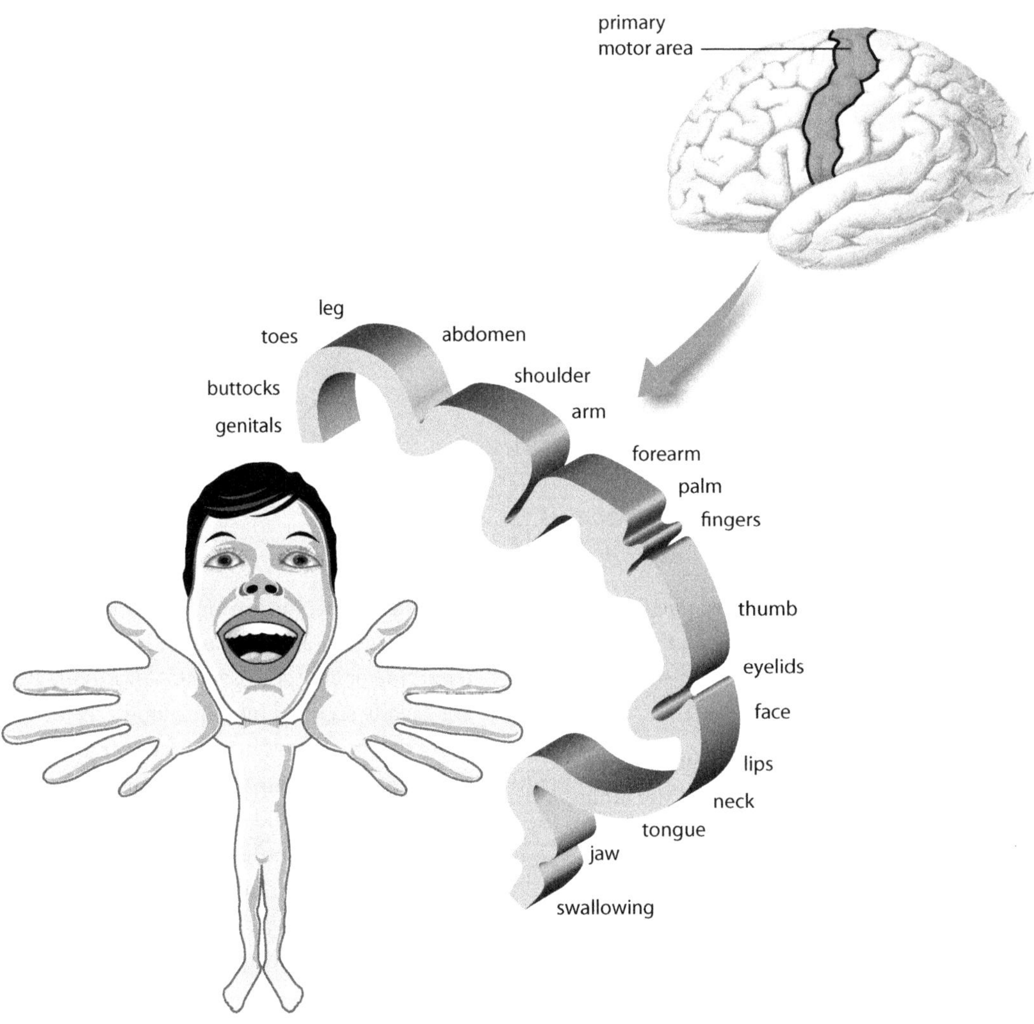

Motor homunculus

Table 6 *The major endocrine glands and their hormones*

Gland	*Hormone*	*Main effects*
Pituitary		
posterior lobe	antidiuretic hormone	stimulates water reabsorption in kidney tubules
(neurohypophysis)	oxytocin	stimulates contraction of uterine muscles during labour; stimulates contraction of cells in the mammary glands during suckling
anterior lobe (adenohypophysis)	follicle-stimulating hormone	stimulates gamete production in the gonads of both males and females; also, secretion of oestrogens from the ovaries of females

Gland	*Hormone*	*Main effects*
	luteinising hormone	females—stimulates ovulation and development of the corpus luteum after ovulation; males—stimulates testosterone secretion by the testes
	thyroid-stimulating hormone	stimulates thyroid gland to produce and release hormones
	adrenocorticotropic hormone	stimulates adrenal cortex to produce and release glucocorticoids, especially cortisol
	growth hormone	stimulates growth, especially of the skeleton; also causes changes in cellular metabolism
	prolactin	involved in milk secretion from mammary glands of females
	melanocyte-stimulating hormone	stimulates production of melanin by skin cells
Thyroid		
	thyroxine	regulates many metabolic processes; also important for growth and development
	calcitonin	lowers blood calcium and phosphate levels by increasing uptake of calcium by bones and inhibiting bone breakdown
Parathyroids		
	parathyroid	raises blood calcium level and reduces blood phosphate
Adrenal glands		
medulla	adrenaline (epinephrine)	prepares the body for fight-or-flight responses by raising blood glucose level, constricting blood vessels in skin, mucous membranes and kidneys and dilating bronchioles in the lungs
	noradrenaline (norepinephrine)	constricts blood vessels in many parts of the body, increases heart rate, increases force of contraction of heart muscle
cortex	mineralocorticoids (e.g. aldosterone)	control fluid and electrolyte balance by causing the kidney to retain sodium ions and excrete potassium ions
	glucocorticoids (e.g. cortisol)	work with other hormones to promote normal carbohydrate metabolism; tend to increase blood glucose level

(continued)

Table 6 *(continued)*

Gland	*Hormone*	*Main effects*
Pancreas		
	insulin	decreases blood glucose level by promoting transfer of glucose from blood into cells; promotes conversion of glucose into glycogen
	glucagon	generally acts in the opposite way to insulin; increases blood sugar level by promoting conversion of glycogen to glucose in the liver
Gonads		
ovaries	oestrogens	promote development and maintenance of female reproductive structures and secondary sexual characteristics
	progesterone	works with oestrogens to prepare the uterine lining for a fertilised egg and the mammary glands for milk secretion
testes	androgens (e.g. testosterone)	promote development and maintenance of male sex organs; stimulate sperm production and development of male secondary sexual characteristics

hormone therapy (HT) an alternative name for HORMONE REPLACEMEN THERAPY

hormone-receptor complex formed when a hormone binds to its specific receptor, either on the plasma membrane or within the cell

hospice a hospital for terminally ill patients with a focus on alleviating suffering; a hospital that provides PALLIATIVE CARE

host **1.** the organism on (or in) which a parasite lives (e.g. humans may be hosts for parasites like head lice or tapeworms) **2.** the recipient of a transplanted organ or a grafted tissue

human pertaining to the human species; a member of the species *Homo sapiens*

human biology the study of all the biological aspects of the human species

human chorionic gonadotropin (hCG) a hormone produced by the developing placenta in a pregnant woman that maintains the corpus luteum; has a similar effect to LUTEINISING HORMONE; also abbreviated as *hCG*; *see also* gonadotropic hormones

human genome the genetic material of an organism; consists of all of the DNA in an individual's CHROMOSOMES as well as that in their MITOCHONDRIA; humans have two genomes, a large chromosomal genome and a much smaller mitochondrial genome

Human Genome Project a project initiated in 1988, and completed in

2003, that aimed to identify all of the approximately 25 000 to 30 000 genes in human DNA; the exact order of the approximately 3 billion base pairs in human DNA was determined

human growth hormone (hGH) secreted by cells in the anterior lobe of the PITUITARY GLAND; essential for normal growth and metabolism; deficiency results in growth retardation or DWARFISM

human immunodeficiency virus (HIV) a virus that causes progressive damage to the body's immune system; transmitted by infected body fluids such as blood and semen; frequently transmitted during unprotected sexual intercourse; infection usually leads to AIDS; variants of the virus have been recognised, such as HTLV-I (human T-lymphotropic virus type I) and HTLV-III

humerus the upper arm bone; the longest and largest bone of the upper limb; also called the *funny bone*; *see also* diagram of skeleton (p. 249)

humidity the water vapour concentration in the air

humoral immunity that part of the immune response in which the body produces antibodies that destroy invading agents (antigens); the antibodies are produced by plasma cells that develop from B lymphocytes (B cells); particularly effective against bacterial and viral infections; also called *antibody-mediated immunity*; *see also* cellular immunity

hunting and gathering the form of existence in which animals were hunted, and vegetables, fruits and nuts were gathered for food; the first definite hunting and gathering groups were *Homo erectus*; in modern hunting and gathering societies up to 70% of the total energy value of the food of a group comes from gathering

Huntington's disease a hereditary disorder; the symptoms, which seldom appear before the age of 40, include occasional involuntary movements of the arms and legs, and writhing movements of the hands, head, trunk and feet; the disease also involves a progressive loss of the ability to think clearly; sometimes called *Huntington's chorea*

hyaline cartilage flexible, supporting CARTILAGE

hybrid offspring that is the result of a mating between individuals of two different genetic constitutions **1.** may be the offspring of a mating between homozygous dominant and homozygous recessive individuals **2.** more often used to describe the offspring of matings between two different species or varieties

hybridisation **1.** the act of producing a HYBRID **2.** may refer to CHEMICAL HYBRIDISATION (or *molecular hybridisation*)

hydatid a cyst that is the larval stage of a tapeworm that lives in the intestine of dogs; the cysts may occur in various tissues of humans, cattle and sheep

hydrocortisone one of a group of hormones known as GLUCOCORTICOIDS

hydrogen bond a weak bond between a hydrogen atom and an oxygen or nitrogen atom; too weak to bind atoms into molecules but very important in holding molecules together or different parts of large molecules together; hydrogen bonds hold the two strands of a DNA molecule together and help to form the three-dimensional structure of proteins;

although individual bonds are weak, large molecules may have several hundred so that together they give considerable stability

hydrogen ion a hydrogen atom that has lost one electron, so that it is positively charged

hydrogen-ion concentration the number of hydrogen ions in a given volume; *see also* pH

hydrolysis a chemical reaction in which large molecules are broken down by the addition of water; the molecule is split into two, one part combining with the OH from the water and the other with the H (e.g. breakdown of a disaccharide into two monosaccharides, breakdown of a dipeptide into two amino acids, breakdown of fats into fatty acids and glycerol)

hydrophilic water-loving; describes substances that are attracted to water, and readily absorbed or dissolved in water

hydrophobic water-hating; describes substances that are repelled by water or that are not readily dissolved in water

hydroxyl ion the negatively charged ion that forms when water dissociates; OH^-; consists of an oxygen and a hydrogen atom; a water molecule that has lost a proton; also called a *hydroxide ion*

hygiene practices that ensure cleanliness and good health

Hylobatidae the family of primates consisting of the so-called lesser apes—the gibbons and siamangs

hymen the fold of tissue covering the external opening of the vagina; the hymen may be ruptured the first time sexual intercourse takes place or as a result of exercise or injury

hyoid bone a horseshoe-shaped bone lying in the soft tissues of the neck that provides an attachment for the base of the tongue; unique in the human skeleton because it does not form a joint with any other bone

hyper- a prefix meaning above, greater than, excessive (e.g. hypersecretion—oversecretion of a gland)

hypercholesterolaemia an abnormally high level of cholesterol in the blood

hyperglycaemia an abnormally high level of sugar in the blood; frequently found in people with diabetes mellitus

hypermetropia a defect of vision such that objects can be clearly seen only at a distance; the image of close objects is focused behind the retina instead of on the retina; corrected by spectacles with convex (converging) lenses for close viewing; also known as *hyperopia*, *long-sightedness* or *far-sightedness*; *see* diagram opposite

hyperopia an alternative name for HYPERMETROPIA

hyperplasia an increase in the number of non-cancerous cells in an organ or tissue, resulting in increased size; caused by an increase in the rate of cell division

hypersecretion oversecretion by a gland; used particularly to apply to secretion of a hormone by an ENDOCRINE GLAND

hypertension abnormally high blood pressure; in adults generally considered to be over 140/90 mm Hg

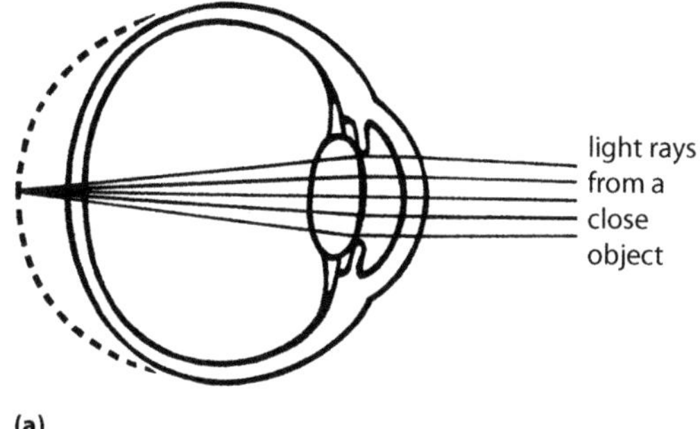

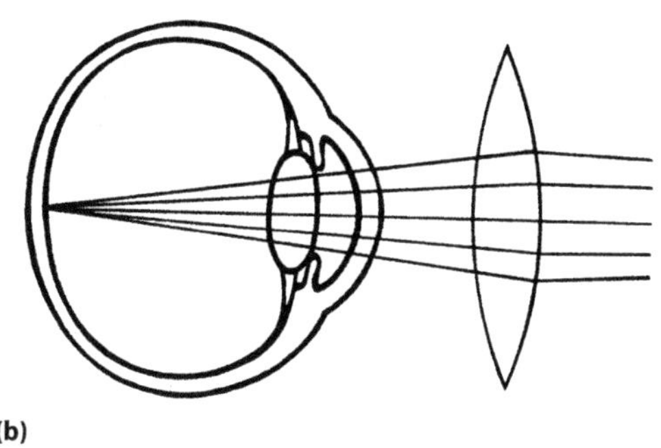

(a) Hypermetropia; (b) Corrected with a convex lens

hyperthermia abnormally high body temperature; the temperature increases above the level required to maintain normal body functions; may lead to collapse and death, particularly in the elderly

hyperthyroidism overactivity of the thyroid gland, resulting in an abnormally high level of thyroid hormones in the blood; *see also* hypothyroidism

hypertonic describes a solution that has a higher concentration of solutes than another solution; often used to describe the concentration of a solution in relation to that of tissue fluid; e.g. hypertonic saline has a higher concentration of salt than normal tissue fluid; *see also* hypotonic, isotonic

hypertrophy the increase in size of a tissue without cell division; the enlargement of the tissue or organ is independent of the body's general growth

hyperventilation extremely rapid or deep breathing; may result in dizziness and even fainting due to the loss of carbon dioxide from the blood

hypnotic a drug that induces sleep

hypo- a prefix meaning under, below, deficient (e.g. hypothalamus—part of the brain that lies below the thalamus)

hypodermis the layer of the skin that attaches it to the underlying tissues or organs; the deeper sublayer of the dermis consisting primarily of loose fibrous connective tissue, blood and lymph vessels and fatty tissue

hypophysis an alternative name for PITUITARY GLAND

hyposecretion undersecretion by a gland; used particularly to apply to the secretion of a hormone by an ENDOCRINE GLAND

hypothalamus the part of the brain lying just below the thalamus and just above, and connected to, the pituitary gland; controls many homeostatic mechanisms through control of the autonomic nervous system; also integrates the functions of the nervous and endocrine systems by controlling secretions of the pituitary gland; *see also* diagram of brain (p. 35)

hypothermia abnormally low body temperature; the temperature drops below the level required to maintain normal body functions

hypothesis (plural *hypotheses*) a possible explanation to account for observations; a hypothesis is normally tested by experiments, the results of which will either support or disprove the hypothesis; when a large body of evidence accumulates to support a hypothesis it may then be called a THEORY

hypothyroidism underactivity of the thyroid gland resulting in a low level of thyroid hormones in the blood; *see also* hyperthyroidism

hypotonic describes a solution that has a lower concentration of solutes than another solution; often used to describe the concentration of a solution in relation to that of tissue fluid; e.g. hypotonic saline has a lower concentration of salt than normal tissue fluid; *see also* hypertonic, isotonic

hypoxia reduced amounts of oxygen in the tissues; may result from reduced levels of haemoglobin in the blood (anaemia) or cardiovascular disorders, such as heart failure

hysterectomy the complete or partial removal of the uterus

I

I band part of the structure of skeletal muscle; *see* sarcomere

ice age a time when vast amounts of ice covered millions of square kilometres of the continents in the Northern Hemisphere; also referred to as a *glacial period* or *glacial age*

ICSI *see* intracytoplasmic sperm injection

Ida a 47 million-year-old fossil primate first described and classified in 2009; *see* DARWINIUS MASILLAE

identical twins TWINS that develop from a single fertilised egg; also called *monozygotic twins*

Ig an abbreviation for IMMUNOGLOBULIN

ileum the last part of the small intestine, between the jejunum and the start of the large intestine; about 3.5 metres in length; only distinguishable from the rest of the small intestine by differences in microscopic structure; *see also* diagram of digestive system (p. 77)

ilium one of the bones that make up the PELVIC GIRDLE

immigration the movement of individuals into a population (e.g. Australia's population growth in recent years has been maintained by immigration)

immovable joint a JOINT in which there is no movement between the bones

immune response a response triggered by foreign substances or micro-organisms entering the body; the response is specific for the particular antigen; it involves production of antibodies to neutralise the antigen and production of lymphocytes called T cells to act against the foreign antigen (infection); also called *immune reaction*

immune system the organs, tissues and cells involved in protecting the body from invading disease-causing organisms and from toxins; includes organs like the lymph nodes, thymus, tonsils and spleen, the cells they produce (such as the various forms of lymphocytes) and the molecules they produce (such as antibodies)

immunisation becoming immune to a disease; can occur naturally as the body fights off the disease or through the introduction of ANTIGENS (in the form of a vaccine) so that immunity is acquired without suffering the illness; exposure to the antigen causes an IMMUNE RESPONSE and the memory of that response is retained so that any future exposure can be dealt with before disease symptoms occur; commonly referred to as *vaccination*

immunity **1.** the ability to resist organisms and chemicals that can damage tissues **2.** resistance to a particular disease by producing a specific type of cell or molecule (antibody) to destroy a particular organism or chemical (antigen); also called *specific immunity*; specific immunity may be active, passive, natural, artificial or acquired or a combination of these; *active immunity* is when the body manufactures antibodies against a foreign antigen; the antibodies deal with the foreign antigen before disease symptoms can occur; active immunity can be acquired naturally by exposure to a disease (e.g. a person is normally immune to chickenpox after having the disease); can also be acquired artificially by vaccination (e.g. vaccination against polio by injection of weakened micro-organisms); *passive immunity* is produced by the introduction of antibodies from an outside source; may be natural, such as when antibodies against an antigen are passed across the placenta from mother to foetus or to an infant in breast milk; may be artificially acquired by injection of antibodies, such as when a person is given antibodies produced by a horse in response to a snake venom; *natural immunity* to a disease occurs without any human intervention; active natural immunity could be acquired by natural contact with an antigen such as when a person has measles or chickenpox and is then immune to the disease; passive natural immunity can be acquired by passage of antibodies from mother to foetus across the placenta or from mother to child in breast milk; *artificial immunity* is produced by artificially giving a person an antigen that triggers the immune response (in which case the immunity is active) or by giving a person antibodies to an infecting antigen (a passive immunity); *acquired immunity* is gained during life as a result of exposure to an antigen; such immunity may be gained either actively or passively, or by natural or artificial means; *see also* immunisation

immunoglobulin (Ig) a group of proteins that act as antibodies; there are 5 classes of immunoglobulins in humans—IgA, IgD, IgE, IgG and IgM; each has a particular chemical structure and a specific role in the immune response; *see also* antibody

immunology the study of the way in which the body resists, or overcomes, infections

immunosuppression the use of drugs to inhibit the immune response; used during and after organ transplants to prevent rejection of the donor organ due to an immune response by the patient; the drug *cyclosporin* is an immunosuppressant used for heart, kidney and liver transplants

immunotherapy treatment of a disease by promoting and supporting the body's immune response to the disease; now used in the treatment of some cancers

impacted fracture a FRACTURE in which the broken ends of bones are forced into each other; *see also* diagram of fractures (p. 104)

impetigo a highly contagious bacterial infection of the skin; more common among children and sometimes called *school sores*

Implanon a contraceptive device that is implanted under a woman's skin; the device prevents ovulation by slowly releasing progestin over a period of three years

implantation the process whereby the early embryo attaches itself to the lining of the mother's uterus, usually 7 to 8 days after fertilisation; the embryo is at the blastocyst stage and the outer layer of cells secretes enzymes that digest and liquefy the lining

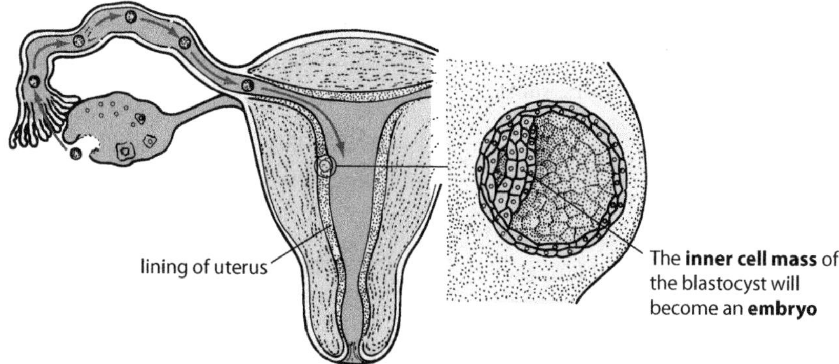

Implantation

of the uterus so that the blastocyst is able to burrow into the lining; *see* diagram above

impotence for a male, inability to achieve an erection of the penis, or to retain an erection long enough to complete sexual intercourse; such a person is said to be *impotent*

impulse also known as a *nerve impulse*; *see* action potential

in situ Latin for 'in place'; a term used to indicate an object in its original position (i.e. in its normal or natural place)

in vitro Latin for 'in glass'; when a procedure is carried out *in vitro*, it is done outside the body in an artificial environment such as a test tube (e.g. in-vitro fertilisation); compare with *in vivo*

in vivo Latin for 'in life'; a process or experiment occurring in the living body; compare with *in vitro*

inactive-*X* hypothesis a hypothesis that suggests that in females only one of the *X*-chromosomes is active in any particular cell; different *X*-chromosomes are thought to be inactivated in different cells, which leads to what is known as a MOSAIC PHENOTYPE in individuals who are heterozygous for alleles carried on the *X*-chromosome; the inactive *X*-chromosome is present in each cell as a BARR BODY; the hypothesis was proposed by Mary Lyon and is sometimes called the *Lyon hypothesis*

inbreeding mating between biologically related individuals

incisor a chisel-shaped tooth; *see* teeth

inclusion a granule or droplet of liquid in the cytoplasm of a cell (e.g. a fat droplet, haemoglobin in red blood cells, melanin in cells of the skin, hair and eyes, glycogen in liver cells)

incomplete dominance the phenotype of a heterozygote being intermediate between that of each homozygote parent (e.g. in snapdragons, if pollen from a white flower fertilises the egg in a red flower, the offspring have pink flowers); *see also* codominance

incubation period the time between infection and the appearance of the symptoms of a disease

incus one of the bones in the middle ear; *see* auditory ossicles

independent assortment the random separation of members of each chromosome pair when gametes are formed at meiosis; results in random combinations of alleles unless the alleles are part of a linkage group (on the same chromosome); that is, different traits are inherited independently of each other (e.g. in pea plants, flower colour is inherited independently of plant stature); also known as Mendel's second law, the principle of independent assortment; also called *genetic reassortment*; *see also* Mendel's laws, gene linkage, diagram below

independent variable the factor under investigation in an experiment; *see* variable

index fossils widely distributed fossils, of limited age range, that are useful for dating the rock strata in which they are found

indicator a dye that changes colour depending on the acidity or alkalinity of a solution; used to determine the pH of a solution

indigenous originating in a particular area or country (e.g. the Australian Aborigines are indigenous to Australia)

indigestion a feeling of pain or discomfort in the stomach or duodenum, caused by overeating or excessive secretion of acid in the stomach

induced abortion the deliberate removal of an embryo or foetus from the uterus before it is capable of survival; *see* abortion

induction of LABOUR; artificially stimulating the beginning of the birth process

infection the invasion of the body by an organism, such as a bacterium or virus, that could cause disease

infectious disease an alternative name for TRANSMISSIBLE DISEASE

infectious mononucleosis an alternative name for GLANDULAR FEVER

inferior describes a structure that is lower in place or position (e.g. the *inferior vena cava*); *see also* superior, diagram (p. 13)

inferior vena cava one of the two veins returning blood to the heart after circulation through the body; *see* vena cava

infertility in males, the inability to fertilise an egg through insufficient production of sperm or obstructed transport of sperm through the sperm duct; in females, disorders that prevent the union of a sperm with an egg or prevent implantation of the fertilised egg, such as blockage of the uterine tubes; sometimes called *sterility*, especially in males

inflammation a defensive response of the body to cell damage by micro-organisms, physical agents or chemical agents; attempts to neutralise and destroy harmful agents at the site of injury, thus preventing spread to other organs; also called the *inflammatory*

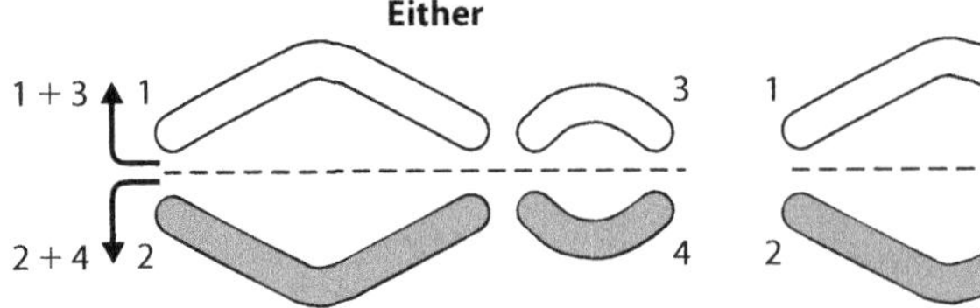

Independent assortment — with just two pairs of homologous chromosomes, four types of gametes are possible: 1 and 3; 1 and 4; 2 and 3; 2 and 4

response; 4 symptoms are characteristic of inflammation—redness, pain, heat and swelling; events of the inflammatory response include:

- vasodilation of small blood vessels in the injured area—this produces the heat and redness
- increased permeability of capillaries in the injured area—loss of fluid from the blood into the tissues produces swelling
- white blood cells leaving the circulation at the site of injury—engulfing of material (phagocytosis) by these cells removes micro-organisms, toxic substances or cell debris
- fibrinogen that has leaked out of the blood being converted to fibrin, which forms a network of threads that wall off the affected area—limits spread to other tissues
- accumulation of pus (a thick fluid that contains living and dead white blood cells and debris from damaged tissue)

see also histamine

inflammatory disease a DISEASE that results in INFLAMMATION (e.g. ARTHRITIS)

inflammatory response an alternative term for INFLAMMATION

influenza a viral infection of the upper respiratory tract; symptoms include chills, fever, headache and muscular aches; commonly called the *flu*

informed consent when a person agrees to a medical procedure, or to be involved in an experiment, after being made aware of all the risks involved

infundibulum **1.** a stalk-like structure that joins the pituitary gland to the hypothalamus; *see* diagram (p. 211) **2.** the funnel-shaped open end of the uterine tube that lies close to the ovary but is not attached to it

ingestion taking in food, liquids or drugs by the mouth

inhalants organic substances that produce effects similar to those of alcohol or anaesthetics when inhaled; vapours inhaled may be from petrol, butane gas, adhesives, dry-cleaning agents or nail polish

inheritance the transmission of genetic material from one generation to the next, i.e from the parents to the offspring

inhibiting factor a factor that slows down release of a hormone; *see* regulating factor

innate behaviour behavioural patterns present at birth and determined basically by hereditary mechanisms (e.g. human infants blind from birth will smile at voices and turn their eyes toward the sound)

innate reflex an automatic response that is inborn and not learned (e.g. the muscles of the iris of the eye changing the diameter of the pupil in response to changing light intensities)

inner cell mass a group of about 30 cells found at one side of a BLASTOCYST; develops into the embryo; also called the *embryoblast*

inner ear the part of the ear containing the organs of hearing and balance; consists of two main sections, a bony labyrinth that is a series of cavities in the temporal bone of the skull, and a membranous labyrinth that lines the bony labyrinth and is separated from it by fluid; the parts of the inner ear are the VESTIBULE, the SEMICIRCULAR CANALS and the COCHLEA; also called the *internal ear*; *see* diagram next page; *see also* diagram of ear (p. 82)

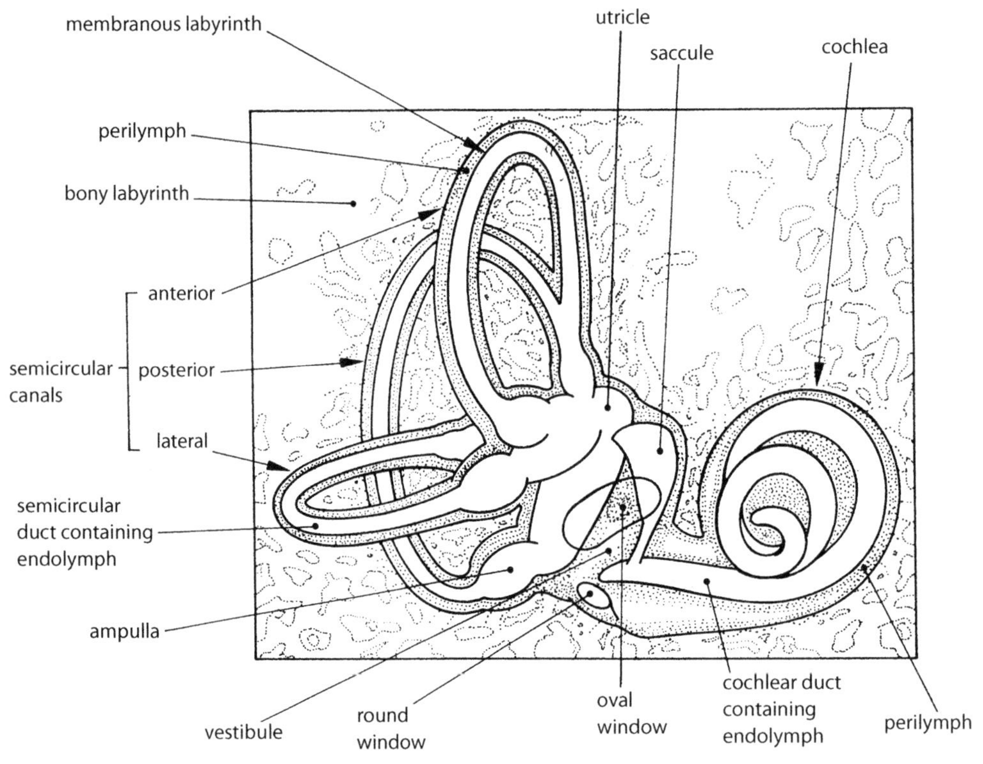

Structures of the inner ear

innervate **1.** to supply nerves to (e.g. the optic nerve innervates the eye) **2.** to stimulate an organ with nerve impulses

inoculation **1.** the introduction of a vaccine to provide immunity against disease (the terms *inoculation*, *immunisation* and *vaccination* tend to be used interchangeably) **2.** the introduction of micro-organisms, such as bacteria or fungi, into a CULTURE MEDIUM

inorganic compound a compound made up of small molecules; usually does not contain carbon, although small molecules that do contain carbon (such as carbon dioxide and carbonate compounds) are also classed as inorganic; *see also* organic compound

insemination the deposition of sperm within the vagina; can be done naturally (by sexual intercourse) or artificially; *see also* artificial insemination

insensible water loss the continual evaporation of water from the body surface and loss of water in exhaled air; so called because a person is unaware of the loss of water; may be up to 600 mL per day

insertion **1.** of a muscle, the end that is fixed to the movable bone—that is, the bone that moves when the muscle contracts (e.g. the insertion of the biceps muscle that bends the arm at the elbow is at the radius—the bone in the forearm); *see also* origin **2.** a mutation caused by the insertion of one or more nucleotides into a DNA molecule

inspiration breathing in; taking air into the lungs; also called *inhalation*

inspiratory centre the region in the MEDULLA OBLONGATA where electrical stimulation results in impulses being sent to the muscles that control the taking of a breath (INSPIRATION)

inspiratory muscles the MUSCLES involved in taking in a breath; the INTERCOSTAL MUSCLES and DIAPHRAGM

inspiratory reserve volume the volume of air that can be forcibly taken into the lungs after taking a normal quiet breath; *see* lung volumes

insula a part of the cerebral cortex that lies deep inside one of the fissures; regarded as a 5th lobe of the cerebral hemisphere

insulin a hormone secreted by the beta cells of the islets of Langerhans of the pancreas; decreases blood sugar level by speeding up transfer of glucose from the blood into cells; also promotes conversion of glucose into glycogen; secretion of the hormone is controlled by a negative feedback system; *see also* glucagon, table of hormones (p. 128)

insulin-dependent diabetes an alternative name for *type 1 diabetes*; *see* diabetes mellitus

integumentary system the skin and structures derived from the skin such as hair, nails, specialised receptors and glands

intelligence the capacity for understanding; the ability to grasp concepts, to reason, to create abstract ideas and to associate ideas

intelligence quotient (IQ) a measure of intelligence; the ratio of mental age (as measured by 'intelligence tests') to chronological age;
IQ = mental age ÷ age in years x 100;
not very reliable and cannot be applied to adults

intelligence test a test that tries to measure a person's inborn intelligence; very unreliable because it is impossible to separate environmental influences from inborn ability; cannot be used to compare individuals from different cultural backgrounds

intelligent design the idea that living things were created by an 'intelligent designer'; a refinement of CREATIONISM

inter- a prefix meaning between (e.g. intercellular fluid—fluid between cells)

intercalated disc an area of thickened material found at right angles to heart muscle fibres; separates the fibres from one another, strengthens the muscle tissue and aids in conduction of nerve impulses from one fibre to another

intercellular fluid an alternative name for TISSUE FLUID

intercostal muscles muscles between the ribs, used in raising or lowering the rib cage during breathing; when breathing in, the external intercostal muscles contract, pulling the ribs upwards and the breastbone outwards; when forcibly breathing out, the internal intercostal muscles contract to pull the ribs downwards

intercostal nerves nerves that take messages to and from the muscles between the ribs

interferon a protein produced by cells infected with viruses; when released from the infected cells it diffuses to uninfected cells, where it triggers responses to inhibit viral replication in those cells; the body's first line of defence against many viruses; unlike antibodies, interferon is not specific to a particular virus

interglacial period a time between ICE AGES

interleukins *interleukin 1*—a protein secreted by macrophages during an immune response; stimulates production of T cells and B cells; also stimulates secretion of interleukin 2; *interleukin 2*—secreted by helper T cells and stimulates production of killer T cells; *see also* immune response, T cells

internal environment the environment within the body

internal respiration an alternative name for CELLULAR RESPIRATION

interneuron an alternative name for CONNECTOR NEURON

internuncial neuron alternative name for a CONNECTOR NEURON

interphase the stage in the life cycle of a cell when it is not dividing; the stage between mitotic divisions; during interphase the DNA duplicates itself ready for the next division; interphase is sometimes called the 'resting phase', but the cell is not resting—it is carrying out all of its normal activities

interstitial cells clusters of cells located between the tubules in a mature testis; secrete the male hormone testosterone

interstitial fluid an alternative name for TISSUE FLUID

intervertebral discs pads of fibrous cartilage that are found between adjacent vertebrae; permit movement of the vertebral column and absorb shock; commonly called *discs*

intestinal glands glands in the mucous membrane of the small and large intestines

intestinal juice the digestive juice secreted by the glands of the small intestine; has a pH of 7.6 (slightly alkaline); contains water, mucus and the enzymes maltase, sucrase, lactase, ribonuclease, deoxyribonuclease and peptidase

intra- a prefix meaning within (e.g. intracellular fluid—fluid within a cell)

intracellular fluid the fluid inside cells; makes up about two-thirds of all body fluid

intracytoplasmic sperm injection a form of IN-VITRO FERTILISATION in which a single sperm is injected into an egg

intrauterine device (IUD) a plastic or metallic device inserted into the uterus that prevents pregnancy; believed to affect the lining of the uterus so that implantation of the embryo cannot occur; sometimes called intrauterine contraceptive device (IUCD); *see* diagram, below

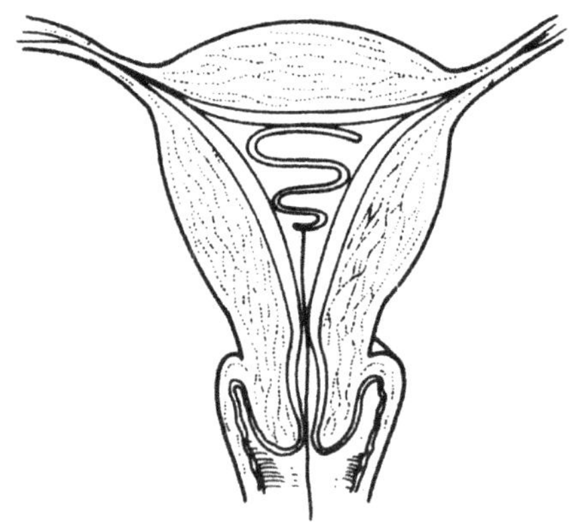

An IUD in place in the uterus

intron a segment of DNA in a gene that is transcribed into messenger RNA but is then removed before the code is used to determine the amino acid sequence in a protein; *see also* exon, protein synthesis

inversion **1.** the movement of the sole of the foot inward at the ankle joint **2.** a type of CHROMOSOMAL MUTATION in which a chromosome segment is turned around end-to-end

in-vitro fertilisation (IVF) a technique for treating certain forms of infertility; involves removing a mature egg from a woman, fertilising it outside her body, then placing the embryo in her uterus; also called *external fertilisation*; babies produced in this way may be called *test-tube babies*; one refinement of the technique, known as *tubal embryo stage transfer* (TEST), involves allowing the fertilised egg to divide, then transferring the embryo to a uterine tube; another refinement, called *pronuclear stage tubal transfer* (PROST), involves transfer of the fertilised egg to the woman's uterine tube; a further variation of the technique, designed to improve on the low success rate of fertilisation outside the woman's body, is known as *gamete intrafallopian transfer* (GIFT), in which the sperm and egg are mixed together immediately after collection from the male and female, then placed into the woman's uterine tubes; another variation is *intracytoplasmic sperm injection*, where a single sperm is injected into an egg

involuntary not under conscious control (e.g. involuntary muscles such as in the wall of the alimentary canal, or in the walls of certain blood vessels, cannot be controlled by conscious effort)

involuntary muscle an alternative name for *smooth muscle*; *see* muscular tissue

iodine an element with the chemical symbol I; an essential component of thyroid hormones

iodine-deficiency disorder a common name for diseases resulting from a deficiency of iodine in the diet—CRETINISM in children and GOITRE in adults

iodine-potassium-iodide a solution of iodine and potassium iodide in water; used as a reagent to check for starch; also used as an antiseptic; may be called *Lugol's solution*

iodised salt table salt to which an iodine compound has been added to help prevent IODINE-DEFICIENCY DISORDER

ion an atom, or group of atoms, that has an electrical charge because of the gain or loss of electrons (e.g. magnesium ion Mg^{++}, ammonium ion NH_4^+, chloride ion Cl^-, sulphate ion SO_4^{--}

ionic bond a chemical bond formed between oppositely charged ions

ionisation an alternative name for DISSOCIATION

ionising radiation radiation that has enough energy to remove an electron from an atom or molecule, thus producing an ion and a free electron (e.g. X-rays, gamma rays, alpha particles)

iris the coloured part of the eyeball; contains muscles able to increase or decrease the diameter of the pupil and thus controls the amount of light reaching the retina; *see also* diagram of eye (p. 94)

irritable bowel syndrome a disorder of the intestine that involves pain, bloating and irregular bowel movements

ischial callosities hardened areas of skin (sitting pads) on the buttocks of Old

World monkeys; also occur to some extent in gibbons

ischium one of the bones that make up the PELVIC GIRDLE

islets of Langerhans clusters of endocrine cells in the PANCREAS

iso- prefix meaning equal; e.g. when two solutions are isotonic, they both have the same concentration of solutes

isoleucine one of the 20 amino acids that are common in proteins; essential in the human diet; *see also* list of amino acids (p. 11)

isometric describes contraction of a muscle without shortening of the muscle; *see also* isotonic (definition 2)

isotonic **1.** describes a solution that has a concentration of solutes equal to that of another solution; often used to describe a concentration equal to that of tissue fluid (e.g. isotonic saline has the same salt concentration as normal tissue fluid); *see also* hypertonic, hypotonic **2.** describes a muscle that shortens while exerting a constant force; *see also* isometric

isotope an alternative form of a chemical element that has the same atomic number (same number of protons and electrons) but a different atomic mass due to the different number of neutrons present in the nucleus; some isotopes (known as *radioisotopes*) are unstable and change into other elements with the emission of certain radiation (e.g. carbon-12 is not radioactive but the isotope carbon-14 decays to nitrogen with the emission of radiation); radioisotopes can be used as tracers to follow the path taken by an element in an organism; some naturally occurring radioisotopes (such as carbon-14) can be used for dating fossils and artefacts

-itis suffix meaning inflammation, e.g. hepatitis is an inflammation of the liver, and conjunctivitis is an inflammation of the eye's conjunctiva

IVF *see* in-vitro fertilisation

J

jaundice a condition in which the liver is unable to remove bilirubin from the blood; may occur because of an increased rate of destruction of red blood cells, blockage of the bile ducts or damage to liver tissue, such as cirrhosis of the liver; bilirubin is one of the products of destruction of red blood cells and some of it enters the bile to be excreted through the alimentary canal; large amounts of bilirubin in the blood give the eyes and skin a yellow colour, one of the symptoms of jaundice; jaundice due to damaged red blood cells is called *haemolytic jaundice*; if due to an obstruction in the bile duct, it is called *obstructive jaundice*; *see also* erythroblastosis

jejunum the middle part of the small intestine, between the duodenum and the ileum; only distinguishable from the rest of the small intestine by differences in microscopic structure; about 2.5 metres long; *see also* diagram of digestive system (p. 77)

joint the junction of two or more bones, or between cartilage and bone; also called an *articulation*; may be *immovable* (*fixed* or *fibrous*), *slightly movable* (*cartilaginous*) or *freely movable* (*synovial*); a fibrous joint is one at which no movement occurs between the bones; the bones are held in place by fibrous tissue, as between the bones of the skull; a cartilaginous joint has only limited movement between the bones, which are held in place by cartilage, as between the ribs and sternum; a SYNOVIAL JOINT has free movement between the bones

joint cavity the space between articulating bones in a SYNOVIAL JOINT

joule a unit of heat energy; the unit used to indicate the amount of energy in foods and the metabolic rate of the body; the former unit of energy, the calorie, equals 4.2 J (approx.); 1 kilojoule (kJ) = 1000 joules (J)

journal a publication, produced at regular intervals, that contains technical and scholarly material; may also be called a *periodical*; some journals, such as *Scientific American* or *Nature*, cover a wide range of subjects; others are restricted to a very narrow area of inquiry (e.g. *Journal of Human Evolution, American Journal of Physical Anthropology*)

junk DNA a sequence of non-coding bases in DNA

Jurassic the middle of the three geological periods in the Mesozoic era, 200–145 million years ago; *see* geological time scale (p. 111)

juxta- prefix meaning 'next to'

juxta-glomerular cells cells in the wall of the blood vessel that delivers blood to the GLOMERULUS of a kidney NEPHRON; these cells secrete the hormone RENIN, which indirectly causes a rise in blood pressure

K

karyotype the number, size and shape of the full set of chromosomes in a cell; usually shown by microscope photographs of metaphase chromosomes arranged in order of size; *see also* diagram below

1 2 3 4 5

6 7 8 9 10 11 12

13 14 15 16 17 18

19 20 21 22 XX

Karotype of a human female

keratin a PROTEIN that replaces the cytoplasm and organelles of the dead cells at the surface layer of the skin in a process called *keratinisation* (or *cornification*); is waterproof, resistant to friction and helps to prevent entry of micro-organisms; also present in hair and nails

ketone bodies substances produced when fats break down in the liver or in fat storage tissue

ketones organic molecules in which a C–O group is located within the molecule rather than at one end (e.g. acetone, CH_3COCH_3)

kidney the principal excretory organ of the human body; filters wastes from the blood and regulates the balance of water and salts in the blood plasma; has an outer layer of tissue (the cortex) and an inner layer (the medulla); the *renal cortex* is the outer, reddish-coloured region of the kidney where glomerular capsules are located; the *renal medulla* is the central, reddish-brown portion of the kidney in which the loop of Henle is located; the medulla contains between 8 and 18 triangular structures called *renal pyramids*; the tip of each renal pyramid is called a *renal papilla*; between each pyramid are extensions of the renal cortex called *renal columns*; the cavity inside the kidney that collects urine before it passes to the ureter is the *renal pelvis*; the *calyces* (singular *calyx*) are cup-like divisions of the kidney pelvis; *minor calyces* drain urine from the renal pyramids and *major calyces* pass urine into the pelvis and out through the ureter; *see* diagram of kidney opposite, diagram of nephron (p. 183), diagram of urinary system (p. 281)

kidney failure any condition in which the kidneys do not function properly

kidney stone a hard, solid mass of chemical materials, usually consisting of calcium oxalate, uric acid and calcium phosphate crystals, that may form in any portion of the urinary tract; also called a *renal calculus*

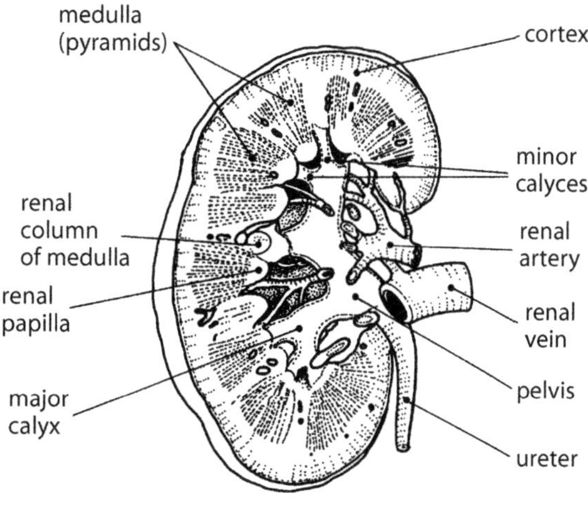

Longitudinal section of a kidney

killer T cell a type of white blood cell that is able to kill damaged or infected cells; also called *cytotoxic T cell*

kilojoule 1000 joules; *see* joule

kinase an enzyme that converts the inactive form of another enzyme into the active form; *enterokinase* is an enzyme secreted by the mucosa of the small intestine; it converts the inactive trypsinogen (part of the pancreatic juice) into the active enzyme trypsin; trypsin is then able to digest proteins

kingdom the first level of biological classification; most biologists recognise two main kingdoms—the plant kingdom and the animal kingdom

kinins a group of polypeptides that cause dilation of arterioles, increase the permeability of blood vessels, attract certain white blood cells and cause pain; *see also* inflammation

Klinefelter's syndrome a genetic condition in which an extra *X*-chromosome is inherited (*XXY*); characterised by small testes that do not produce sperm, breast enlargement and sparse body hair; *see also* trisomy

knapping the process of making stone tools by chipping rocks

knuckle-walking the form of semi-erect quadrupedal locomotion used by chimpanzees and gorillas; characterised by all 4 limbs touching the ground, with the weight of the arms resting on the knuckles of the hands

Krebs cycle a series of chemical reactions that take place in the mitochondria of cells; pyruvic acid from the anaerobic phase of respiration is broken down to carbon dioxide; the energy released is transferred to carrier molecules and through the electron transport chain to ATP; also known as the *citric acid cycle* or the *tricarboxylic acid (TCA) cycle*

kuru an infectious disease, once found in areas of New Guinea, that affected the central nervous system; spread by eating the brain of an infected person

kwashiorkor a type of PROTEIN-ENERGY MALNUTRITION

kya an abbreviation meaning 'thousand years ago' (e.g. 25 kya means 25 000 years ago)

L

labia folds of skin that surround the vaginal opening of a female; the *labia majora* are the outside folds, which are covered with pubic hair and contain fat storage tissue, oil glands and sweat glands; the *labia minora* are inside the two larger folds, have no pubic hair and contain no fat and few sweat glands but numerous oil glands

labour the events that accompany childbirth; divided into three stages—*dilation*, when the neck of the uterus (cervix) gets wider; *expulsion*, when the baby is pushed out of the uterus and through the vagina; *placental stage*, when the placenta and membranes (afterbirth) are expelled from the uterus

labour pains pains that correspond to contractions of the uterus during childbirth

labyrinth the inner or internal ear, made up of the semicircular canals, the utricle and saccule and the cochlear duct; consists of an outer *bony labyrinth*, the hollows through the temporal bone of the skull that form the tubes of the inner ear, and an inner *membranous labyrinth* that follows the same shape as the bony labyrinth; between the bony and membranous labyrinths is a fluid called *perilymph*, and the membranous labyrinth is filled with *endolymph*; *see also* diagram of structures of the ear (p. 82) and diagram of inner ear (p. 140)

lachrymal gland an alternative name for TEAR GLAND

lacrimal gland an alternative name for TEAR GLAND

lactation the secretion and ejection of milk by the mammary glands; lactogenic hormone, from the anterior lobe of the pituitary gland, promotes lactation; the sucking action of the infant stimulates release of the hormone oxytocin from the posterior lobe of the pituitary and milk is passed into the ducts of the mammary glands, a process known as *milk letdown*

lacteal a lymph capillary in each villus of the small intestine; fat from digested food is absorbed into the lacteals and eventually reaches the blood by way of the lymph system

lactic acid a product of anaerobic respiration of glucose in muscles; build-up of lactic acid in muscle cells inhibits the action of certain enzymes, thereby contributing to muscle fatigue; lactic acid is converted to pyruvic acid, mainly in the liver, when oxygen is available; *see also* oxygen debt, anaerobic respiration

lactogenic hormone an alternative name for the hormone PROLACTIN

lactose a polysaccharide sugar consisting of a molecule of glucose and a molecule of galactose; occurs only in milk and is sometimes called *milk sugar*

lactose intolerance inability to digest LACTOSE due to a deficiency of the enzyme lactase; symptoms may include stomach cramps, bloating, flatulence and diarrhoea

lacuna (plural *lacunae*) a space in the matrix of bone or cartilage that is occupied by a cell; *see also* diagram of microscopic structure of bone (p. 34)

Laetoli footprints a 23-metre-long trail of footprints left in volcanic ash 3.5 million years ago at Laetoli, south of Olduvai Gorge in East Africa; the footprints were made by three hominids who had acquired an upright, bipedal, free-striding gait; they clearly show that the hominids had a well-developed arch to the foot and no divergence of the big toe

lamella (plural *lamellae*) one of the concentric rings of matrix around the Haversian canals of compact bone; *see also* diagram of microscopic structure of bone (p. 34)

lamellated corpuscles an alternative name for PACINIAN CORPUSCLES

land bridge a term used to describe the situation when low sea-levels in the past resulted in the joining of land masses that are today separated by sea (e.g. Alaska and Siberia are separated by the Bering Strait but in past ice ages they were joined by a land bridge)

language communication by use of a system of symbols; should not be confused with animal communication, which is by vocalisations (calls), scents, facial expressions, gestures and postures

lanugo fine downy hair that covers the foetus during a late stage in its development; normally lost before birth

large intestine the part of the intestine between the small intestine and the anus; made up of the CAECUM, COLON and RECTUM; *see also* diagram of digestive system (p. 77)

laryngitis redness, pain and swelling (inflammation) of the mucous membranes of the larynx; swelling of the vocal folds prevents them from vibrating freely, causing hoarseness or loss of voice; caused by an infection or by irritants such as cigarette smoke

larynx a short air passage that connects the back of the throat (pharynx) with the windpipe; also called the *voice box*; surrounded by 9 pieces of cartilage and containing two mucous membranes, the VOCAL FOLDS; the edges of the vocal folds, the VOCAL CORDS, vibrate to produce sound; the bulge of the larynx in the neck is called the *Adam's apple* and is formed by the *thyroid cartilage*; other cartilages are the *epiglottis*, which closes the larynx when swallowing; the *cricoid cartilage*, a ring attached to the top of the trachea; a pair of *arytenoid cartilages*, which can move the vocal folds; *see also* diagram of respiratory system (p. 236)

latent learning an alternative term for COGNITIVE LEARNING

lateral adjective meaning side; e.g. lateral (side) view of an organ

laxative a drug or medicinal preparation that will relieve constipation and bring about defecation; *see also* cathartic

learning a relatively permanent change in behaviour that results from practice or experience

legionnaires' disease a form of pneumonia that can be fatal; caused by the bacterium *Legionella pneumophila*; the bacterium occurs in soil and has often been found in air-conditioning systems; so named because the bacterium was found in the air-conditioning ducts of a Philadelphia hotel after an outbreak of the disease at a conference of the American Legion in 1976

Leishman's stain a stain used in the preparation of microscope slides of blood; stains the red cells pink, and the nuclei of white cells purple

lemur one of the so-called primitive primates; a prosimian found today on the island of Madagascar; lemurs include both nocturnal and diurnal species

Lemuriformes one of three infraorders of prosimians; includes lemurs and sifakas (superfamily Lemuroidea) and the aye-aye (superfamily Daubentonioidea)

lens a biconvex, transparent structure in the eyeball that bends light rays to focus on the retina; composed of layers of protein fibres that are arranged like the layers of an onion; *see also* diagram of eye (p. 94)

leprosy a chronic bacterial infection that affects the skin and nerves; if untreated, damage to skin, nerves, eyes and limbs may be permanent; also called *Hansen's disease*

leptin a hormone produced by fat storage tissue; helps to regulate appetite and metabolism

leptotene the first stage of prophase I of MEIOSIS, during which the chromosomes become visible as thin threads

lesion any localised, abnormal change in a tissue (e.g. a cut or scratch, a fracture, an ulcer or abscess, a wart or pimple)

lesser ape one of the members of the family Hylobatidae (i.e. a gibbon or a siamang); *see* ape

less-industrialised country (LIC) a country that is just beginning to industrialise; *see* developing country; *see also* more-industrialised country

lethal allele an ALLELE that results in the premature death of an individual who inherits it; if the allele is dominant, death of the heterozygote occurs; if recessive (lethal recessive), it causes death in the homozygote (e.g. the allele for sickle-cell anaemia is a lethal recessive, causing death in persons who inherit the allele from both parents); also referred to as a *lethal gene*

lethal recessive *see* lethal allele

leucine one of the 20 amino acids that are common in proteins; essential in the human diet; *see also* list of amino acids (p. 11)

leuco- a prefix meaning white (e.g. leucocyte—white blood cell)

leucocytes non-pigmented cells in the blood, also known as *white blood cells*; there are about 5000–9000 per cubic millimetre of blood, but they also occur in other tissues such as in the spleen, kidney, lymph nodes and thymus; they protect against infection by ingesting invading micro-organisms and by producing antibodies that react specifically with foreign proteins; also spelt *leukocytes*; types of leucocytes include *granulocytes*—white blood cells that have granules in the cytoplasm and a lobed nucleus; granulocytes make up about 70% of all white blood cells; *agranulocytes—white* blood cells that have non-granular cytoplasm and a large spherical nucleus; agranulocytes make up the other 30% of white cells and are of two types, monocytes and lymphocytes; *monocytes* (or *macrocytes*) are the largest of the white blood cells; formed in the red bone marrow; capable of a flowing movement and can leave the capillaries to enter the tissues, where they develop into large cells called macrophages; macrophages move to sites of invasion by bacteria or foreign materials, where they ingest particles by PHAGOCYTOSIS; *lymphocytes* are a type of white blood cell found in lymph nodes and lymph; play an important role in the immune system, being involved in the production of antibodies; lymphocytes are formed in the bone marrow and differentiate into either B lymphocytes (B CELLS) or T lymphocytes (T CELLS); *see also* diagram showing the types of white blood cells below

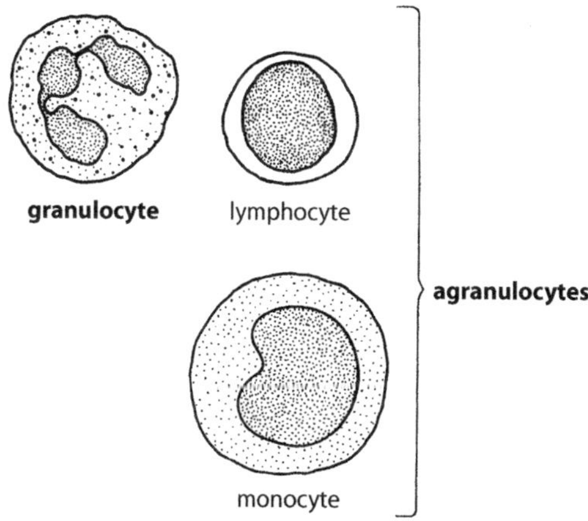

Leucocytes

leukaemia a cancer of the blood-forming tissues in which there may be excess of white cells in the blood or overproduction of immature white cells in the bone marrow, preventing normal production of red cells and platelets

libido the sexual drive or urge; may be either conscious or unconscious

lice (singular *louse*) tiny insects that live as external parasites on the skin; cause intense itching

life expectancy the expected life span of a person; calculated using statistical evidence; will depend on the person's sex, year of birth and age; life expectancies are published in LIFE TABLES

life table a table that provides an estimate of the chance that an individual has of dying by a certain age; based on statistical evidence; life table analysis is used to estimate life expectancy (e.g. the life expectancy at birth of an Australian male born in 2007 was 79 years, but if that person lived to age 79 he would still have a life expectancy of several more years)

lifestyle a person's way of life; expressed through activities, interests, attitudes and values

lifestyle disease a disease that is caused, or made worse, by a person's lifestyle; usually related to lack of exercise or unbalanced diet

ligament an elastic connective tissue that joins bones, composed of densely packed elastic fibres; *see also* diagram of structure of synovial joint (p. 264)

ligand a molecule that triggers a signal by binding to a receptor

ligase enzymes enzymes that bind molecules together; in particular, enzymes that join fragments of DNA molecules together

ligation in recombinant DNA technology, the process of joining together short strands of DNA during replication

limbic system an area in the cerebrum and diencephalon of the brain that is concerned with certain aspects of behaviour and emotions

limiting factor **1.** for a population or a species, any factor in the physical or living environment that restricts population growth or distribution of a species; a common limiting factor in the physical environment for many species is the availability of water; a limiting factor from the living environment could be the amount of interspecies competition **2.** for a physiological process, any factor that limits the rate of the process, such as low concentration of one of the reactants or accumulation of one of the products

line graph a GRAPH in which the relationship between two variables is shown by a line joining points; *see* illustration (p. 116)

linguistic anthropology a branch of ANTHROPOLOGY that focuses on the nature of human language and its relationship to culture

linguistic evidence evidence from the study of languages; used to determine relationships between different groups of humans

linguistics the study of languages; evidence from linguistics may be used in anthropology to determine relationships between different groups of humans

linkage the tendency for certain genes to be inherited as a group; *see* gene linkage

lipase an enzyme involved in the chemical digestion of fat to fatty acids and glycerol; a part of the pancreatic juice, where it is known as *pancreatic lipase*

lipid a type of organic compound that contains the elements carbon, hydrogen and oxygen; usually insoluble in water; simple lipids include fats, oils and waxes; more complex ones are steroids, phospholipids and glycolipids; lipids are part of the structure of cell membranes and are stored in adipose tissue as an energy reserve

lipogenesis the production of fats from glucose or amino acids; occurs in the liver when the diet contains more carbohydrate than is needed for energy or more protein than can be used by the cells

lipolysis breakdown of fats (lipids) into fatty acids and glycerol

liposome an artificial VESICLE made out of material similar to the cell membrane; used to deliver drugs directly into cells when treating cancer and some other diseases

listeriosis an infectious disease caused by the bacterium *Lysteria*; dangerous to infants, the elderly, and people with weakened immune systems; contracted by eating infected food; relatively rare

literature review a survey of all the publications relating to a particular area of research

lithics artefacts made of stone; stone tools

liver a large organ located in the upper abdomen below the diaphragm; has many functions, including secretion of bile, storage of glycogen and other materials, conversion of nutrient molecules, maintenance of blood glucose levels, breakdown of many substances (such as hormones and alcohol and other drugs), breakdown of worn out red and white blood cells; *see* position of liver on diagram of digestive system (p. 77)

living-floor a former land surface that was once occupied by a hominid; frequently in a cave and often preserved in sedimentary rock; may be represented by a narrow layer in the material deposited in the cave; contains tools, fossilised remains of food, coprolites (fossilised faeces) and sometimes even hominid fossils; also called *occupational layer*

lobe a rounded segment of an organ or body part, e.g. lobes of the liver, lung or breast

lobule a small lobe or a subdivision of a lobe, e.g. lobes of the female breast are divided into lobules

local hormone a HORMONE that acts on cells in the area where it is produced

lock-and-key model a model that explains the specific nature of molecular 'recognition' in biological systems; the specific action of enzymes is explained because, according to the model, the shape of the active site of the enzyme molecule is complementary to the shape of the substrate molecule, so that each enzyme is able to combine with only one particular substrate, much like a key being specific to a particular lock; the model can also be applied to neurotransmitters and their receptors at synapses and nerve-muscle junctions; it also applies to the way in which a particular antibody combines with a specific antigen; *see* diagram next page

locus (plural *loci*) the point along the length of a chromosome where the gene for a particular trait is located

long bone any bone that is longer than it is wide; contains a marrow cavity; examples are the humerus, femur, ulna and tibia

long sight an alternative name for HYPERMETROPIA

longevity how long a person lives

longitudinal related to the long axis of a structure

longitudinal arch the arch of the foot running from front to back; *see* arches

longitudinal fissure a deep cleft between the CEREBRAL HEMISPHERES of the brain

longitudinal muscle smooth muscle with fibres arranged lengthwise along an organ (e.g. the wall of the alimentary canal has a layer of longitudinal muscle)

longitudinal section a SECTION through the long axis of a tissue, organ or other structure

longitudinal studies research that investigates the same group of people over an extended period of time; useful for investigating factors contributing to diseases like cancer and cardiovascular disease; *see also* cross-sectional studies

loop of Henle the U-shaped section of the kidney tubule; plays a major role in the reabsorption of water and salts from the liquid filtered out of the blood; also known as the *loop of the nephron*; the part of the loop that extends away from the proximal convoluted tubule into the medulla of the kidney is called the *descending limb*; the part that extends towards the cortex and the distal convoluted tubule is called the *ascending limb*; *see also* diagram of kidney nephron (p. 183)

loop of the nephron an alternative name for LOOP OF HENLE

loose connective tissue an alternative name for AREOLAR TISSUE

loris a small, solitary, nocturnal prosimian found today in Asia and Africa

low power the lowest magnification on a light microscope

lower motor neuron a nerve cell that carries impulses from the spinal cord to a muscle; carries impulses from the upper

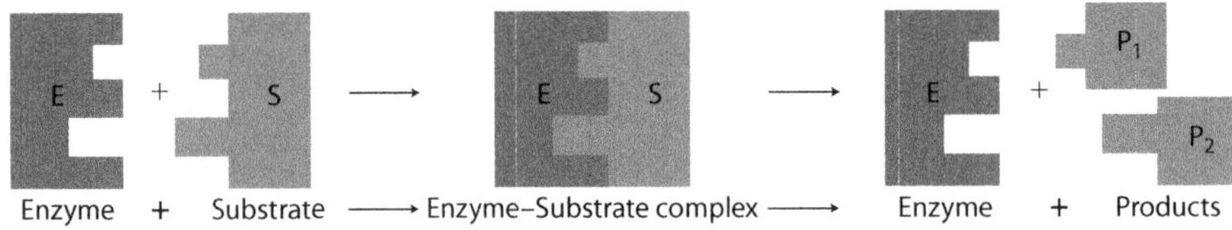

Lock-and-key model

motor neurons to the muscles; *see also* upper motor neuron

lower oesophageal sphincter an alternative name for CARDIAC SPHINCTER

LSD an abbreviation for LYSERGIC ACID DIETHYLAMIDE

Lucy the nickname given to a 40% complete skeleton of a female *Australopithecus afarensis*; found in the Hadar region of Ethiopia; dated at 3.18 million years ago; an important find because although the angle of the pelvic bones indicated that the specimen was not fully human, the structure and angle of the head of the femur, the angle of the knee joint, and the nature of the bones of the foot, all indicated bipedal locomotion; named after the song 'Lucy in the Sky with Diamonds', popular at the time the fossils were discovered in 1973

lumbago FIBROSITIS of muscles in the small of the back

lumbar vertebrae the vertebrae of the lower back; *see* vertebral column

lumen the space within an artery, vein, intestine or tube

lung one of a pair of organs for gas exchange occupying the chest cavity; each lung is supplied with an air tube, the bronchus, that divides into tiny tubes called bronchioles that end in clusters of alveoli (air sacs); *see also* diagram of respiratory system (p. 236)

lung cancer cancer of the epithelial tissue of the lung; often due to constant irritation from inhaled cigarette smoke or other pollutants; also called *carcinoma of the lung*

lung volumes a measure of various aspects of lung capacity; lung volumes (or *pulmonary volumes*) include the following: *tidal volume* (or *tidal air*)—the volume of air breathed in and out in one breath; average tidal volume for adults when at rest is 500 mL; *inspiratory reserve volume*—the volume of air that can be forcibly taken into the lungs after taking a normal quiet breath; *expiratory reserve volume*—the volume of air that can be forced out of the lungs after a normal exhalation; *vital capacity*—the maximum volume of air that can be forcibly breathed out after taking as much air as possible into the lungs; average for young adult males about 4700 mL, for females about 3400 mL; vital capacity is the sum of the inspiratory reserve volume, the tidal volume and the expiratory reserve volume; *residual volume* (or *residual air*)—the volume of air left in the lungs after a person has forced out as much air as possible; in adult females, about 1000 mL; in males, about 1200 mL; *dead space* (or *dead air volume*)—the volume of air in that part of the respiratory system where gas exchange does not occur (windpipe and other air passages); about 150 mL in adults; *respiratory minute volume* (or *minute volume of respiration*)—the volume of air taken into the lungs in one minute;

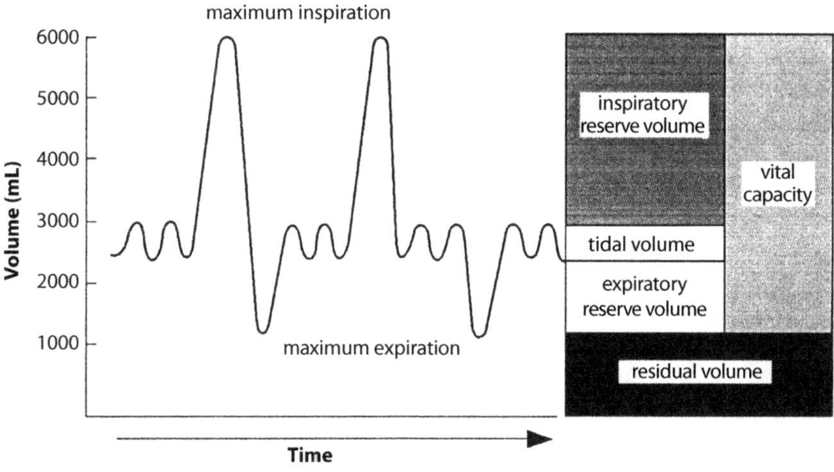

Lung volumes

calculated by multiplying tidal volume by breathing rate; average about 6000 mL per minute; *see* diagram above

luteinising hormone (LH) a hormone secreted by the anterior lobe of the pituitary gland; in females, stimulates ovulation and development of the corpus luteum after ovulation; in males, stimulates testosterone secretion by the testes; secretion is controlled by a releasing factor from the hypothalamus; *see* table of hormones (p. 128)

luteotrophic hormone (LTH) an alternative name for PROLACTIN

lymph the colourless fluid in the lymphatic system; comes from fluid between the cells that has seeped out of the blood capillaries; has a similar composition to blood plasma but with fewer proteins; contains white blood cells, mostly lymphocytes; is carried by lymph vessels to veins at the base of the neck, where it re-enters the blood stream; *see also* diagram of lymphatic system on next page

lymph capillary a microscopic vessel that carries lymph; closed at one end; slightly larger and more permeable than blood capillaries

lymph gland an alternative name for LYMPH NODE

lymph node an oval or bean-shaped structure located along a lymphatic vessel; has several lymph vessels leading into it and one or two leading out; involved in protection against infection by filtering foreign cells and substances from the lymph and producing lymphocytes; contains macrophages and lymphocytes, which destroy foreign substances; also called *lymphatic node* or *lymph gland*; *see also* diagram of lymph node next page left

lymph system an alternative name for LYMPHATIC SYSTEM

lymphatic an alternative name for LYMPHATIC VESSEL

lymphatic sinuses irregular channels in lymph nodes; *see also* sinus

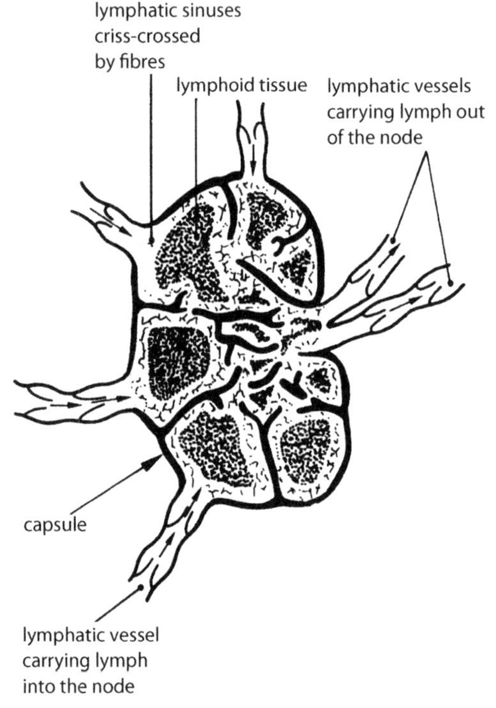

Lymph node

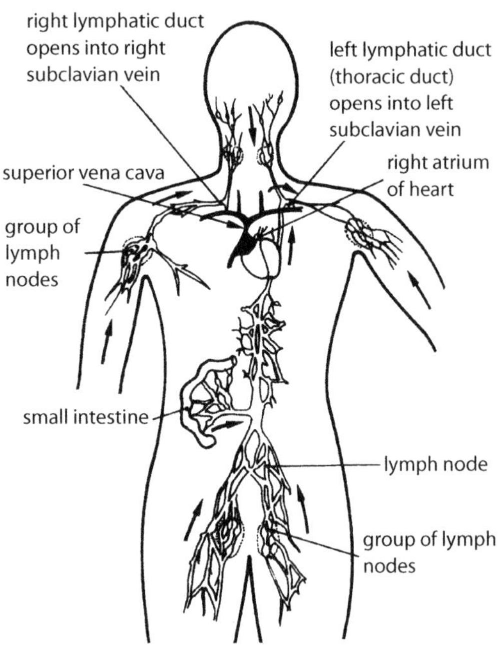

Main vessels of the lymphatic system

lymphatic system a network of vessels that drains excess fluid from the tissues and returns it to the blood; also transports fats from the alimentary canal to the blood and is involved in protecting the body from foreign cells and substances (*see* lymph node); sometimes called the *lymph system*; *see* diagram of lymphatic system right

lymphatic tissue an alternative name for LYMPHOID TISSUE

lymphatic vessel a large vessel that collects lymph from the lymph capillaries; lymphatic vessels join up and eventually return lymph to the blood; the lymph is squeezed along by contractions of the skeletal muscles, with valves preventing backflow; also called *lymph vessels* or *lymphatics*; *see also* diagram of lymphatic system above

lymphocyte a type of white blood cell; produced in lymph nodes and other lymphoid tissue; involved in production of antibodies (proteins that inactivate antigens); plays an important part in the immune response; *see also* B cells, T cells

lymphoid tissue tissue containing many lymphocytes and macrophages; found mostly in the lymph nodes but also in the bone marrow, tonsils, spleen and thymus; also called *lymphatic tissue*

lymphokines proteins secreted by T cells (e.g. interleukin 2, which stimulates the production of killer T cells); other lymphokines attract macrophages to the site of infection, prevent macrophages from migrating away from the site of infection and intensify the effect of killer T cells; *see also* T cells

lymphoma cancer of tissues associated with the lymphatic system (e.g. Hodgkin's disease, which is a cancer arising in the lymph nodes, and Burkitt's lymphoma)

Lyon hypothesis an alternative name for INACTIVE-*X* HYPOTHESIS

lysergic acid diethylamide (LSD) an hallucinogenic drug that occurs naturally in a fungus that grows on grain; most of that used is made in laboratories; the hallucinogenic effect is unpredictable—it can be pleasant or frightening

lysine one of the 20 amino acids that are common in proteins; essential in the human diet; *see also* list of amino acids (p. 11)

lyso-, -lysis prefix or suffix meaning breaking or destruction (e.g. a *lysosome* is an ORGANELLE containing enzymes that break down large molecules, and *haemolysis* is the destruction of red-blood-cell membranes)

lysosome a structure in the cytoplasm of a cell; consists of a membrane enclosing enzymes that can break down large molecules; cells that actively engulf materials have many lysosomes for the breakdown of the consumed matter; *see also* diagram of cell structure (p. 44)

lysozyme an enzyme that breaks down bacterial cell walls; provides protection against invasion of the body by bacteria; found in tears, saliva, perspiration and nasal secretions

M

M phase the part of the CELL CYCLE when nuclear division (mitosis) occurs

macro- a prefix meaning large or great (e.g. macromolecules—very large molecules, such as those of proteins, lipids and nucleic acids)

macrocytic anaemia a type of ANAEMIA characterised by abnormally large red blood cells

macroevolution **1.** major evolutionary changes in organisms; changes that lead to the formation of new genera and families; changes taking place above the species level **2.** the study of long-term evolutionary change, focusing on biological evolution over many generations and on the origin of taxonomic groups above the species level; *see also* microevolution

macromolecule a large molecule; molecular mass is several thousand or more (e.g. fats, nucleic acids, polysaccharides, proteins)

macrophage a cell derived from a monocyte (a type of white blood cell) that is capable of engulfing and destroying cell debris and micro-organisms; some, called *wandering macrophages*, leave the blood and migrate to places where an infection has occurred; others, called *fixed macrophages*, enter organs such as the liver, lungs, spleen, brain, lymph nodes and bone marrow, where they remain to consume and digest worn out cells and cell debris

macroscopic able to be seen with the naked eye

macroscopic anatomy an alternative name for GROSS ANATOMY

macula lutea an alternative name for the YELLOW SPOT of the retina of the eye

magnetic resonance imaging (MRI) a method of obtaining images of body organs; the patient is surrounded by a magnetic field that causes hydrogen and some other atomic nuclei to become aligned; adjusting the strength of the field makes it possible to determine the amount of energy absorbed by the various nuclei; a computer program is used to plot the distribution of the nuclei and provide an image of body organs; formerly known as *nuclear magnetic resonance* or *nuclear magnetic resonance tomography*

magnification a measure of the apparent increase in size of an object when seen through a microscope or when presented as a photograph; the magnification of a microscope is calculated by multiplying the magnification of the objective lens by the magnification of the ocular lens; when photographs, and sometimes drawings, of microscopic objects are published, the magnification is quoted

malaria an infectious disease caused by the parasite *Plasmodium vivax*, a single-celled organism carried from one person to another by the *Anopheles* mosquito; the *Plasmodium* cells pass into the human bloodstream and feed in the red blood cells; when they burst out of the red blood cells the infected person suffers cycles of fever and chills

malignant neoplasm a growth capable of spreading to other parts of the body; *see* neoplasm

malignant tumour a growth capable of spreading to other parts of the body; *see* neoplasm

malleus one of the bones of the middle ear; *see* auditory ossicles

malnutrition poor nutrition, either from too much or too little nutrition, or the improper balance of nutrients; frequently used to refer to a deficiency in one or more nutrients, usually protein

Malpighian layer an alternative name for the *stratum germinativum*, a layer of the EPIDERMIS of the skin

Malthusian doctrine a term that refers to the ideas put forward by Thomas Malthus (1766–1834) on human populations; Malthus, in an article called *An Essay on the Principle of Population*, pointed out that the human population tends to increase at a rate that outstrips food production; these ideas were a major influence on Charles Darwin when he was formulating his theory of evolution through NATURAL SELECTION

maltose a disaccharide sugar made up of two glucose molecules chemically joined; also called *malt sugar*

mammal a member of the class MAMMALIA

Mammalia a class within the phylum Chordata, subphylum Vertebrata; the class to which humans belong; members of the class are known as *mammals*; all mammals are warm blooded, have hair, secrete milk from mammary glands, have teeth of four different types (incisors, canines, premolars and molars) and almost all have young that are born alive and not hatched from eggs

mammary gland a modified sweat gland of female mammals that secretes milk for the nourishment of the young; in humans, mammary glands are referred to as breasts and begin to become prominent during puberty

mammary papilla an alternative name for the NIPPLE of the breast

man in biological terms, the common name for a human; a member of the species *Homo sapiens*; frequently used to refer to the male person in the human species

mandible the bone that forms the lower jaw; fossil mandibles of human ancestors are important in determining the course of human evolution, as the mandible is a good indication of the shape of the rest of the head, and the teeth it contains are useful in indicating diet; *see also* diagram of skeleton (p. 249)

manipulated variable an alternative name for the *independent variable* in an experiment; *see* variable

marasmus a type of PROTEIN-ENERGY MALNUTRITION

mariculture food production using marine organisms; *see* aquaculture

marijuana a drug derived from dried parts of the Indian hemp plant; *see* cannabis; also spelt *marihuana*

marker proteins in biochemical analysis, the presence of marker proteins in a body fluid helps to diagnose disease

marrow the soft tissue that fills the spaces within bones; *red bone marrow* (or *red marrow*) occurs in almost all the bones of a foetus but in an adult is present only in certain bones such as the ribs, vertebrae, sternum (breastbone) and pelvis; blood cell production takes place in the red bone marrow; *yellow bone marrow* (or *yellow marrow*) is the region of a bone where fat is stored; most bones of adults contain yellow bone marrow

marrow cavity a cavity in the shaft (diaphysis) of a LONG BONE; used to store fat; flat bones have no marrow cavity; also called *medullary cavity*

mass extinction many species becoming extinct at roughly the same time (e.g. many of the large reptiles, including the dinosaurs, became extinct around 65 million years ago)

mast cell a type of cell found in loose connective tissue, especially along blood vessels; probably derived from BASOPHILS; produces heparin to prevent blood clotting in the vessels; also thought to produce histamine

mastectomy removal of the breast, usually due to cancer

mastication chewing

mastitis a condition in which the lactating breast becomes engorged with milk, hard, reddened, tender and painful; may be due to bacterial infection

matrix intercellular material; the material between the cells of a tissue; connective tissues have a lot of matrix; the matrix of blood is the liquid plasma; the matrix of bone is the hard material that gives the bone its rigidity

maxilla the bone that forms the upper jaw; *see also* diagram of skeleton (p. 249)

mean alternative term for AVERAGE; more correctly called the *arithmetic mean*

measles an infectious disease caused by the *Morbilla* virus; occurs more commonly in children; characterised by red spots on the skin, an increased temperature, cough and sore throat; an attenuated live virus vaccine is available for children

mechanical digestion the mechanical breakdown of food into small particles; mechanical digestion is begun by the teeth, continued by the churning action of the stomach and completed by movements of the small intestine; the bile salts mechanically digest fats by breaking them into small droplets

mediastinum the part of the chest cavity between the lungs; contains the heart, trachea and oesophagus

medulla the central part of an organ (e.g. the adrenal medulla, the kidney medulla)

medulla oblongata the part of the brain that joins the spinal cord; contains all the fibres that carry messages between the brain and spinal cord; also contains centres that regulate heartbeat, breathing, blood vessel diameter, coughing, sneezing, swallowing and vomiting; also known as the *medulla*; *see also* diagram of brain (p. 35)

medullary cavity an alternative name for MARROW CAVITY

meiosis the type of division of the nucleus of a cell that results in cells with half the usual chromosome number; sometimes called *reduction division*; occurs in the ovaries during production of eggs and in the testes during production of sperm; meiosis involves two nuclear divisions with only one duplication of chromosomes so that four haploid cells are produced from one diploid cell; the phases of meiosis are given the same names for each division as those for mitosis—prophase, metaphase, anaphase and telophase; in prophase of the first meiotic division, *prophase I* (which can be subdivided into the stages LEPTOTENE, ZYGOTENE, PACHYTENE, DIPLOTENE and DIAKINESIS), the chromosomes become visible with a light microscope; each chromosome is already duplicated and is in the form of a pair of chromatids; the nuclear membrane breaks down, nucleolus disappears and the spindle forms; like chromosomes pair off and crossing over of chromatids may occur; during *metaphase I* the pairs of chromosomes line up across the centre of the spindle; during *anaphase I* the like chromosomes separate and are drawn to opposite ends of the cell; in *telophase I* the chromosomes group at opposite ends of the cell, two daughter nuclei form and the cytoplasm may divide to form two daughter cells; this first meiotic division results in the formation of nuclei in which there is only one of each type of chromosome; the second meiotic division begins with *prophase II*, in which two new spindles develop at right angles to the first; in *metaphase II* the chromosomes (each consisting of a pair of chromatids) align themselves across the centre of the spindle; in *anaphase II* the centromeres divide so that the chromatids become daughter chromosomes and the daughter chromosomes are pulled apart towards opposite ends of the spindle; meiosis is completed with *telophase II*, in which the daughter chromosomes group at opposite poles of the cell, nuclear membranes form around them and the cytoplasm divides so that four haploid cells have been produced from the original diploid cell; *see also* mitosis and diagram of meiosis next page

Meissner's corpuscles receptors for touch; found in the surface layers of the skin; most numerous in fingertips, the palms of the hands and soles of the feet; also called *corpuscles of touch*; *see* diagram of skin structure (p. 250)

melan- a prefix meaning black (e.g. melanocyte—a cell that contains a dark pigment)

melanin a yellow-black pigment produced by special skin cells (melanocytes); the amount of melanin determines skin colour; protects the body from harmful ultraviolet radiation; also found in hair and in parts of the eye

melanocyte-stimulating hormone (MSH) a hormone produced by the anterior lobe of the pituitary gland; increases skin pigmentation by stimulating production of the pigment melanin by pigment-producing cells (melanocytes); secretion is controlled by a regulating factor from the hypothalamus; *see also* table of hormones (p. 128)

melanocytes special cells within the epidermis of the skin that synthesise the pigment melanin from an amino acid called tyrosine

melanoma a malignant tumour of the skin that develops from pigment-producing cells, often from a mole; the risk of melanoma is increased by frequent exposure

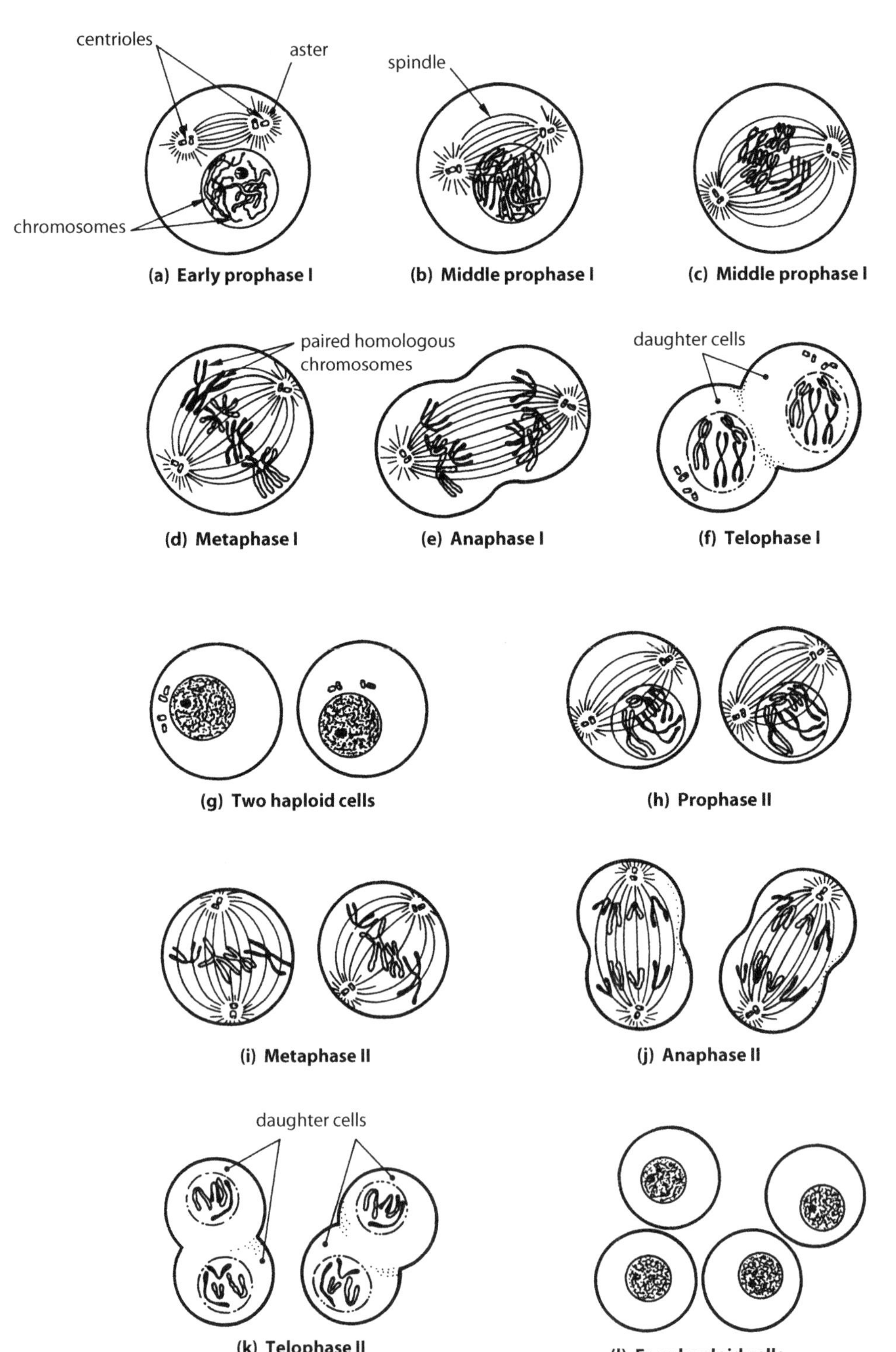

(a) Early prophase I **(b) Middle prophase I** **(c) Middle prophase I**

(d) Metaphase I **(e) Anaphase I** **(f) Telophase I**

(g) Two haploid cells **(h) Prophase II**

(i) Metaphase II **(j) Anaphase II**

(k) Telophase II **(l) Four haploid cells**

Phases of meiosis

to the ultraviolet rays of the sun; *see also* neoplasm

melanosome an ORGANELLE that contains the pigment melanin; found in MELANOCYTES

membrane channel a channel formed by a protein molecule; allows water, ions or other solutes to pass across the cell membrane

membrane potential the electrical voltage across the membrane of a cell; usually used to describe the voltage across the membrane of an unstimulated nerve cell, between –70 and –85 millivolts; brought about by uneven distribution of ions on either side of the membrane; also called a *resting potential* or *resting membrane potential*; the difference in electrical charge on either side of the nerve cell membrane is maintained by the *sodium–potassium pump*, a mechanism in the membrane of a nerve cell that uses energy to transport sodium ions out of the cell and potassium ions into the cell; *see also* action potential, polarised membrane

membrane receptor *see* receptor protein

membranous labyrinth part of the inner ear; *see* labyrinth

memory cell a cell formed by the immune system when exposed to a foreign antigen, which is then able to recognise an antigen to which the body has previously been exposed and bring about the immune response much more rapidly; *see also* B cell, T cell, anamnestic response

menarche the first menstrual bleeding in a female; usually occurs between the ages of 11 and 15 but can be earlier or later; occurs as a part of the period of development known as PUBERTY

Mendelian genetics a branch of genetics concerned with the patterns and processes of INHERITANCE; named after Gregor Mendel, the first scientist to work out many of the principles involved; the practical application of Mendel's laws

Mendelian laws an alternative name for MENDEL'S LAW

Mendel's laws two laws of genetics proposed by Gregor Mendel in 1866; the first law, the *law* (or *principle*) *of segregation*, states that for each characteristic of an organism there are two 'factors' that determine the characteristic (we now call these factors *genes*); when eggs and sperm are produced these factors (genes) separate so that only one goes into each sex cell; at fertilisation the new individual receives one form of each gene from the male parent and one form of each gene from the female parent; the second law, the *law* (or *principle*) *of independent assortment*, states that the factors (genes) separate independently of one another; that is, when the members of a pair of genes separate at the formation of the sex cells, the genes for one characteristic separate independently of those for other characteristics; it is now known that this law is not true for genes that are on the same chromosome—those genes are inherited as a linked group; *see also* gene linkage

meningeal layers alternative name for MENINGES

meninges (singular *meninx*) membranes covering the whole of the central nervous system—the brain and spinal cord; consist of three layers: outer *dura mater*, composed of fibrous connective tissue; middle *arachnoid* (or *arachnoid mater*); inner *pia mater*, composed of loose connective tissue that closely follows the surface of the brain and spinal cord; *see* diagram below

meningitis inflammation of the MENINGES; may be caused by bacteria (bacterial meningitis), viruses (viral meningitis) or amoebae (AMOEBIC MENINGITIS)

meniscus (plural *menisci*) a cartilaginous disc found in some synovial joints, such as the knee joint, that divides the joint cavity into two parts; allows bones of different shapes to fit together, helps maintain stability of the joint and directs synovial fluid to areas of greatest friction; also called *articular disc*; *see also* diagram of structure of synovial joint (p. 264)

menopause the time when a woman ceases to menstruate; the termination of the menstrual cycle; in human females, menopause usually begins between the ages of 40 and 50, with menstrual cycles becoming less and less frequent; menopause is part of a period in the female life cycle called the CLIMACTERIC; *see also* andropause

menorrhagia excessive menstrual flow; *see also* menstruation

menses an alternative name for MENSTRUATION

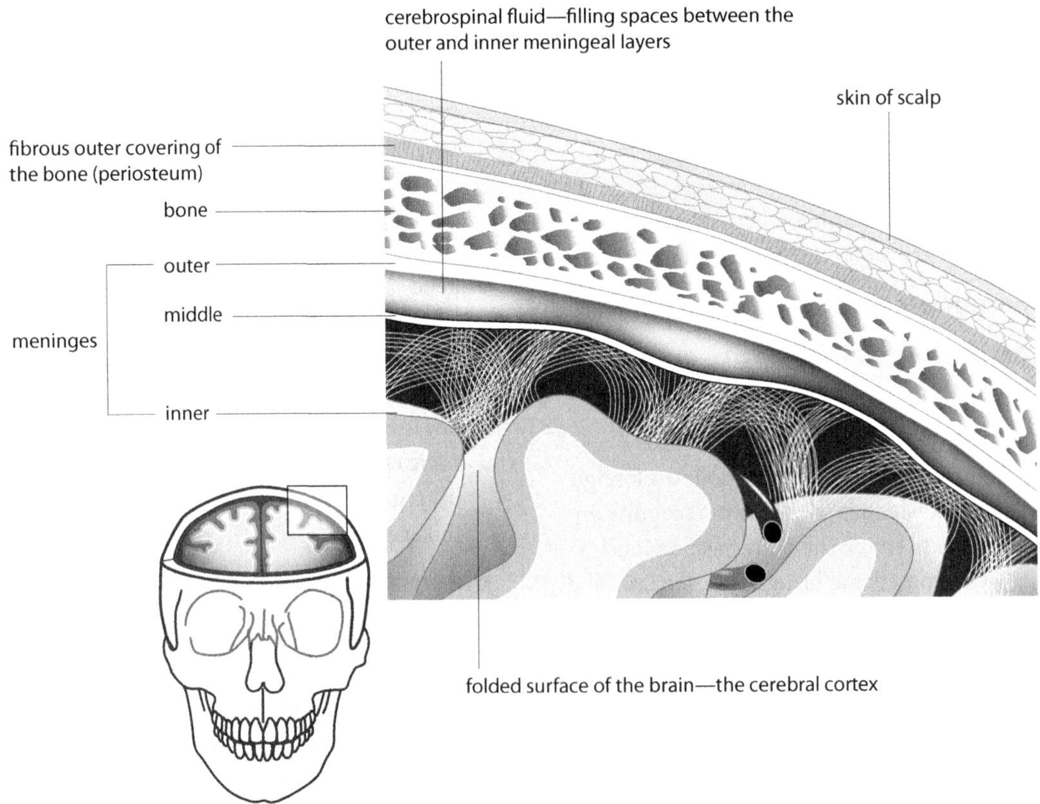

Meninges

menstrual cycle the regular series of changes, controlled by hormones, that take place in the lining of the uterus (endometrium) of a non-pregnant female; occurs in humans, apes and Old World monkeys; each month the uterine lining thickens in preparation to receive an embryo, but if fertilisation does not occur the lining is shed with consequent loss of blood and tissue through the vagina (menstruation); the uterine lining then rebuilds until the start of the next cycle, when it breaks down again if pregnancy has not occurred; average length of the cycle in humans is about 28 days but may be from 24 to 35 days; menstrual cycles usually begin between the ages of 11 and 15, and continue until menopause; *see also* menstruation, ovarian cycle, menopause, illustration of menstrual cycle below

menstruation the periodic discharge of blood and tissue fluid due to the breakdown of the lining of the uterus; total volume is about 30 mL; the menstrual flow passes through the cervix and the vagina to the exterior; it lasts for about the first 5 days of the menstrual cycle and is caused by a sudden reduction in the secretion of oestrogens and progesterone; also known as *menses*, a *period, menstrual period* or the *menstrual phase* of the menstrual cycle

mes-, meso- a prefix meaning middle (e.g. mesoderm—the middle layer of the three layers of primary tissues in an embryo)

mesencephalon the middle part of the brain; also called the MIDBRAIN

mesenchyme connective tissue in an embryo from which all other connective tissues develop; lies beneath the skin and along developing bones of the embryo

mesenteric artery the artery that carries blood to the intestines; *see also* diagram of major arteries (p. 20)

mesentery a membrane attached to the internal wall of the abdomen that keeps the internal organs such as the stomach, intestines and spleen in position; contains the nerve, blood and lymph supply to the abdominal organs

mesocephalic having a medium-sized head; *see* cephalic index

mesoderm the middle layer of the PRIMARY GERM LAYERS

mesomorph a human body type; *see* somatotype

mesothelioma cancer of the membrane that lines the outside of the lungs or the abdominal tissues; usually caused by exposure to asbestos fibres

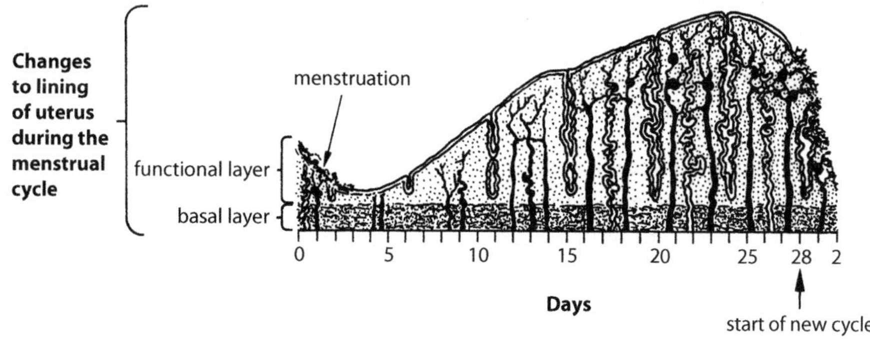

Menstrual cycle

mesothelium the layer of squamous epithelium that covers the outside of SEROUS MEMBRANES

Mesozoic one of the major subdivisions of geological time; the 5th era, 251–65 million years ago; also known as the 'age of reptiles', as these were the dominant animals of the time, occupying most of the earth's surface; the first mammals and birds also appeared during the Mesozoic era; *see* geological time scale (p. 111)

messenger RNA (mRNA) a form of RNA; *see* ribonucleic acid

meta-analysis combining the results of several similar investigations to obtain an overall result

metabolic pathway the series of reactions (or steps) that make up a metabolic process (e.g. glycolysis, the anaerobic phase of cellular respiration in which glucose is broken down to pyruvic acid, consists of about 9 separate reactions making up a metabolic pathway)

metabolic rate the rate at which energy is released by the breakdown of food; *see also* basal metabolic rate

metabolic waste any waste produced by chemical reactions within the cells; in humans the main metabolic wastes are urea, uric acid, carbon dioxide and water

metabolic water water formed as a product of CELLULAR RESPIRATION

metabolism all the chemical reactions occurring in the body (if referring to metabolism of the organism) or in a cell (if referring to cellular metabolism); includes anabolism and catabolism; *anabolism* is the joining of small molecules to make larger molecules, which requires energy (e.g. the build-up of proteins from amino acids, glycogen from glucose and fats from fatty acids and glycerol); also referred to as *anabolic processes* or *synthesis*; synthesis is not confined to living organisms—it can also be done in the laboratory; *catabolism* is the breaking down of large molecules into smaller molecules, which releases energy (e.g. the reactions of respiration where molecules like glucose are broken down to carbon dioxide and water); also referred to as *catabolic processes*; sometimes spelt *katabolism*

metacarpals the 5 bones that make up the palm of the hand; collectively the bones are known as the *metacarpus*; *see also* diagram of skeleton (p. 249)

metaphase the 2nd phase of MITOSIS and of each of the two divisions of MEIOSIS

metastasis (plural *metastases*) the spreading of disease from one part of the body to another; in the case of CANCER, spreading from the primary growth to other organs; *see also* neoplasm

metatarsals the 5 bones between the ankle and the toes of the foot; collectively the bones are known as the *metatarsus*; *see also* diagram of skeleton (p. 249)

metencephalon part of the brain composed of the CEREBELLUM and the PONS VAROLII

methadone a laboratory-made drug with similar effects to those derived from opium; used to treat heroin dependence

methamphetamine a drug that stimulates the central nervous system; common names are *meth* and *speed*; in crystal form the drug is often called *ice*

methionine one of the 20 amino acids that are common in proteins; essential in the human diet; *see also* list of amino acids (p. 11)

methylene blue a stain used to examine some tissues with a microscope; colours the nucleus of cells dark blue and the cytoplasm pale blue

micro- **1.** a prefix meaning small (e.g. microfilament—a very small filament) **2.** in the international system of units, one-millionth (e.g. micrometre—one-millionth of a metre)

microbe a microscopic organism; sometimes used to describe micro-organisms in general and sometimes restricted to pathogenic micro-organisms

microbiology the study of micro-organisms such as bacteria, protozoa, viruses and some fungi

microevolution **1.** small evolutionary changes, resulting from adaptation; changes within a species rather than development of new species (e.g. development of strains of bacteria that are resistant to penicillin) **2.** evolutionary change that takes place over a short period of time; the study of microevolution focuses on changes in allele frequencies from one generation to the next

microfilaments fine filaments of protein present in the cytoplasm of most cells that also form part of the CYTOSKELETON; usually made of actin, although in muscle cells also of myosin; able to contract and therefore involved in processes requiring cell movement, such as PHAGOCYTOSIS

micrometre a very small unit of length, equal to one-millionth of a metre or one-thousandth of a millimetre; often used as the unit of measurement in microbiology; also called a *micron*; symbol μm

micron *see* micrometre

micro-organism an organism that can only be seen with the aid of a microscope; may be a virus, bacterium, protozoan or one of the microscopic algae or fungi; may also be referred to as a *microbe*

microphage a small phagocytic white blood cell; *see also* phagocyte

microscope slide a piece of glass, usually rectangular, on which a specimen is placed for viewing with a light microscope; commonly called a *slide*

microsurgery any surgical procedure carried out with the aid of a microscope

microtome an instrument for cutting very thin sections (slices) of tissue; the sections can then be mounted on a microscope slide for examination

microtubule-organising centre an alternative name for CENTROSOME

microtubules straight hollow rods within the cytoplasm of a cell that form part of the CYTOSKELETON; they provide shape and support for the cell and serve as tracks along which organelles can move in the cell; they are also the main components of CILIA and FLAGELLA; in many cells they radiate from a CENTROSOME; *see also* centrioles

microvilli (singular *microvillus*) microscopic projections from the membrane of cells lining the small intestine; increase surface area for absorption of nutrients; *see also* diagram of microvilli next page top

micturition the act of expelling urine from the urinary bladder; also called *urination*

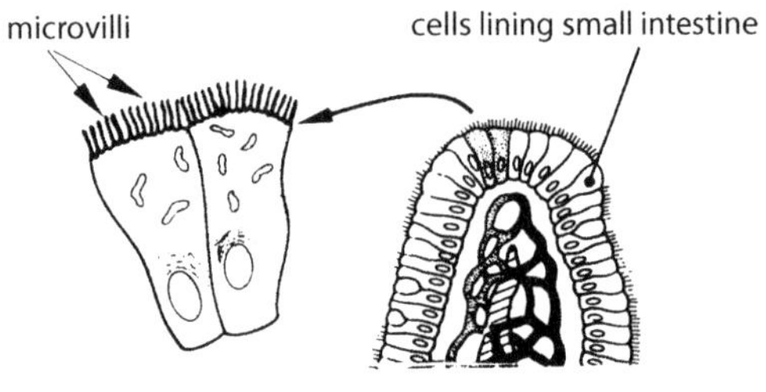

Microvilli

midbrain the part of the brain between the pons and the diencephalon; contains nerve fibres that connect the cerebrum with lower parts of the brain and with the spinal cord; also has centres that control reflex movements of the head and eyeballs; also called the *mesencephalon*; *see also* diagram of brain (p. 35)

middle ear a hollow in one of the bones of the skull separated from the outer ear by the eardrum and from the inner ear by a thin layer of bone containing the oval and round windows; connected to the throat by the Eustachian tube; contains the three small bones (auditory ossicles) that transfer vibrations from the eardrum to the inner ear; *see* diagram of ear (p. 82)

milk letdown the passage of milk into the ducts of the mammary glands; *see* lactation; and diagram below

milk teeth an alternative name for *deciduous teeth*; *see* teeth

milli a prefix meaning one-thousandth (e.g. millimetre—one-thousandth of a metre)

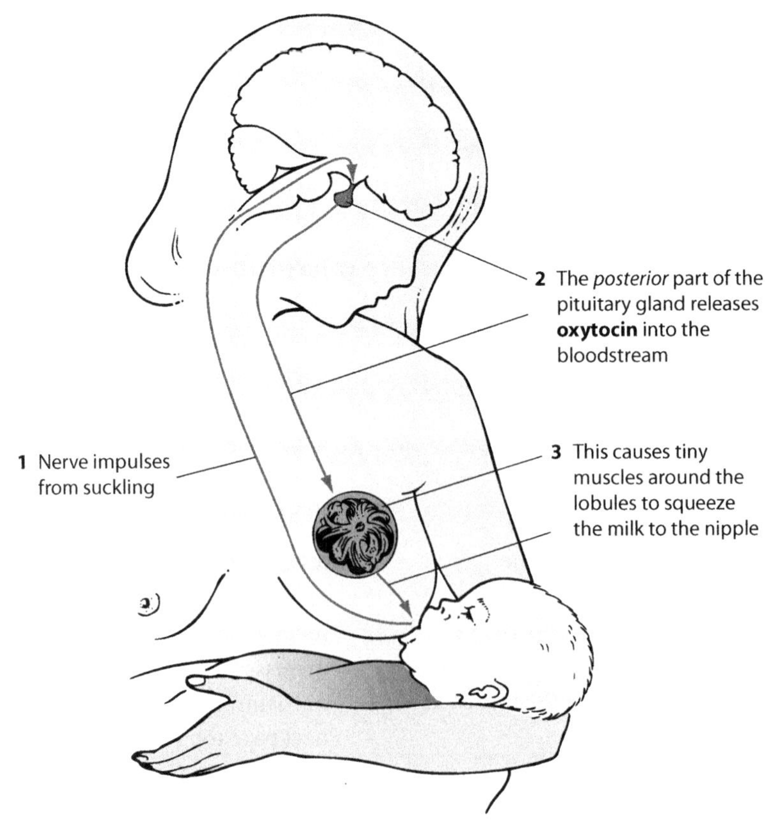

Milk letdown

Millon's test a test for the presence of protein in food, tissue or other organic material; Millon's reagent, which contains mercuric nitrate and nitrite, reacts with proteins to produce a white precipitate that turns red on heating; because of the danger of mercury compounds the BIURET TEST is now preferred as a test for protein

Minamata disease nervous disorders resulting from methyl mercury poisoning; named after Minamata Bay in Japan, where the disease was first documented

mineral an inorganic substance that may be essential for normal body functioning (e.g. iron, calcium, phosphorus, potassium); *see* table of mineral nutrients next page

mineralocorticoids a group of hormones secreted by the adrenal cortex; involved in homeostasis of water, sodium and potassium; most mineralocorticoid activity is due to ALDOSTERONE

mini-pill a contraceptive pill that contains only progestin; *see also* combined pill

mini-stroke a common name for a TRANSIENT ISCHAEMIC ATTACK; *see also* stroke

minute volume of respiration an alternative term for RESPIRATORY MINUTE VOLUME

Miocene one of five epochs in the Tertiary period of geological time; the 4th epoch of the Cainozoic era, between the Oligocene and the Pliocene, dating roughly 23–5.3 million years ago; the first apes are thought to have evolved during the Miocene; *see* geological time scale (p. 111)

miscarriage the natural expulsion of the foetus from the uterus before it is capable of survival; also called *spontaneous abortion*

missing link a term used to describe a creature that is part modern human and part modern ape; the term was first used in Darwin's time after the publication of his book *The Descent of Man*; it has little scientific value, as it implies that modern humans have evolved from modern apes; in reality, modern apes and modern humans have both evolved from a common ancestor; the discovery of Piltdown man, a fossil with a large skull and an ape-like jaw, was thought to be the so-called 'missing link'; since its discovery early in this century, Piltdown man has been shown to be an elaborate hoax

mitochondrial disorder any disorder that affects the functions of the MITOCHONDRIA, or a disorder resulting from changes in mitochondrial DNA

mitochondrial DNA (mtDNA) DNA that is found in the mitochondria of the cells rather than in the nucleus; mitochondrial DNA is inherited only through the mother, as eggs contain mitochondria but sperm do not, so the mtDNA is not shuffled during reproduction; the only factor affecting mtDNA evolution over long periods of time is MUTATION, thus mtDNA can be used to trace the ancestry of a species back through females to some original source; using this technique it has been estimated that the mtDNA for all modern women has come from a single woman, often called 'Eve', who lived in Africa about 140 000 years ago

mitochondrial Eve the name given to a woman who lived in Africa about 140 000 years ago; by tracing MITOCHONDRIAL DNA, researchers have decided that she was the maternal ancestor of all humans living today

Table 7 *Minerals vital for normal body functioning*

Mineral	*Importance*
Calcium	Required for formation of bones and teeth, blood clotting, normal skeletal and cardiac muscle activity, normal nerve function, movements of cells, chromosome movement during mitosis and meiosis, glycogen metabolism, synthesis and release of neurotransmitters
Phosphorous	Required for formation of bones and teeth; a component of compounds involved in storage and transfer of energy (such as ATP); a component of nucleic acids (RNA, DNA), and of many enzymes
Sodium	Major influence on distribution of water through osmosis; important in functioning of muscles and nerves; the main positively charged ion in the extracellular fluid
Potassium	Important for transmission of nerve impulses and muscle contraction; the main positively charged ion in the intracellular fluid
Iron	Part of the compound haemoglobin that carries oxygen in the blood; also part of myoglobin in muscles; a component of the cytochromes that are involved in the electron transport system
Iodine	Essential for synthesis of thyroid hormones that regulate metabolic rate
Copper	Necessary for synthesis of haemoglobin and the pigment melanin
Zinc	A part of the enzymes carbonic anhydrase (involved in carbon dioxide metabolism), and carboxypeptidase (involved in protein digestion); essential for normal growth, wound healing, taste, appetite and sperm production
Chlorine	Important in acid-base balance of the blood, water balance in the tissues and a part of the hydrochloric acid formed in the stomach
Magnesium	Important in normal muscle and nerve function; a part of many coenzymes; also has a role in bone formation
Sulphur	A part of many hormones and vitamins and therefore involved in regulation of many body activities
Fluorine	A component of bones and teeth; inhibits dental decay
Manganese	Necessary for urea formation, haemoglobin synthesis, activation of some enzymes, growth, reproduction and lactation
Cobalt	A part of vitamin B_{12} that is needed for normal maturation of red blood cells
Chromium	Improves production and effects of insulin
Selenium	Prevents breakage of chromosomes; may play a part in preventing some birth defects

mitochondrial Eve model a model of human evolution; also called the AFRICAN ORIGIN MODEL

mitochondrion (plural *mitochondria*) a structure in the cytoplasm of a cell in which the aerobic stage of respiration occurs; consists of a double membrane with the inner membrane folded into projections called *cristae*; *see also* diagram of cell structure (p. 44)

mitosis the process of division of the nucleus of a cell in which the two daughter nuclei have the same number and type of chromosomes as the parent nucleus; the term mitosis is often used loosely to mean cell division; the process is continuous but for ease of explanation is divided into the stages prophase, metaphase, anaphase and telophase; the first phase of mitosis is *prophase*, during which chromosomes become visible with a light microscope, the nuclear membrane breaks down, nucleolus disappears and the spindle forms; during the second phase, *metaphase*, the chromosomes (pairs of chromatids) line up across the centre of the spindle; during *anaphase* the centromeres divide and the daughter chromosomes are pulled apart and are drawn to opposite ends of the cell; during the final phase of mitosis, *telophase*, the chromosomes group at opposite ends of the cell, two daughter nuclei form, and the cytoplasm then divides to form two daughter cells; the cells then enter *interphase*, that stage in the life cycle of a cell when it is not dividing; the stage between mitotic divisions; during interphase the DNA duplicates itself ready for the next division; *see also* cell cycle, diagram of mitosis on next page

mixed nerve a nerve containing both sensory and motor fibres; that is, a nerve with both afferent and efferent nerve cells; all spinal nerves and some cranial nerves are mixed nerves

mL the abbreviation for millilitre

MMR vaccine a vaccine used for immunisation against measles, mumps and rubella; contains live attenuated viruses

mobility aids aids designed to assist walking (e.g. walking sticks, crutches and walking frames)

model a simplified representation of an idea or process; may be a diagram or flow chart, a simplified description of a complex situation, or a physical model such as a model of a cell; examples are the stimulus–response model, and the lock-and-key model of enzyme action

modelling learning by imitation; a form of COGNITIVE LEARNING

modulator a control centre involved in homeostasis (e.g. the hypothalamus is the modulator for regulation of body temperature)

molar a tooth with broad crowns and rounded cusps for crushing or grinding food; *see* teeth

molecular biology the study of the carbohydrates, lipids, proteins, nucleic acids and other large molecules found in cells

molecular dating a method of investigating the sequence and timing of the evolution of closely related groups that diverged from a common ancestor; comparison of the molecular differences between the species is used for what is known as a *molecular clock*; the fossil record is used as a means of checking the accuracy of the dates; DNA sequences and amino acid sequences in proteins are among the molecules compared; *see* chemical hybridisation and DNA fingerprinting for descriptions of techniques for comparing DNA sequences

molecular mass the sum of the atomic masses of all the atoms in a molecule; also called *molecular weight*

molecular motors a motor converts energy into movement; molecular motors convert energy from CELLULAR RESPIRATION into movement of a cell or part of the cell; myosin, the protein responsible for muscle contraction, is a molecular motor

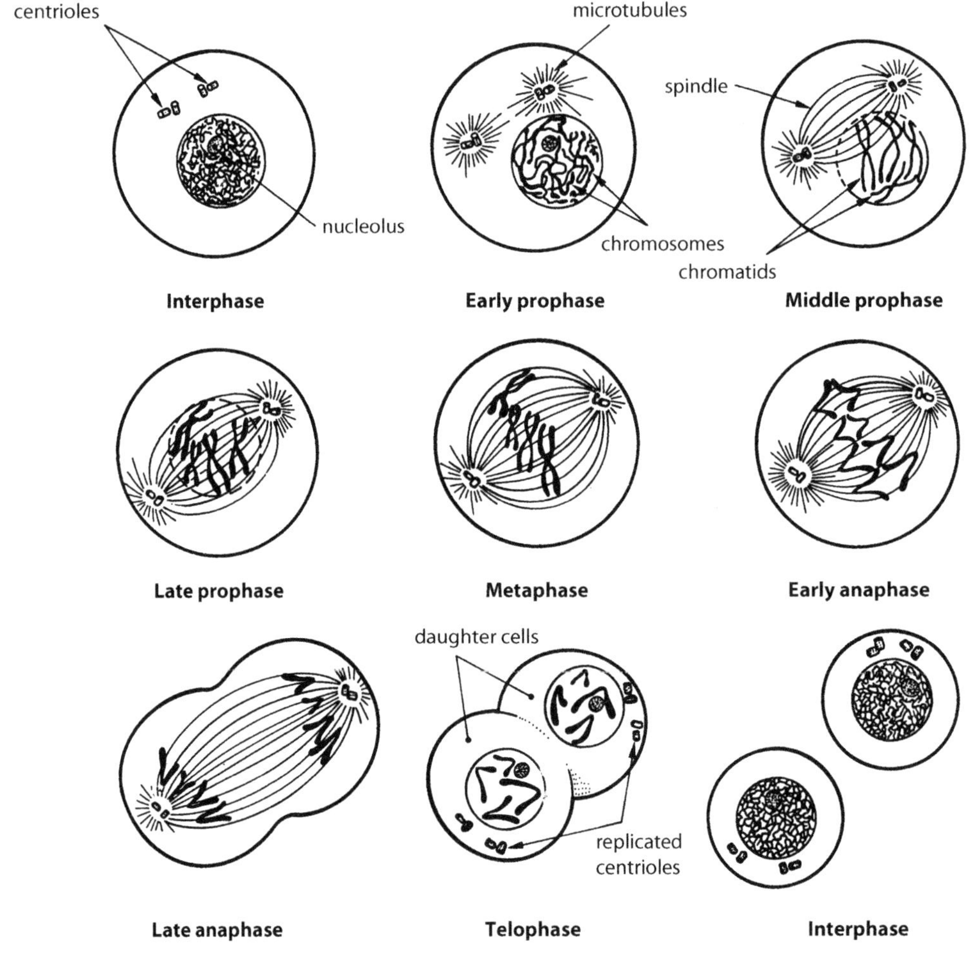

Phases of mitosis

molecule two or more atoms chemically combined; the smallest particle of an element or compound that can exist independently (e.g. the oxygen molecule, O_2, and the water molecule, H_2O)

monkey a member of the order Primates; the main feature that distinguishes monkeys from apes is the presence of a tail; monkeys are classified as NEW WORLD MONKEYS, which inhabit Central and South America (e.g. marmosets, howler monkeys and spider monkeys) and OLD WORLD MONKEYS, which inhabit Africa and Asia (e.g. macaques, baboons, mandrills and colobus monkeys); *see also* classification of primates (p. 219)

mono- a prefix meaning one (e.g. monosaccharide—a carbohydrate made up of one simple sugar)

monoacylglycerol an alternative name for a *monoglyceride*; *see* fat

monoclonal antibody an antibody produced in the laboratory by a group of genetically identical cells (a clone); such an antibody is very pure because it is produced by a group of cells all derived from one original cell; using this technique, antibodies against specific cancer cells have been produced

monocyte the largest of the white blood cells; *see* leucocytes

monogamy an exclusive sexual bond between an adult male and an adult female for a long period of time; for humans it usually means marriage to only one person at a time

monogenic disorder an inherited disease controlled by a single gene (e.g. cystic fibrosis is caused by a recessive ALLELE of one gene)

monogenic inheritance the type of inheritance in which the characteristics are controlled by only one pair of alleles; *see also* polygenic inheritance

monoglyceride a FAT with molecules that have only one fatty acid

monohybrid an organism that has two different alleles for the one gene; is HETEROZYGOUS for a particular gene (e.g. has one allele for ability to roll the tongue (*R*) and one for non-rolling (*r*), *or* has one allele for normal pigmentation (*A*) and one for albinism (*a*)); *see also* dihybrid

monohybrid cross a mating between individuals in which only one pair of contrasting characteristics is being considered (e.g. ability to roll the tongue versus non-rolling *or* albinism versus normal pigment production); *see also* dihybrid cross

monomer a relatively simple compound that can combine with other monomers to form a long chain known as a POLYMER; glucose is a monomer that can be combined with other glucose molecules to form glycogen; amino acids are monomers that can combine to form proteins

monosaccharide a CARBOHYDRATE with molecules containing 3–7 carbon atoms

monosomy the general name for a condition where only one chromosome (rather than the normal pair) is present in body cells (e.g. Turner's syndrome, in which individuals have only one X-chromosome and no second X-chromosome or Y-chromosome); *partial monosomy* is when only part of a chromosome is missing; *see also* trisomy

monounsaturated fatty acid a FATTY ACID that could contain one more hydrogen atom; *see also* fat

monozygotic twins TWINS that develop from a single fertilised egg; also called *identical twins*

more-industrialised country (MIC) a country with an economy based on industry; *see also* developed country, less-industrialised country

morning sickness nausea and vomiting that affects many women during pregnancy; often occurs in the morning but can occur throughout the day

morning-after pill an emergency contraceptive pill that is taken 24 to 72 hours after sexual intercourse; consists of a large dose of oestrogen that prevents IMPLANTATION of an embryo

morphine a drug obtained from the resin of the seed pod of the OPIUM poppy

morphogenesis the change in size and shape of organs, or the whole organism, during development

morphology the study of body form and structure of organisms, including that of embryos and fossils

mortality death; mortality, fertility and migration are the three main determiners of population size and growth

mortality rate an alternative term for DEATH RATE

morula a very early stage in embryonic development; a solid mass of cells produced as a result of the successive division of the fertilised egg; about the same size as the original egg; this stage is reached a few days after fertilisation

mosaic evolution the concept that major evolutionary changes tend to take place in stages, not all at once (e.g. human evolution shows a mosaic pattern in that small canine teeth, large brains and tool use did not all evolve at the same time)

mosaic phenotype possessing cells that may have a number of different genetic make-ups and that therefore show differing characteristics; may be caused by an inactive *X*-chromosome in persons who are heterozygous for alleles carried on the *X*-chromosome; *see also* inactive-*X* hypothesis

mother–infant group a type of social structure in which the basic social group consists of a mother and her dependent offspring (e.g. orang-utans have this type of social structure)

motile able to move (e.g. sperm cells are motile)

motion-sickness a feeling of dizziness or nausea caused by the motion of a vehicle (e.g. a car, ship or plane)

motor a term that describes the activity of an effector organ—a muscle or a gland (e.g. the motor area of the brain controls muscular movement); *see* following entries

motor areas parts of the cerebral cortex that control muscular movement (e.g. the primary motor area controls specific muscles, the motor speech area translates thoughts into speech, the premotor area controls complex learned motor activities such as writing or playing a musical instrument); *see also* diagram of motor areas below

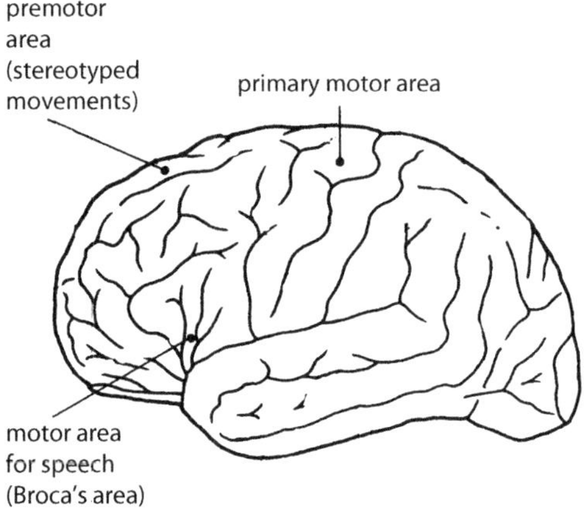

Motor areas of the cerebral cortex

motor association area the part of the CEREBRUM where behaviour is planned; nerve cells in this area plan the sequence and amount of muscle contraction required for actions like walking, riding or speaking

motor development the development of the ability to co-ordinate the muscle groups of the body

motor division an alternative name for the efferent division of the peripheral NERVOUS SYSTEM

motor end plate an alternative name for NEUROMUSCULAR JUNCTION

motor neuron a nerve cell that carries messages from the brain or spinal cord to effector organs (muscles and glands); also called *effector neuron*; *see also* diagram of motor neuron (p. 187)

motor skills a learned series of movements that result in a smooth, coordinated action; *gross motor skills* involve large muscles, such as in walking; *fine motor skills* involve small muscles, such as in threading a needle

motor speech area an alternative name for BROCA'S AREA of the brain

motor unit a motor nerve cell and all the muscle fibres stimulated by that cell

mountain sickness an alternative name for ALTITUDE SICKNESS

mRNA an abbreviation for *messenger* RNA; *see* ribonucleic acid, protein synthesis

mtDNA an abbreviation for MITOCHONDRIAL DNA

mucilage a sticky substance produced by some micro-organisms

mucosa a mucous membrane; in particular, the mucous membrane that forms the internal lining of the alimentary canal

mucous membrane a membrane that secretes mucus; the membrane that forms the lining of the alimentary canal, the reproductive tract, the respiratory system and the urinary tract; may also be called *mucosa*

mucus a thick, slimy and sticky fluid secreted by mucous glands and by mucous membranes; mucus functions as a lubricant (such as in saliva to lubricate food when swallowing) and as a sticky material to trap particles (such as in the windpipe to trap dust and other matter)

multi- a prefix meaning much or many (e.g. multinucleate—having many nuclei)

multiallelic inheritance when more than two alleles are involved in the inheritance of a characteristic (e.g. inheritance of ABO blood group where an individual has any two of three possible alleles); *see also* multiple alleles

multifactorial disease a disease that results from the combined effects of a number of genetic and environmental factors

multimale group a type of social structure in which the main social group is made up of several adult males, several adult females and their offspring; the most common form of social structure found in primates; *see also* one-male group, mother–infant group

multiple alleles a group of alleles (more than two) that can influence the development of a characteristic; *see also* multiallelic inheritance

multiple drug resistance resistance of bacteria to a number of antibiotics

multiple sclerosis a disease in which there is gradual destruction of the myelin sheaths of nerve cells in the central nervous system; the sheaths become hard scars called scleroses; transmission of nerve impulses is disrupted, resulting in symptoms such as impaired muscle co-ordination

multipolar neuron a nerve cell with one axon and many dendrites; the most common type of neuron; most neurons of the brain and spinal cord are multipolar; *see also* bipolar neuron; diagram of connecting neuron (p. 187)

multipotent stem cells *see* stem cells

multiregional evolution model one of the models put forward to explain the evolution of modern humans; states that the change from archaic to modern forms of *Homo sapiens* took place in all regions of the Old World, although not necessarily at the same time; *see also* African origin model

mumps a disease caused by infection with a paramyxovirus; characterised by the inflammation and enlargement of the salivary glands behind the ear (parotid glands), and sometimes by inflammation of the testes or ovaries; accompanied by fever and extreme pain during swallowing; more common in children; spread by droplet infection; an attenuated virus vaccine is available

murmur an unusual heart sound; may be of no consequence, but may indicate malfunction of a heart valve

muscle a tissue or organ capable of shortening in length; skeletal muscles are attached to the bones by tendons; smooth muscle occurs in the walls of internal organs; cardiac muscle is the muscle of the heart; *see also* muscular tissue, diagrams of muscular system (pp. 178, 179)

muscle fibres the long cylindrical cells that make up skeletal muscles; also called *myofibres*; each has several nuclei and is surrounded by a plasma membrane called the *sarcolemma*; each contains smaller cylindrical structures, the *myofibrils*, running lengthwise within the cytoplasm; these are made up of even smaller structures, the *myofilaments*; myofilaments are of two types—*thick myofilaments* are made of a protein called *myosin*, and *thin myofilaments* are made of the protein *actin*; when a muscle fibre contracts, the myofilaments slide past one another; the overlapping arrangement of the myofilaments gives the muscle its striated appearance when viewed with a microscope; *see* diagram of muscle fibres right, sliding filament theory, sarcomere, diagram of structure of sarcomere (p. 241)

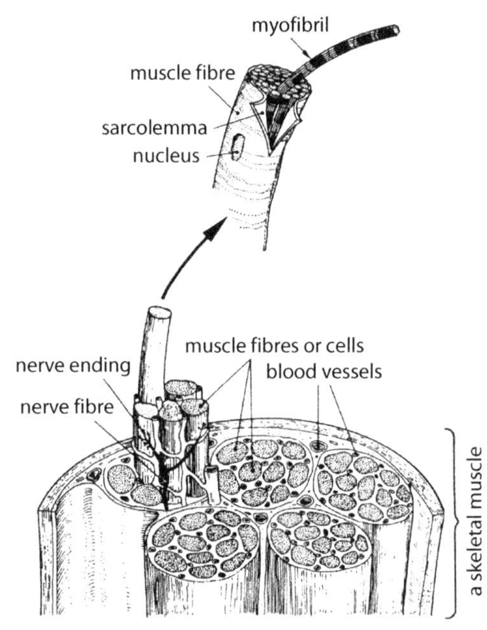

Muscle fibres

muscle spindles stretch receptors in the skeletal muscles; provide the brain with feedback needed to regulate tension in skeletal muscles; particularly abundant in muscles that require fine control; a type of *proprioceptor*

muscle tone the partial contraction of skeletal muscles; at any particular time some fibres in the muscle are contracted and some relaxed; does not produce fatigue because the contracted fibres are continually changing; tone is important in maintaining body posture; also called *tonus*

muscular dystrophy a general term referring to a number of diseases in which there is progressive wasting of muscles and eventual death; one form is called *Duchenne muscular dystrophy*—an inherited wasting disease of the leg muscles and later the arms, shoulders and chest—which first becomes evident between 1 and 6 years, with death usually occurring by the late teenage years; Duchenne muscular dystrophy is controlled by a gene carried on the X-chromosome, and is therefore much more common in boys than girls (*see also* sex linkage); another form is *facioscapulohumeral muscular dystrophy*, which affects boys and girls equally—it initially affects the muscles of the face, shoulder blades and upper arms

muscular system the body system made up of the skeletal muscles; *see* diagrams on pages 178–179

muscular tissue tissue made up of cells that are specialised for contraction; provides movement, maintains posture and produces heat; specialised into three types—skeletal muscle, smooth muscle and cardiac muscle; *skeletal muscle* is joined to the bones and is under voluntary control; made up of thousands of long cylindrical cells or fibres; also called *voluntary muscle, striated muscle* or *striped muscle*; *see also* muscle fibres, diagram of structure of skeletal muscle (p. 241); *smooth muscle* is not under conscious control; it is found in the walls of internal organs such as the stomach, intestines, uterus and bladder; called smooth because its cells do not have the striated appearance of skeletal and cardiac muscle; also called *non-striated muscle, plain muscle* or *involuntary muscle*; *cardiac muscle* forms the wall of the heart; contracts to pump the blood; similar in structure to skeletal muscle but the fibres branch and rejoin, forming a network; *see* diagram of muscular tissue right

muscularis layers of muscle responsible for movement of the digestive tract

mutagen *see* mutagenic agent

mutagenic agent any agent that increases the rate at which mutations occur (e.g. *X*-rays, mustard gas, ultraviolet radiation); also called a *mutagen*

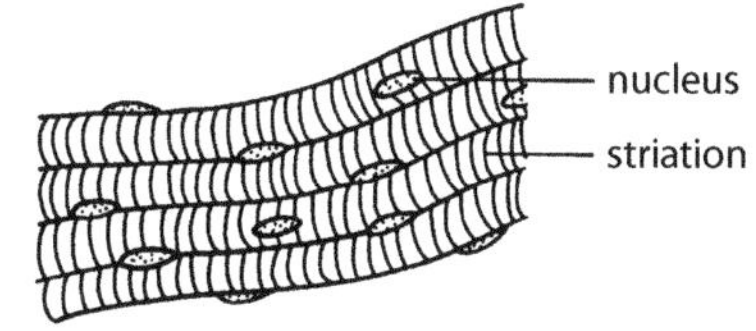

smooth muscle cell
nucleus
(b)

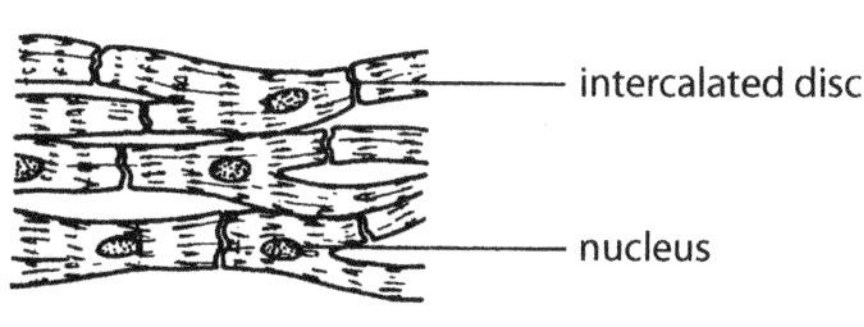

Muscular tissue (a) skeletal; (b) smooth; (c) cardiac

mutant **1.** an organism with a characteristic resulting from a mutation **2.** a gene that has undergone mutation

mutation a new variation, resembling neither parent, that occurs quite suddenly and purely by chance; caused by a permanent change in a gene or chromosome(an altered sequence of bases in the DNA molecule); if the change is present in a gamete or gamete-forming cell it is called a *germ-line mutation* and may be inherited by offspring; a *somatic mutation*, a change that occurs in a body cell, is not passed on to offspring; a *point mutation* or *gene mutation* is an alteration to a single gene, thus producing a new allele—the characteristic normally produced by that gene is changed or destroyed; a CHROMOSOMAL MUTATION is a change to the structure and/or number of chromosomes—in offspring, the characteristics normally produced by the affected chromosome are changed or destroyed; new characteristics formed as

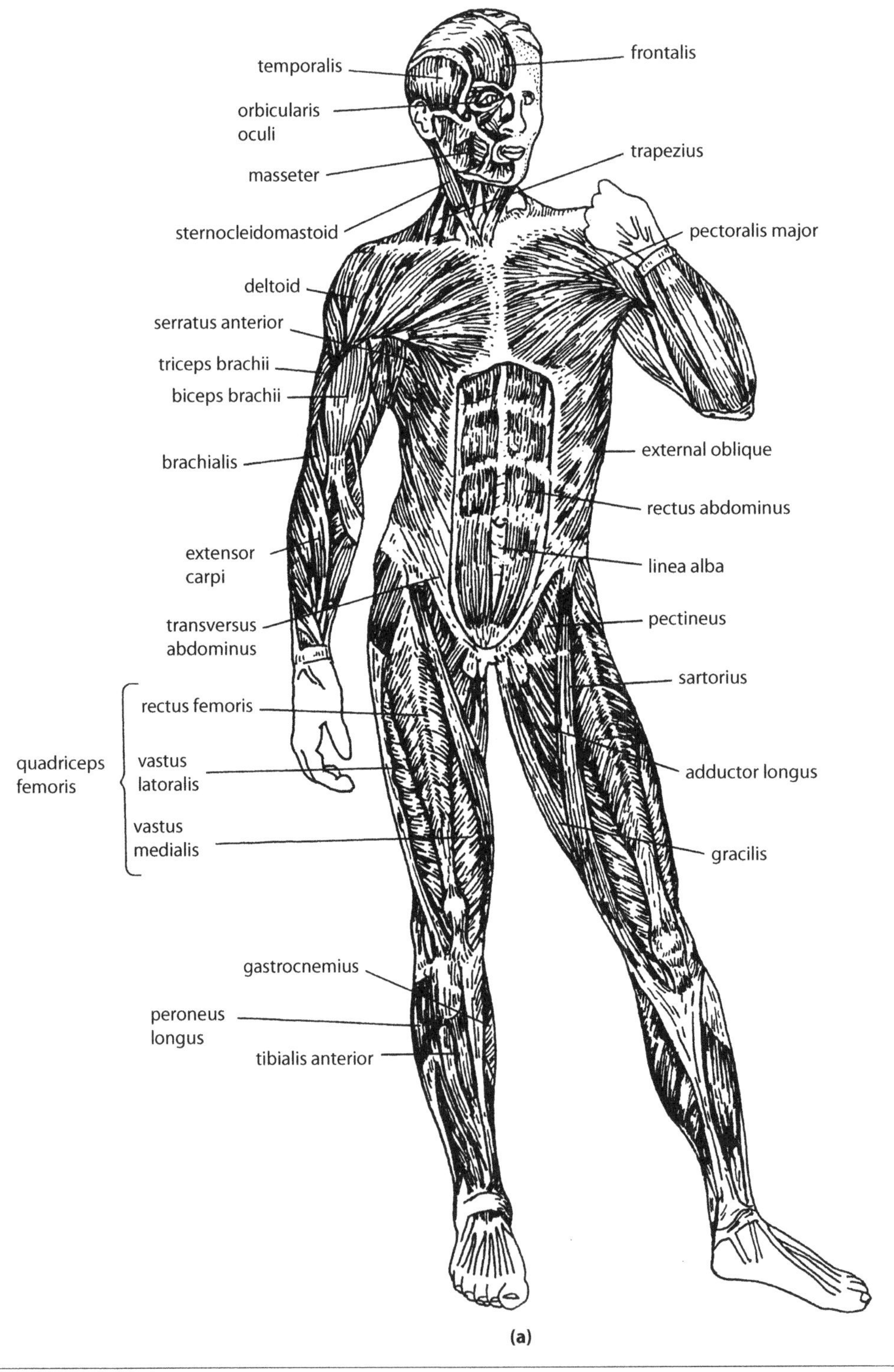

Muscular system (front)

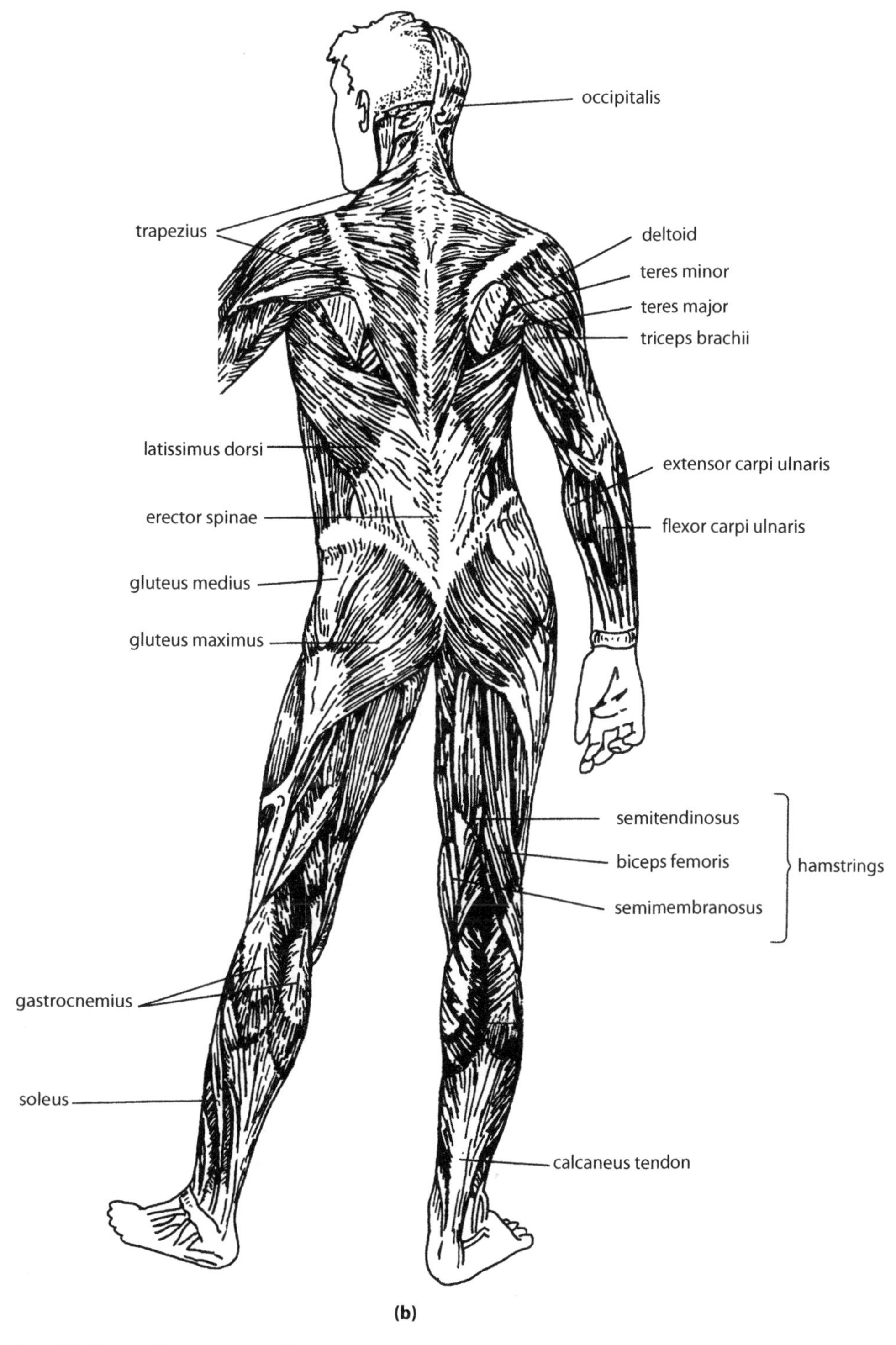

Muscular system (rear)

a result of mutations other than somatic mutations can be inherited by subsequent generations; most are harmful, but some may be advantageous and are thus favoured by natural selection

mutualism a relationship in which two participating species live together in such a way that both benefit (e.g. the association between yellow baboons and impalas; the excellent vision of the baboon combined with the keen sense of hearing and smell of the impala make it very difficult for a predator to approach unnoticed; each species recognises the other's call of alarm)

mya an abbreviation for 'million years ago' (e.g. 2.5 mya means 2.5 million years ago)

myasthenia gravis a condition characterised by extreme weakness of skeletal muscles; thought to be due to an abnormal immune response that produces antibodies directed against acetylcholine receptors; the receptors are on the muscle fibres at the neuromuscular junction, and so the muscle fibres do not respond to the acetylcholine released from the nerve endings; may be fatal but in some patients can be stabilised with drugs

myelencephalon part of the brain that forms the MEDULLA OBLONGATA

myelin sheath a white, fatty sheath that surrounds some nerve fibres; increases the speed of conduction of a nerve impulse and insulates the fibre from adjacent fibres; the myelin sheath is formed by flattened cells called *Schwann cells* (or *neurolemmocytes*) that wind around the axon many times, pushing the cytoplasm and nucleus of the Schwann cell to the outside layer; the myelin sheath is the inner portion of the structure, consisting of 20–30 layers of cell membrane; the outer layer, consisting of the nucleus and cytoplasm of the Schwann cell, is called the *neurilemma* (or *sheath of Schwann*); *see also* diagram of nerve cell (p. 187)

myelinated fibre a nerve fibre (axon or dendron) that has a myelin sheath

myelitis inflammation of the spinal cord; most frequently caused by the invasion of the nervous tissue by viruses, but may be due to other agents, such as bacteria and fungi; may result in a variety of symptoms, including paralysis and coma and, in extreme cases, death

myo- a prefix meaning muscle (e.g. myofibre—a muscle fibre)

myocardial infarction the medical term for HEART ATTACK

myocardium the middle layer of the heart wall, made up of heart (cardiac) muscle

myofibres an alternative name for MUSCLE FIBRES

myofibrils cylindrical structures running lengthwise within the cytoplasm of skeletal MUSCLE FIBRES

myofilaments structures that make up the myofibrils of skeletal MUSCLE FIBRES

myoglobin a protein similar to haemoglobin that is found in muscle fibres; increases the rate of diffusion of oxygen into the fibres and is able to combine with oxygen to provide the muscle with small stores of oxygen

myopia a defect of vision such that objects can be clearly seen only when they are close to the eye; the image of distant objects is focused in front of the retina instead of onto the retina; corrected by spectacles with concave (diverging) lenses for distant viewing; also called *short-sightedness* or *near-sightedness*; *see also* diagram opposite

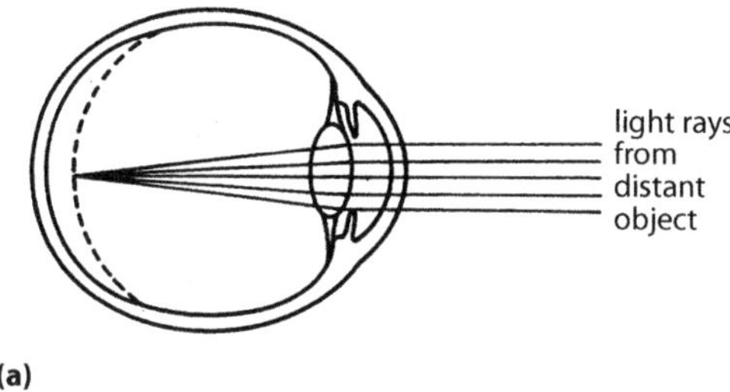

(a)

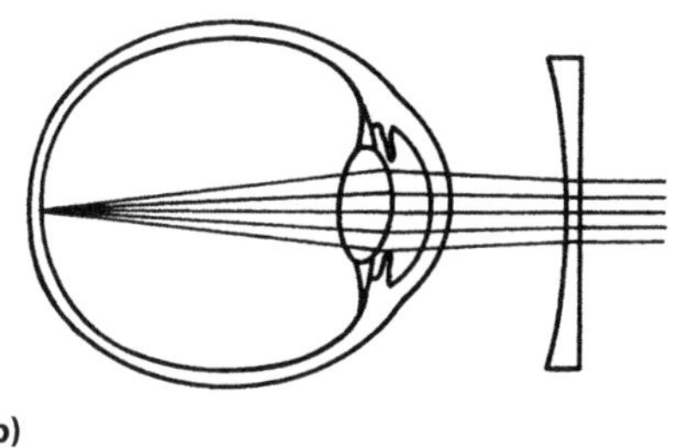

(b)

(a) Myopia; (b) corrected with a concave lens

myosin a protein of which myofilaments of muscle are made; *see* muscle fibres

myxoedema a condition that occurs if there is undersecretion of the hormone thyroxine from the thyroid gland during adulthood; the face becomes swollen and puffy, heart rate is slow, body temperature is low and the person generally lacks energy

myxovirus a group of RNA viruses that have a tendency to make red blood cells stick together; many produce diseases in humans such as influenza, measles and mumps

N

NAD an abbreviation for NICOTINAMIDE ADENINE DINUCLEOTIDE

nail a thin, horny plate, consisting of modified cells that develop from the outer layer of the skin; forms a protective covering on the upper surface of the ends of the fingers and toes; replaced the claw in primate evolution; the sharp, curved claws were reduced and modified to flat plates (the nails), which gave support to the terminal pads that were so important in maintaining a grip on a branch; although possession of nails rather than claws is considered to be a primate characteristic, some of the prosimians (e.g. tarsiers) have claws on some of their digits; *see* diagram below

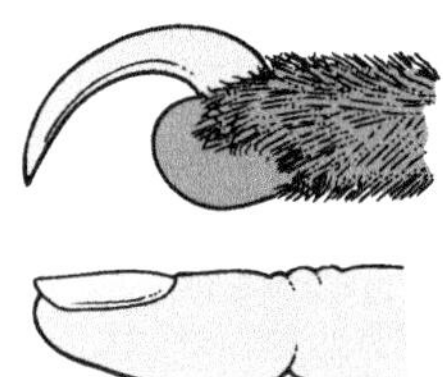

The difference between claw and nail

nano- prefix meaning very small; if used in measurement, it means *one billionth* or 10^{-9} (e.g. a nanometre is one billionth of a metre or one thousandth of a micrometre)

nanometre an extremely small unit of length; one nanometre is one thousand millionths (10^{-9}) of a metre; used to measure structures viewed through an electron microscope; 1000 nanometres = 1 micrometre; symbol *nm*; *see also* micrometre

narcosis a state of stupor or unconsciousness that has been induced by a drug; *narcotics* are drugs that produce such effects; alcohol, heroin, methadone, morphine and opium are all narcotics because they all produce unconsciousness or loss of control over bodily functions when used in high doses; at lower doses narcotics can be used as sedatives or to relieve pain

narcotic a drug that causes NARCOSIS

nasal cavity a cavity that opens at the nostrils on the outside and into the throat (pharynx) on the inside; divided into two parts by a vertical partition of bone and cartilage; in the upper nasal cavity, three shelves (or projections) increase the internal surface area so that air breathed in can be filtered, warmed and moistened before passing to the lungs; also contains receptors for smell

natality alternative term for BIRTH RATE; also called *natality rate*

natural immunity IMMUNITY to a disease that occurs without any human intervention

natural selection the process by which a population or species becomes better adapted to its environment; first proposed by Charles Darwin and Alfred Russel Wallace in 1858;

individuals with favourable characteristics are able to compete successfully in a particular environment; they are more likely to survive and produce a greater number of offspring than those with less favourable characteristics; after many generations the whole population will have the desirable characteristics; in terms of genetics, natural selection in a population can be described as phenotypes that have a high survival rate tending to pass their genes on to succeeding generations

nausea the discomfort, or sick feeling, that precedes VOMITING

Neanderthal people formerly called *Neanderthal man*; *see Homo neanderthalensis*

neck of a tooth, the constricted part where the tooth meets the gum; between the crown and the root; *see also* diagram of tooth (p. 267)

negative feedback where the response to a stimulus brings about a change that is opposite to, or reduces the effect of, the original stimulus; *see* feedback system

nematodes unsegmented worms, many of which are parasitic; also called *roundworms*

neo- a prefix meaning new (e.g. neoplasm—new growth)

neonatal during the first 4 weeks of a baby's life (after birth)

neonate a newborn infant; up to about 4 weeks old

neoplasm any abnormal new growth of tissue; a *malignant neoplasm* (or *malignant tumour*) is a growth that is capable of spreading to other parts of the body and thus causing secondary growths; a *benign neoplasm* (or *benign tumour*) is a growth that does not spread

neoteny the retention of juvenile features in adults; thought to be important in human evolution because adult humans have many characteristics that are present in juvenile apes, such as little body hair, thin skull bones, small jaws and high ratio of brain weight to body weight

nephron the functional unit of the kidney; consists of the glomerular capsule with its glomerulus and a very long tubule; nephrons filter blood at the glomerulus and, as the filtrate passes along the tubule, add some wastes and excess substances to the filtrate and reabsorb some useful substances from the filtrate; the liquid remaining after these processes is urine; there are estimated to be over a million nephrons in each kidney; *see* diagram below

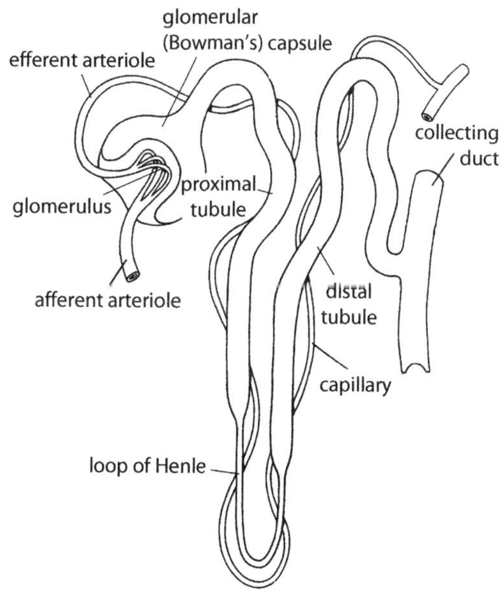

Nephron

nephron loop an alternative name for the LOOP OF HENLE

nephron tubule *see* nephron

nerve a bundle of NERVE FIBRES held together by connective tissue, located outside the central nervous system

nerve cell a common name for NEURON

nerve deafness a form of hearing loss; *see* deafness

nerve fibre any projection from the body of a nerve cell; usually refers to an axon with its associated coverings; *see also* diagram of nerve cell (p. 187)

nerve impulse also called an ACTION POTENTIAL

nerve tract a bundle of nerve fibres in the central nervous system; *see* tract

nervous system the body system involved with control and co-ordination of the body; consists of the brain, spinal cord, nerves and sensory receptors; the *central nervous system* (*CNS*) is the brain and spinal cord; all nervous impulses that stimulate the functioning of muscles and glands originate in the CNS; information from receptors must be sent to the CNS if any interpretation and response is required; the *peripheral nervous system* consists of the 12 pairs of cranial nerves and 31 pairs of spinal nerves that connect the parts of the body with the central nervous system; the peripheral nervous system is divided into afferent and efferent divisions; the *afferent division* (or *sensory division*) contains nerve fibres that carry impulses into the brain and spinal cord from receptors; the *efferent division* (or *motor division*) contains nerve fibres that carry impulses out of the brain and spinal cord; the part of the efferent division of the peripheral nervous system that carries nerve impulses from the brain and spinal cord to skeletal muscles is called the *somatic nervous system* (or *somatic division*); the part of the efferent division of the peripheral nervous system that carries nerve impulses from the brain and spinal cord to smooth muscle, cardiac muscle and glands is the *autonomic nervous system* (or *autonomic division*); it controls involuntary activities of the body such as heartbeat, movements of the alimentary canal and changes in the diameter of the pupil of the eye; the autonomic nervous system is further divided into *sympathetic* and *parasympathetic divisions*; these two divisions commonly oppose each other; *see* table of effects of sympathetic and parasympathetic stimulation (p. 186), diagram showing relationship of parts of nervous system opposite

nervous tissue the nerve cells that make up the nervous system; tissue that initiates and carries messages

net diffusion the net movement of ions or molecules from a higher to a lower concentration; *see* diffusion

neural arch part of a vertebra that forms the VERTEBRAL CANAL

neural disease a disease of the nervous system

neural tube defects a group of conditions affecting the brain or spinal cord of newborn babies; include SPINA BIFIDA and ANENCEPHALY

neurilemma a sheath around a nerve fibre; *see* myelin sheath

neuro- a prefix meaning related to a nerve or the nervous system (e.g. neuron—a nerve cell, neurologist—one who studies the nervous system)

neurodegenerative disease a disease in which nerve cells of the brain and spinal cord are gradually lost; examples are Alzheimer's disease, multiple sclerosis and Parkinson's disease

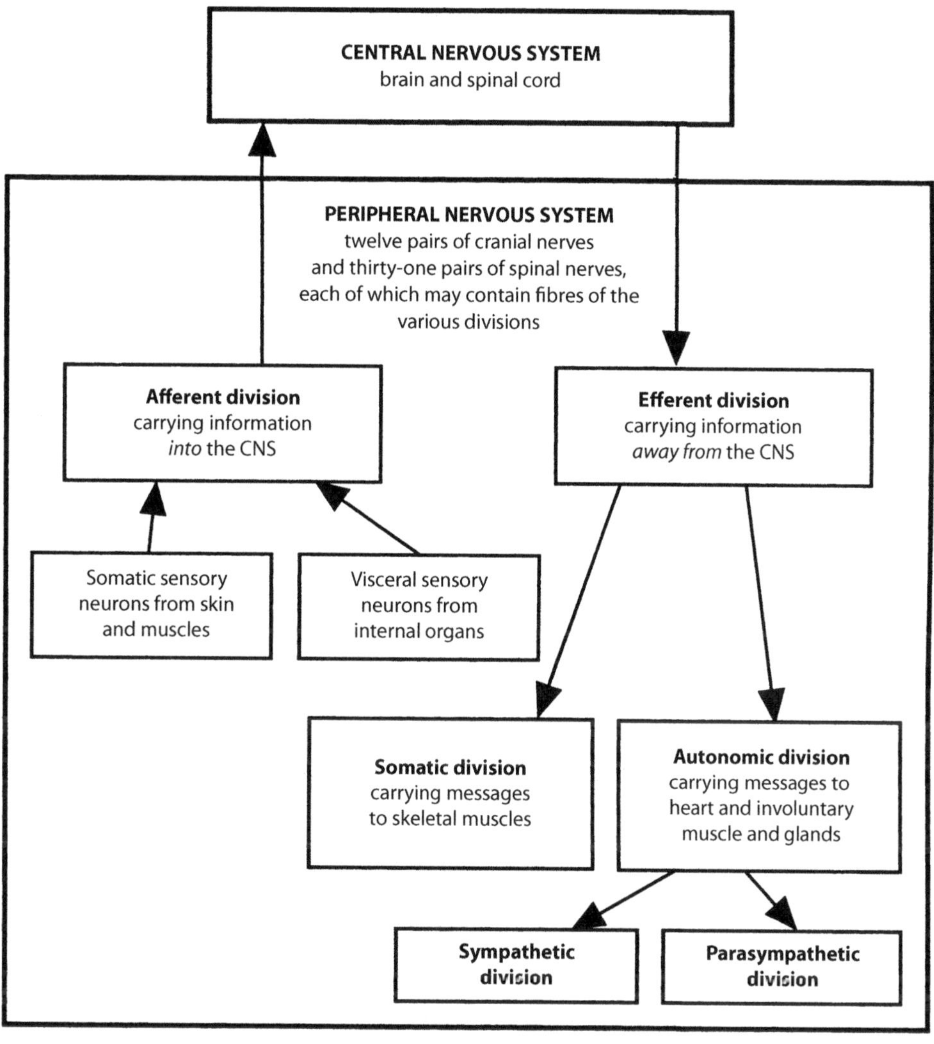

Organisation of the nervous system

neuroendocrine refers to secretions from the nervous system into the blood and from the nervous system to adjacent cells; *see also* neurosecretion, neurohormone

neuroendocrine cells an alternative name for NEUROSECRETORY CELLS

neurofibril node an alternative name for NODE OF RANVIER

neurofibromatosis an inherited disease in which growths occur on nerve cells; the growths may be harmless but may be serious if they press on nerves or other tissues

neuroglia cells in the central nervous system that provide support and protection for the nerve cells; they are smaller than nerve cells and 5–10 times more numerous; also called *glia* or *glial cells*

neurohormone a chemical produced by neurosecretory cells; functions as a hormone, producing specific effects in

Table 8 *The effects of the autonomic nervous system*

Structure	*Effects of sympathetic stimulation*	*Effects of parasympathetic stimulation*
Heart	Increases rate and strength of contraction	Decreases rate and strength of contraction
Lungs	Dilates bronchioles	Constricts bronchioles
Glands		
Salivary	Decreases secretion	Stimulates secretion
Sweat	Local stimulation of secretion	General stimulation of secretion
Gastric and intestinal	Inhibits secretion	Stimulates secretion
Pancreas	Inhibits secretion of enzymes and insulin; stimulates secretion of glucagons	Stimulates secretion of enzymes and insulin
Stomach and intestine	Decreases peristalsis	Increases peristalsis
Liver	Increases release of glucose; decreases bile secretion	Decreases release of glucose; increases bile secretion
Eye		
Iris	Dilation of pupil	Constriction of pupil
Ciliary muscle	Relaxation for distant vision	Contraction for near vision
Arterioles		
Skin	Constriction	No effect for most
Salivary glands	Constriction	Dilation
Skeletal muscle	Constriction or dilation depending on location	No effect
External genitalia	Constriction	Dilation
Kidney	Decreased urine production; secretion of renin	No effect
Urinary bladder	Relaxation of muscles of wall	Contraction of muscles of wall
Muscles attached to hair follicles	Contraction—erection of hairs	No effect
Spleen	Contraction—stored blood passed into circulation	No effect

some other part of the body; *see also* neurosecretion

neurohypophysis the posterior lobe of the PITUITARY GLAND

neurolemmocytes an alternative name for *Schwann cells*; *see* myelin sheath

neurology the study of the nerves and the nervous system

neuromuscular junction the junction between branches of a motor nerve cell and a muscle fibre; also called the *motor end plate*; *see also* diagram, opposite left

neuron an individual nerve cell; the basic unit of the nervous system; conducts nerve impulses from one part of the body to another; consists of three distinct parts: the cell body that contains the nucleus and cellular organelles, branched extensions of the cytoplasm (dendrites) that carry nerve impulses towards the cell body, and a single extension of cytoplasm (the axon) that carries impulses away from the cell body; also called a *nerve cell*; also spelt *neurone*; *sensory neurons* carry messages from receptors into the brain and spinal cord; *motor neurons* carry impulses from the brain and spinal cord to the muscles and glands; *connector neurons* link sensory and motor neurons in the brain and spinal cord; *see* diagrams, right

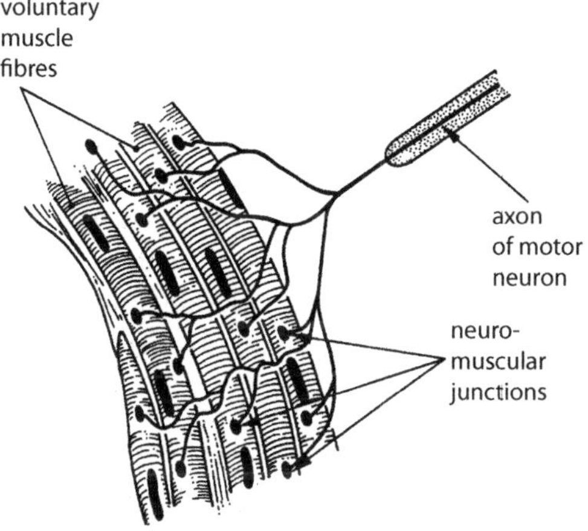

Neuromuscular junctions

neuropeptides polypeptides that occur in the brain and modulate the activity of neurotransmitters (e.g. endorphins, enkephalins)

neurophysiology the study of the functioning of the nervous system

neurosecretion secretion of hormones by nerve cells; there are *neurosecretory cells* with cell bodies in the hypothalamus and axons that pass to the posterior lobe of the pituitary gland; the hormones oxytocin and antidiuretic hormone are secreted by neurosecretory cells and stored in the posterior pituitary until required

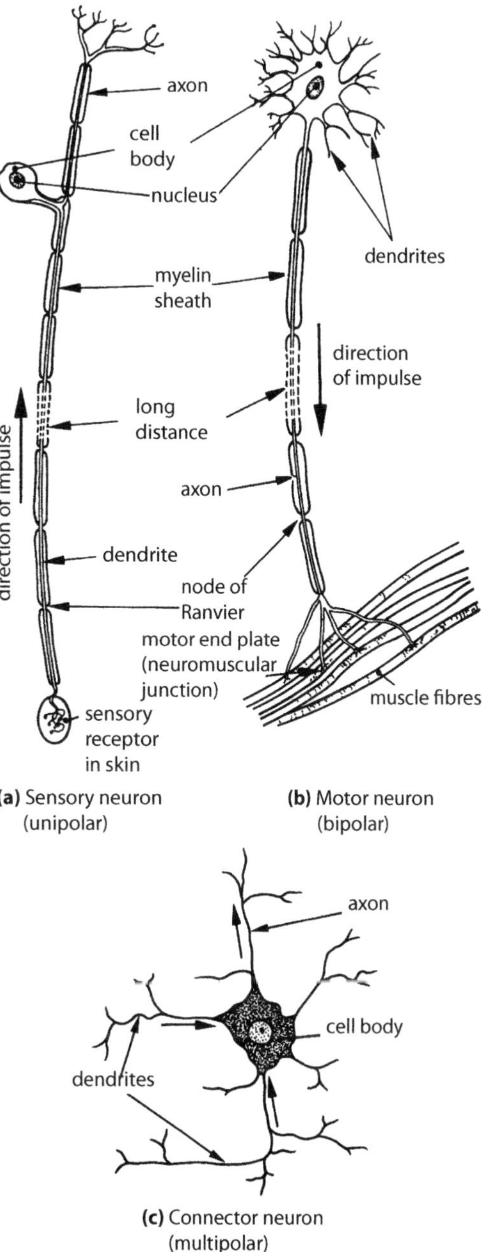

(a) Sensory neuron; (b) motor neuron; (c) connector neuron

neurosecretory cells nerve cells that secrete hormones; *see* neurosecretion

neurosis (plural *neuroses*) an emotional disorder characterised by some impairment of thinking or judgment; anxiety is the main symptom

neurotransmitter a chemical messenger that diffuses across the small gap (synapse) between branches of adjacent nerve cells or from the branches of nerve cells to a muscle fibre or glandular cell; in effect it carries the nerve impulse across the synapse; neurotransmitters can excite or inhibit the nerve cells beyond the synapse (postsynaptic neurons); the best known neurotransmitter is *acetylcholine*; *see also* dopamine, serotonin, diagram of transmission across a synapse (p. 263)

neutral neither acidic nor basic; having a pH of 7; *see* pH scale (p. 208)

neutron one of the particles that make up the nucleus of an atom; has no electrical charge and its mass is similar to that of a proton

neutrophil the most common type of white blood cell; characterised by granular material in the cytoplasm; important in combating bacterial infections; also called *a polymorph*

New World monkeys primates that belong to the parvorder Platyrrhini; live in southern Mexico, Central America and South America; almost exclusively tree-living; distinguished from the Old World monkeys by having broader noses, more sideways-facing nostrils with a broader division between the nostrils, three premolars on each side of each jaw (instead of two), and many have a prehensile tail (e.g. marmosets, tamarins, howler monkeys, spider monkeys, capuchins, night monkeys)

niacin an alternative name for the vitamin NICOTINIC ACID

nicotinamide adenine dinucleotide (NAD) a co-enzyme involved as a hydrogen carrier in the Krebs cycle and other oxidation-reduction reactions; during the Krebs cycle, which is part of aerobic respiration, the co-enzymes NAD and FAD are reduced to become $NADH_2$ and $FADH_2$; they now contain the energy from the breakdown of glucose; in the electron transport chain the energy in the two co-enzymes is transferred to ATP

nicotine the main drug contained in tobacco; the addictive ingredient of tobacco

nicotinic acid one of the B complex of vitamins; water soluble; good sources are meat, fish, liver, yeast, whole grains, peas and beans; an essential part of a co-enzyme involved with energy-producing reactions; deficiency causes *pellagra*, a disease characterised by diarrhoea, skin disorders and mental disturbances; also called *niacin*

nictitating membrane a transparent fold of skin (third eyelid) that protects the eyes of birds and reptiles; in humans it occurs as a VESTIGIAL ORGAN in the corner of the eye

night blindness poor vision in dim light although vision is normal in bright light; caused by a deficiency of vitamin A in the diet

nipple the pigmented, wrinkled projection on the surface of the mammary gland that is the location of the milk-secreting ducts; also called *mammary papilla*

nitric oxide a gas released in minute quantities by many cells; acts as a chemical messenger, triggering a number of responses; a major function is to cause relaxation of muscle cells in artery walls, so keeping arteries dilated; also functions as a neurotransmitter

nitrogenous waste nitrogen compounds, excreted by the kidneys, that are the waste products from the breakdown of amino acids; nitrogenous wastes include urea, uric acid, creatinine and a small amount of ammonia

NMR tomography an abbreviation for *nuclear magnetic resonance*; now known as MAGNETIC RESONANCE IMAGING

nocturnal describes an animal that is active during the night; tarsiers and lorises are prosimians that are nocturnal; of the anthropoid primates, the only one that is nocturnal is the night monkey; *see also* diurnal

node of Ranvier a gap in the myelin sheath of a nerve fibre; nerve impulses jump from one node to the next, increasing the speed of conduction; also called *neurofibril node*; *see* diagram of nerve cell (p .187)

non-disjunction the failure of a normal pair of chromosomes to separate during the first division of meiosis; results in one of the daughter cells receiving an extra chromosome while the other daughter cell is lacking in that chromosome; in humans, Down's syndrome or TRISOMY-21 is a result of non-disjunction of the 21st pair of chromosomes; KLINEFELTER'S SYNDROME and TURNER'S SYNDROME also result from non-disjunction

non-insulin-dependent diabetes also known as *type 2 diabetes*; *see* diabetes mellitus

non-opioid analgesic a pain-killing drug that does not come from or contain opium or morphine (e.g. paracetamol and aspirin)

non-random mating any pattern of choice of mate that is not totally random; influences the distribution of genotype and phenotype frequencies in a population; non-random mating is very common in human populations (e.g. it is highly likely that a person will choose a partner based on such characteristics as skin colour, height, weight, hair colour, religion or educational level rather than make a purely random choice); *see also* assortative mating

non-rapid eye movement (NREM) sleep one of two types of normal SLEEP

non-self describes foreign substances that are not part of an individual's usual body chemistry; *see* self

non-self antigen any compound foreign to the body that triggers an immune response

non-specific defence defence against disease that works against any type of infecting micro-organism and does not depend on previous exposure to the PATHOGEN; examples are external barriers (like the skin) and the inflammatory response (*see* inflammation)

non-specific genital infection a sexually transmitted disease caused by *chlamydia*

non-steroidal anti-inflammatory drugs (NSAIDs) drugs that reduce inflammation and pain but do not contain steroids (e.g. aspirin, ibuprofen, indomethacin and naproxen)

non-striated muscle an alternative name for *smooth muscle*; see muscular tissue

nonsense codon an alternative name for TERMINATION CODON

noradrenaline a hormone secreted by the adrenal medulla that has a similar action to stimulation by the sympathetic division of the autonomic nervous system; is also an important neurotransmitter in the sympathetic nervous system and central nervous system;

also called *norepinephrine*; *see also* table of effects of sympathetic stimulation (p. 186), table of hormones (p. 128)

norepinephrine an alternative name for NORADRENALINE

normal saline a salt solution with a concentration approximately equal to that of body fluids; used in intravenous drips to replace lost fluid or for patients who cannot take fluid by mouth

notochord the flexible internal rod that runs along the back of an animal; animals possessing a notochord at some period during their life are known as chordates (members of the phylum Chordata); marks the location of the backbone in embryos of vertebrates

NREM sleep an abbreviation for *non-rapid eye movement sleep*; *see* sleep

nuchal crest a bony crest on the skull of some primates at the edge of the area where the neck muscles are attached; runs vertically from the large opening through which the spinal cord passes (the foramen magnum) to a small bump just under the back of the skull (the external occipital protruberance); serves to increase the area for attachment of the neck muscles to the bone at the back of the skull (the occipital bone); prominent in heavy-jawed primates like the gorilla and some of the robust australopithecines

nuclear envelope an alternative name for NUCLEAR MEMBRANE

nuclear family a family consisting of the father, mother and unmarried children living together in the same home; *see also* extended family

nuclear magnetic resonance a former term for MAGNETIC RESONANCE IMAGING

nuclear medicine the branch of medicine that is involved in the use of radioactive isotopes for diagnosis of disease (e.g. use of radioactive iodine as a tracer when studying functioning of the thyroid gland) or in the treatment of disease (e.g. use of radiotherapy to treat cancer)

nuclear membrane the membrane around the nucleus of a cell; also called the *nuclear envelope*; consists of a double membrane with a space between the two layers; numerous openings in the membrane, known as *nuclear pores*, are connected to the endoplasmic reticulum in the cytoplasm and allow movement of materials between the nucleus and the cytoplasm; *see also* diagram of cell structure (p. 44)

nuclear pore an opening in the NUCLEAR MEMBRANE

nuclease an enzyme that breaks down nucleotides into pentose sugar and a nitrogen base; *deoxyribonuclease*, found in both pancreatic juice and intestinal juice, breaks down DNA; *ribonuclease*, also found in digestive juices, breaks down RNA

nucleic acid an organic compound made up of chains of nucleotides; there are two main types—DNA and RNA; *see* deoxyribonucleic acid, ribonucleic acid, diagram of DNA structure (p. 72)

nucleolus (plural *nucleoli*) a small spherical body within the nucleus of a cell; each nucleus contains one or more nucleoli; composed of protein; makes and stores RNA; *see also* diagram of cell structure (p. 44)

nucleoplasm jelly-like material inside the nucleus of a cell; *see also* diagram of cell structure (p. 44)

nucleoside a nitrogenous base (adenine, cytosine, guanine, thymine or uracil) attached to a 5-carbon (pentose) sugar such as ribose or deoxyribose; a nucleoside plus a phosphate group is a nucleotide, one of the subunits of a nucleic acid; adenosine is a nucleoside

nucleosome a subunit of a chromosome; consists of a core of small proteins called histones with part of a DNA molecule wrapped around it

nucleotide a basic structural unit of a nucleic acid; consists of a simple (pentose) sugar, a phosphate group and a nitrogen base; the base may be adenine, guanine, cytosine, thymine or uracil; *see also* deoxyribonucleic acid, ribonucleic acid, nucleotide diagram below

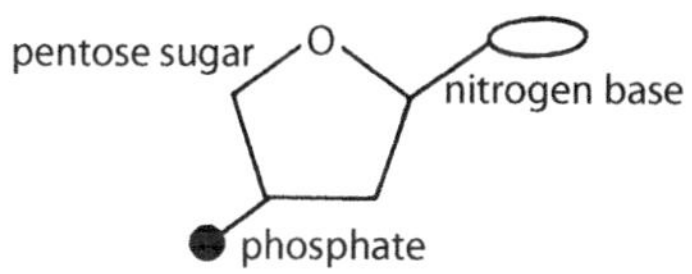

Nucleotide

nucleus (plural *nuclei*) **1.** of a cell, a structure within the cell that contains the hereditary material; the genes in the chromosomes of the nucleus control the structure of proteins produced by the cell; *see also* diagram of cell structure (p. 44) **2.** of an atom, the central part, made up of protons and neutrons

null hypothesis a hypothesis which states that the independent variable has no effect on the dependent variable (e.g. when testing a possible new drug, the null hypothesis would be that the new drug has no effect)

nutrient any substance in food that provides energy, is essential for growth or assists in the functioning of the body

nutrition the food required for normal body functioning or the processing of food within the cells

NuvaRing the trade name for a flexible plastic ring impregnated with female hormones; the ring is placed in the vagina and prevents conception by releasing small amounts of hormone over a period of three months

O

obesity being excessively overweight; more than 20% over normal weight for the person's height; such a person is said to be *obese*

objective lens the microscope lens closest to the stage of a microscope; most microscopes have a revolving nosepiece on which objectives of differing magnifications are mounted

objectivity making judgments based on evidence and not being influenced by emotions or prejudice

obligatory reabsorption reabsorption of water in the kidney tubule due to osmosis; the amount of water absorbed in this way is determined by osmotic pressure, as the tubules have no control over osmosis; *see also* facultative reabsorption

observation in science, anything that you see, smell, hear, taste or touch

obsessive-compulsive disorder an anxiety disorder in which the affected person feels compelled to constantly repeat trivial and meaningless actions

obstetrics the branch of medicine that deals with the care of the mother and baby before, during and after birth

occipital pertaining to the back of the head; *see also* occipital bone, occipital bun

occipital bone the bone making up the back of the skull

occipital bun a slight protrusion at the rear of the skull; a feature often found in Neanderthal fossils

occipital lobe one of the four lobes of each cerebral hemisphere of the brain; occurs towards the rear of the brain; *see* diagram of lobes of the cerebrum (p. 105)

occlusion closure or blockage (e.g. a coronary occlusion is a blockage of one of the arteries supplying the heart muscle)

ocular the microscope lens through which one looks; also called the *eyepiece*

oedema the swelling of parts of the body resulting from the accumulation of fluid in the tissues; increase in the amount of tissue fluid (e.g. swelling of the feet and ankles after standing or sitting still for a long time); also spelt *edema*

oesophagus the tube that carries food from the throat to the stomach; *see also* diagram of digestive system (p. 77)

oestradiol one of the steroid hormones that belongs to the group referred to as oestrogens; produced by the OVARIAN FOLLICLE; chemical formula $C_{18}H_{24}O_2$

oestriol one of the steroid hormones that belongs to the group referred to as oestrogens; produced by the OVARIAN FOLLICLE; chemical formula $C_{18}H_{24}O_3$

oestrogen replacement therapy an alternative name for HORMONE REPLACEMENT THERAPY

oestrogens the general name for female sex hormones; produced mainly in the ovaries and placenta; concerned with development and maintenance of female reproductive structures and secondary sexual characteristics; include oestrone, oestriol and oestradiol; *see* table of hormones (p. 128)

oestrone one of the steroid hormones that belongs to the group referred to as oestrogens; produced by the OVARIAN FOLLICLE; chemical formula $C_{18}H_{22}O_2$

oestrus the time, in most species of mammal, when a female will sexually accept a mate; many primate species, including humans and apes, do not have a period of oestrus and are able to mate at any time

oil gland an alternative name for SEBACEOUS GLAND

oil immersion a technique used with a light microscope when high magnifications (up to 1000x) are required; a small drop of 'immersion oil' is placed in the small gap between the microscope slide and a special oil immersion objective lens; the oil increases the amount of light passing through the lens

Old World monkeys primates that belong to the parvorder Catarrhini and superfamily Cercopithecoidea; live in Africa and Asia; many are ground-living, such as the baboons; distinguished from the New World monkeys by having narrower noses, down-facing nostrils, a thinner division between the nostrils, two premolars on each side of each jaw (instead of three) and often the presence of buttock pads (ischial callosities) (e.g. colobus monkeys, macaques, mandrills, patas monkeys, proboscis monkeys)

Oldowan tools very simple tools made by removing several flakes from a stone; the flakes removed could also be used as cutting tools; the stone tool culture of *Homo habilis*; sometimes called *pebble tools*; *see also* diagram of Oldowan tools below

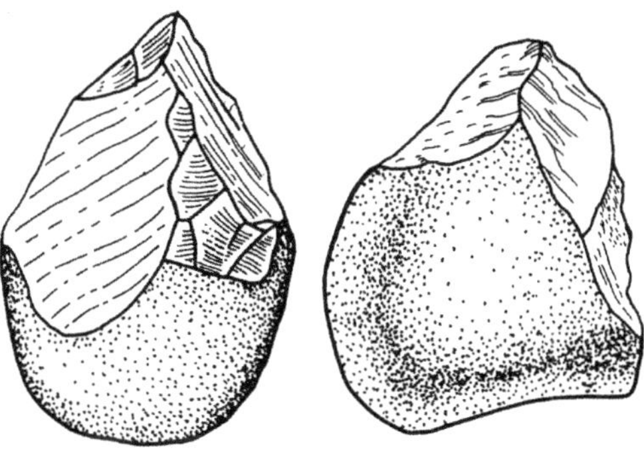

Oldowan tools

olestra a substance made from sucrose that tastes and behaves like fat but cannot be absorbed by the body; also known by brand name Olean; food cooked using olestra contains much less energy than similar food cooked in oil and is therefore less fattening; approved for use in snack foods by the US Food and Drug Administration in January 1996 but not approved in Australia; long-term use may cause health problems because olestra absorbs the fat-soluble vitamins and carotenoids from other foods before the body has a chance to absorb them

olfaction the sense of smell

olfactory cells nerve cells in the nasal cavity; each has a branch that ends in 6–8 cilia called *olfactory hairs*; the hairs react to odours in the air breathed in and stimulate the olfactory cells

olfactory hairs cilia that project from OLFACTORY CELLS

olfactory receptor an alternative name for OLFACTORY CELL

Oligocene the 3rd epoch of the Cainozoic era of geological time; between the Eocene and Miocene epochs; dating roughly 34–23 million years ago; the first anthropoids appeared during the Oligocene; *see* geological time scale (p. 111)

omega-3 fats a group of polyunsaturated fats, found mainly in fish and in canola, walnut and soya bean oils; omega-3 fats reduce the likelihood of clots in the blood vessels and thus reduce the risk of strokes and heart attacks

omnivore an animal that feeds on both plants and animals; such an animal is said to be *omnivorous* (e.g. chimpanzees)

oncogene a gene that is able to convert a normal cell into a cancerous cell; a slightly altered version of a normal gene called a *proto-oncogene*; some proto-oncogenes are activated to become oncogenes by mutations caused by carcinogens; others are activated by certain viruses

oncology the branch of medical science that deals with the study of tumours

one-egg twins an alternative name for *identical twins*; *see* twins

one gene–one enzyme hypothesis an idea, first proposed in 1941, that each gene is responsible for directing the synthesis of a single enzyme; now replaced by the ONE GENE–ONE POLYPEPTIDE HYPOTHESIS

one gene–one polypeptide hypothesis the idea that each gene is responsible for directing the synthesis of a single polypeptide; a refinement of the *one gene–one enzyme hypothesis* that was necessary because, although almost all enzymes are proteins, not all proteins are enzymes; this hypothesis accounts for the synthesis of structural proteins like keratin in hair and hormones that are proteins such as insulin

one-male group a type of social structure in which the primary social group is made up of a single adult male, several adult females and their offspring (e.g. a gorilla group, although occasionally one or more young adult males may join the group); *see also* multimale group, mother–infant group

ontogeny the growth and development of an organism from the fertilised egg to the mature adult

oocyte a cell that undergoes meiosis to produce an egg cell; *primary oocytes* are diploid cells in the ovary that divide by meiosis; from the first meiotic division two cells are produced; one, the *secondary oocyte*, receives most of the cytoplasm and divides again to produce the egg; the other, called a *polar body*, has no function and disintegrates; *see also* diagram of oogenesis (p. 107)

oogenesis the formation of egg cells; *see also* gametogenesis and diagram (p. 107)

oophorectomy the removal of the ovaries by surgery

ootid one of the four haploid cells produced by meiotic division of an oocyte; the ootid receives most of the cytoplasm from the original oocyte and matures to become the egg; the other 3 cells, the polar bodies, disintegrate; *see also* diagram of oogenesis (p. 107)

operant conditioning a learning situation in which a new response is learned to a stimulus that remains the same; *see* conditioning

operator gene a gene that activates production of messenger RNA in nearby STRUCTURAL GENES; part of an OPERON

operon a segment of DNA that includes an OPERATOR GENE, a PROMOTER GENE and one or more STRUCTURAL GENES; functions as a unit to produce messenger RNA

ophthalmic relating to the eye (e.g. the ophthalmic artery carries blood to tissues of the eye and eye socket)

opiate any drug derived from opium (e.g. morphine, codeine, heroin)

opioid analgesics pain-killing drugs that act like morphine; made from opium derivatives or related synthetic chemicals; examples are codeine, morphine and oxycodone

opioids any drug either derived from opium or with similar effects to opium derivatives (e.g. pethidine and methadone are made synthetically but have similar effects to some opium derivatives and are therefore referred to as opioids)

opium a drug obtained from the resin that oozes out of the cut seed pod of the opium poppy; *morphine* and *codeine* can be separated from the opium resin and both are used medically as painkillers; *heroin*, made from codeine or morphine by a chemical process, may be used to relieve severe pain or may be used illegally for the feeling of well-being it gives

opposability the ability of the fleshy tip of the first digit to touch the corresponding tip of each of the other digits (e.g. humans have an *opposable thumb*, but the big toe is not opposable)

opposable thumb *see* opposability

opsonins a group of proteins that coat particles foreign to the body, such as micro-organisms, making them more susceptible to phagocytosis; a process called *opsonisation*

opsonisation making foreign particles susceptible to phagocytosis; *see* opsonins

optic chiasma the point under the hypothalamus of the brain where the optic nerves from the left and right eye meet and cross over; because of the crossing over, nerve impulses from the left eye are interpreted on the right side of the brain and vice versa

optic disc an alternative name for BLIND SPOT

optic nerve the 2nd cranial nerve; one of a pair of nerves that carries nerve impulses from the eye; *see also* table of cranial nerves (p. 65), diagram of eye (p. 94)

orang-utan one of the anthropoid apes, *Pongo pygmaeus*; there are two subspecies—one from the island of Borneo, and the other living on the island of Sumatra; large, vegetarian apes with reddish-brown hair, living in tropical rainforests; adult males usually live by themselves, the basic social group being that of mother and infant; orang-utan translates from Malay as 'man of the forest'

orbit the bony cavity of the skull that holds the eyeball; the eye socket; *see* diagram of skeleton (p. 249)

order a category in the classification of organisms; the level of classification between class and family; each order contains one or more families; humans belong to the order Primates

Ordovician one of the 7 periods in the Palaeozoic era of geological time; between the Cambrian and Silurian periods; 488–444 million years ago; the early vertebrates, including jawless fish, first appeared during the Ordovician period; *see* geological time scale (p. 111)

organ a structural and functional unit of the body, made up of one or more tissues (e.g. the heart, kidney, liver, eye, a bone or muscle)

organ of Corti the organ of hearing; also called the *spiral organ*; located inside the cochlear duct, which is inside the cochlea of the inner ear; consists of supporting cells and hair cells; long hair-like extensions from the hair cells protrude into the liquid of the cochlear duct and the bases of the hair cells link up with nerve fibres; projecting over the hair cells and in contact with the ends of the hairs is a flexible jelly-like membrane called the *tectorial membrane*; the membrane that supports the base of the hair cells is called the *basilar membrane*; movement of fluid in the cochlear duct results in movement of the hairs, so that the original sound wave is converted into nerve impulses, which are transmitted to the brain; *see also* diagram of ear (p. 82)

organ system a group of organs that work together to perform a particular task, such as the digestive system or the circulatory system

organelle a membranous structure within a cell; specialised for a particular function within the cell (e.g. mitochondrion, Golgi apparatus, endoplasmic reticulum); *see also* diagram of cell structure (p. 44)

organic compound a compound consisting of large molecules and always containing carbon and hydrogen; such compounds present in the human body include carbohydrates, proteins, lipids, nucleic acids, adenosine triphosphate (ATP) and many others; *see also* inorganic compound

organic matter matter that has come from a living or dead organism; in organic chemistry, organic matter is any large molecule containing carbon atoms

organism an individual living thing

orgasm the climax of sexual stimulation; the period of greatest intensity of pleasure during sexual intercourse; intense emotional and physical feelings associated with ejaculation of semen from the penis in the male, and with contraction of muscles around the uterus and vagina of the female; also called the *climax*

origin of a muscle, the end that is fixed to the stationary bone; the end that remains stationary when the muscle contracts (e.g. the origin of the biceps muscle that bends the arm at the elbow is at the scapula—the shoulder blade); *see also* insertion

origin of life the beginning of life on the earth; evidence to date suggests that life began on the earth through a process of chemical evolution; experiments have shown that amino acids, and other combinations of molecules that are present in simple cells, can be produced by discharging electrical energy into a mixture of chemicals much like those in the atmosphere of the ancient earth; it has been suggested that for such reactions (leading to the beginnings of life) the energy source could have been the sun and lightning

Orrorin tugenensis a newly described species of hominin that lived in eastern Africa between 6.1 and 5.8 million years BP; named after the Tugen Hills of Kenya, the site where it was found in 2000; fossils include fragmentary arm and thigh bones, lower jaws, and teeth; the thigh bones resemble those of *Australopithecus*, suggesting that it had a similar walking style to these early hominins even though it existed almost 3 million years earlier; limbs indicate it was probably adapted to both bipedality and tree climbing

orthognathism a flat facial profile; the jaw does not project; the opposite of PROGNATHISM

osmolarity a measure of the quantity of dissolved substances in urine or some other liquid

osmoreceptors receptors in the hypothalamus that detect changes in the osmotic pressure of the blood (changes in the concentration of dissolved substances in the blood)

osmosis the diffusion of water molecules through a differentially permeable membrane from an area of high water concentration (and low concentration of dissolved substances) to an area of lower concentration of water molecules (and higher concentration of dissolved substances)

osmotic pressure the pressure due to osmosis; usually measured as the pressure that would be required to prevent osmosis; *see also* illustration of osmotic pressure below

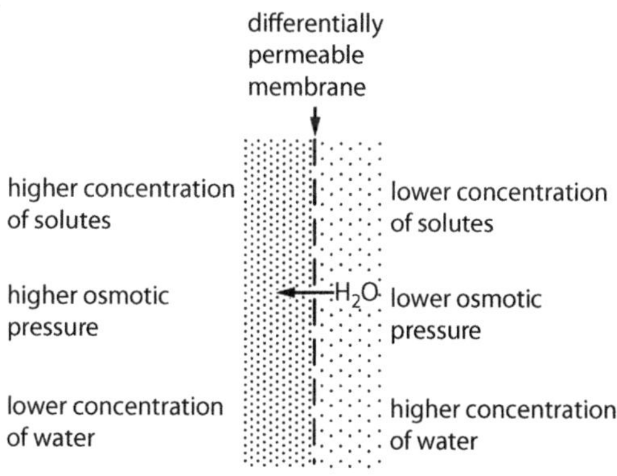

Osmotic pressure

ossicle one of the bones in the middle ear; *see* auditory ossicles

ossification the process by which bone is formed in the skeleton; *see* osteoblast

osteo-, osseo-, oss- prefixes meaning bone (e.g. osteoma—bone cancer, ossification—the formation of bone)

osteoarthritis a form of ARTHRITIS

osteoblast a bone-forming cell; secretes substances, partly composed of fibres, into the spaces between cells, and then deposits calcium salts to harden the material; eventually becomes trapped in the matrix it has deposited and loses the ability to deposit bony material; is then called an *osteocyte*; *see also* remodelling

osteoclast a cell that breaks down bone during REMODELLING

osteocyte a mature bone cell; *see* osteoblast

osteomalacia loss of calcium and phosphorous from bones of adults due to deficiency of vitamin D; bones most affected are those of the pelvis, legs and spine

osteomyelitis the term used to describe all the infectious diseases of bone; the most common micro-organisms to infect bone are bacteria called *Staphylococcus aureus*

osteon an alternative name for HAVERSIAN SYSTEM

osteoporosis a disorder characterised by a reduction in the amount of bone; the bones of the skeleton become porous, fragile and easily broken; more common in middle-aged or elderly people, especially females after menopause; up to middle age, sex hormones maintain bone by stimulating osteoblasts to form new bone; decreased sex hormone secretion causes osteoblasts to become less active and the mass of bone to decline

otoliths particles of calcium carbonate in the utricle and saccule of the inner ear; involved in sensing the position of the head; when the head moves (with respect to gravity), they move and stimulate hair cells in the utricle and saccule; *see* diagram, below

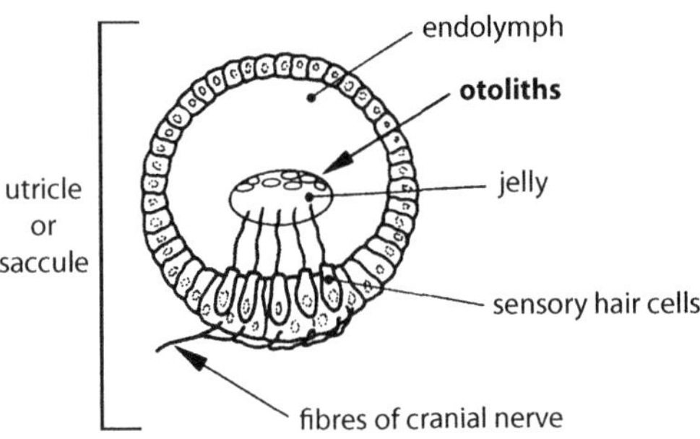

Otoliths

out of Africa hypothesis an alternative name for AFRICAN ORIGIN MODEL

outer ear that part of the ear consisting of the pinna, the ear canal and the eardrum; also called the *external ear*; *see* diagram of ear (p. 82)

outlier in the results of a study, a figure that lies well outside the general trend of the data; often ignored when calculating means or plotting graphs of experimental data

ova mature eggs; plural of ovum; *see also* diagram of oogenesis (p. 107)

oval window a small, membrane-covered opening between the middle and inner ear, directly above the round window; one of the three small bones in the middle ear, the stirrup, is connected to the oval window; vibrations of the stirrup make the oval window (*fenestra ovalis* or *fenestra vestibuli*) vibrate, and waves of vibration then occur in the liquid in the cochlea; as the oval window moves inwards, another opening just below it, the *round window* (*fenestra rotunda* or *fenestra cochlea*), moves outwards to compensate; *see also* cochlea, diagram of ear (p. 82)

ovarian cycle a series of changes that occur in the ovary approximately every 28 days, associated with the maturation of an egg; events of the ovarian cycle correlate with those of the menstrual cycle; during the first few days of the cycle, while menstruation is occurring, a number of primary follicles develop into secondary follicles; one follicle, known as the *Graafian follicle*, then matures ready for release of the egg (ovulation), which occurs on day 14 of the 28-day cycle; after ovulation the Graafian follicle becomes

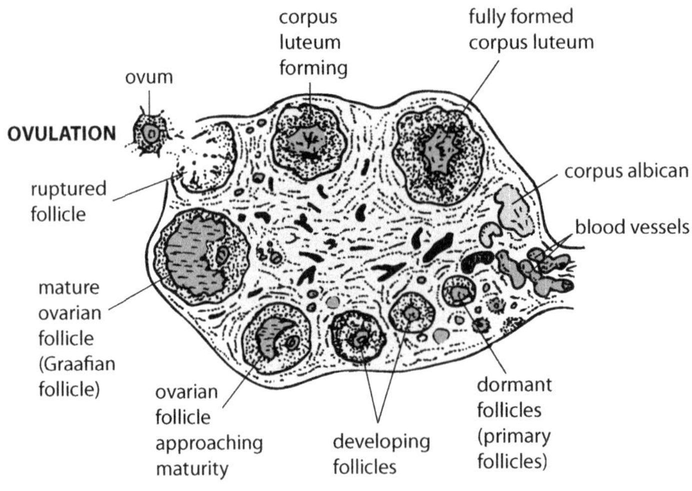

Ovarian cycle

the corpus luteum; if fertilisation does not occur, the corpus luteum degenerates and the cycle begins again; *see also* menstrual cycle, diagram previous page bottom

ovarian cyst a fluid-filled sac in the ovary or on the surface of the ovary; may disappear without treatment or may cause pain and require surgical removal

ovarian follicle a group of cells that surrounds a developing oocyte in an ovary; *see also* Graafian follicle and diagram, previous page

ovary the organs in which the female gametes, the ova (or eggs), and the female hormones (oestrogens and progesterone) are produced; *see also* diagram of female reproductive system (p. 235)

overweight the condition of having more body fat than the optimum for good health; defined as a BODY MASS INDEX of 25 to 29.9

oviduct an alternative name for UTERINE TUBE

ovulation the release of the egg from the Graafian follicle; usually occurs on day 14 of a typical 28-day ovarian cycle; release is initiated by a surge of luteinising hormone from the pituitary gland; after release the egg is swept into the uterine tube by the action of cilia; *see* diagram, previous page

ovum (plural *ova*) the mature egg cell; *see also* diagram of oogenesis (p. 107)

oxidation the removal of electrons and hydrogen ions from a molecule; *see* oxidation–reduction reactions

oxidation–reduction reactions *oxidation* is the removal of electrons and hydrogen ions from a molecule; when oxidation occurs, the hydrogen ions are transferred to another compound by co-enzymes; *reduction* is the addition of electrons and hydrogen ions to a molecule; in a cell, the two types of reaction are always coupled together so that as one molecule is oxidised another is reduced; also called *redox reactions*

oxidative phosphorylation the process by which ADP is combined with phosphate to form ATP in the electron transport system of aerobic respiration

oxygen debt extra oxygen required after exercise, in addition to the normal amount required at rest; during vigorous exercise, muscles respire anaerobically (without oxygen) and lactic acid accumulates in the muscles; after exercise, oxygen is required to remove the lactic acid, and so heavy breathing continues for some time to pay back the 'debt'; the term *recovery oxygen* is preferred by some authorities

oxygenated containing a lot of oxygen (e.g. oxygenated blood is blood that has just been through the capillaries of the lungs and contains a relatively high proportion of oxygen); blood is oxygenated in the lungs; *see also* deoxygenated

oxygenated blood blood containing a lot of oxygen; blood that has just passed through the lungs

oxyhaemoglobin a HAEMOGLOBIN molecule combined with oxygen

oxyntic cell an alternative name for PARIETAL CELL

oxytocin a hormone produced by the hypothalamus and released by the posterior lobe of the pituitary gland; stimulates contraction of the muscles of the uterus during labour; also stimulates contraction of cells in the mammary glands during suckling; *see* table of hormones (p. 128), neurosecretion

P

pacemaker a group of specialised cells in the wall of the right atrium of the heart; regulates heartbeat by sending out nerve impulses over the heart muscle; the inbuilt rhythm of the pacemaker can be changed by nerve impulses from the autonomic nervous system or by certain hormones such as adrenaline; also called the *sinoatrial node* or *SA node*; the term pacemaker is also used to describe an artificial device that generates and delivers electrical signals to the heart to maintain a regular heartbeat; the pacemaker causes the atria to contract and also activates the ATRIOVENTRICULAR (AV) NODE; from the AV node a bundle of special muscle fibres, the *atrioventricular bundle* (or *bundle of His*), carries the stimulus to the ventricles, causing them to contract; *see* diagram of pacemaker below

pachytene the 3rd stage of prophase I of MEIOSIS; during pachytene the chromosomes thicken and each is seen to be a pair of chromatids; CROSSING OVER between chromatids can occur at this stage

Pacinian corpuscles pressure receptors located in the skin; deeper in the skin than touch receptors; also called *lamellated corpuscles*; *see also* diagram of skin (p. 250)

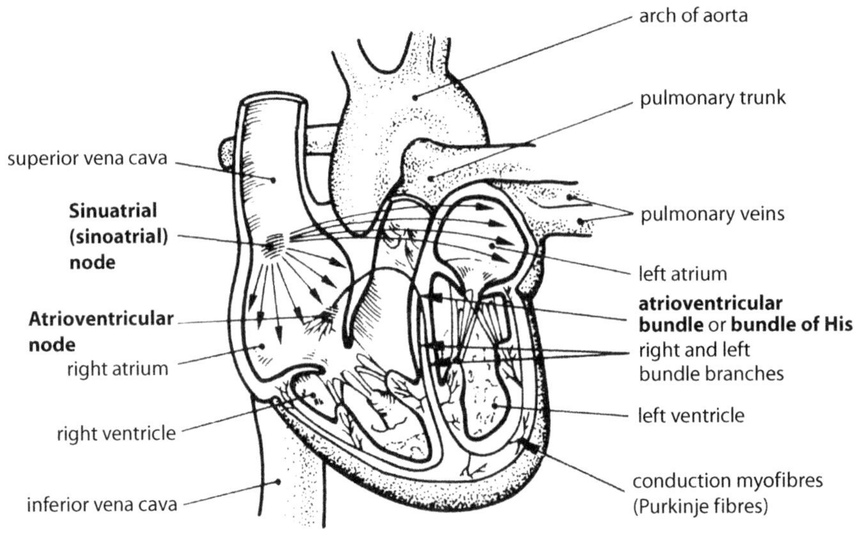

Location of pacemaker

paediatrician a medical practitioner specialising in PAEDIATRICS

paediatrics the field of medicine specialising in the diagnosis and treatment of diseases of children from birth to adolescence; a doctor specialising in this field is called a *paediatrician*

pain an unpleasant or distressing feeling resulting from stimulation of pain receptors; pain has a protective function because it is associated with injury or disease; sudden pain often triggers a reflex response (e.g. withdrawing the hand from a painful stimulus)

pair-bond a relationship established between a male and a female for breeding purposes; the relationship lasts at least until any offspring achieve independence, but in some cases (such as with gibbons) the pairing may last for life

palaeo- a prefix meaning ancient, from the Greek *palaios* (e.g. palaeobiology—the study of extinct organisms)

palaeoanthropology the study of human evolution through examination of fossil remains; research in this area concentrates on determining who the ancestors of modern humans were, when and how they evolved, and why they evolved; palaeoanthropologists work closely with archaeologists to reconstruct the behaviours of the ancestors of modern humans

palaeobiology the biology of extinct plants and animals

Palaeocene the earliest of 7 epochs in the Cainozoic era of geological time; 65–56 million years ago; the primate-like mammals lived during this period; *see* geological time scale (p. 111)

palaeoecology the study of the relationship between extinct organisms and the environment in which they lived

palaeomagnetism refers to the direction of the earth's magnetic field in past times; *see* geomagnetic dating

palaeontology the study of fossils; that is, the study of the remains, or traces, of ancient organisms preserved in rocks

palaeopathology the study of disease in prehistoric populations based on the analysis of skeletal remains and archaeological evidence

palaeospecies species identified from fossil remains and named on the basis of their physical similarities to other species (e.g. *Homo erectus* is an extinct species that has similarities to *Homo sapiens* but also has distinct differences); the point in time at which one species evolves into another cannot be determined, and so palaeospecies that form an evolutionary line are usually separated by long periods of time; often two such species are separated by a period of time for which no fossils have been found; when 'missing links' are subsequently discovered, the classification may have to be revised

Palaeozoic the 4th era of geological time, dating roughly 620–251 million years ago; the word means 'ancient life' and this era was so named because it is the first era to contain abundant fossils; the first vertebrates appeared during this era, including the reptiles and mammal-like reptiles; *see also* geological time scale (p. 111)

palate the roof of the mouth, which separates the mouth from the nasal cavity

palliative care nursing care that serves to relieve or alleviate pain and suffering without curing; the type of care that may be given to the terminally ill; may be given at home or in a HOSPICE

palynology the study of fossilised spores, pollen grains, micro-organisms and microscopic fragments of organisms; useful in determining the ages of sedimentary rocks

pancreas a gland lying between the stomach and duodenum; is both an exocrine and an endocrine gland; secretes pancreatic juice from the exocrine cells; clusters of endocrine cells in the pancreas called *islets of Langerhans* (or *pancreatic islets*) secrete hormones; the islets contain three types of cells—*alpha cells* secrete the hormone GLUCAGON, *beta cells* secrete the hormone INSULIN and *delta cells* secrete GROWTH HORMONE inhibiting factor; *see* diagram of digestive system (p. 77)

pancreatic amylase an enzyme found in pancreatic juice; breaks down starch into disaccharides

pancreatic duct a tube that carries pancreatic juice into the duodenum; the bile duct joins into the pancreatic duct just before it enters the duodenum; *see also* diagram of digestive system (p. 77)

pancreatic islets an alternative name for *islets of Langerhans* of the PANCREAS

pancreatic juice the alkaline liquid secreted by the exocrine cells of the pancreas; contains digestive enzymes including amylases, proteases (trypsin), lipase, ribonuclease and deoxyribonuclease

pancreatic lipase an enzyme found in pancreatic juice that breaks down fats into fatty acids and glycerol; *see also* lipase

pancreatic protease an enzyme found in pancreatic juice that breaks down proteins into small chains of amino acids; also called *trypsin*; is secreted in the inactive form *trypsinogen*, which is converted to trypsin on contact with an enzyme secreted by the lining of the intestine

pandemic describes a disease that occurs over a large geographic range, such as over an entire continent or even the whole world (e.g. malaria is a pandemic disease)

panmixis an alternative term for RANDOM MATING

pannus abnormal tissue produced in rheumatoid ARTHRITIS

pantothenic acid one of the B complex of vitamins (vitamin B_5); water soluble; stored in the liver and kidneys; good sources are liver, kidney, green vegetables, cereals; deficiency probably causes poor muscle co-ordination and nervous disorders

Pap smear an abbreviation for PAPANICOLAOU TEST

Papanicolaou test a test for cancer of the cervix in females; a few cells are scraped from the neck of the uterus (cervix), smeared onto a microscope slide and examined microscopically for any signs of abnormality; allows cancerous cells to be detected before any symptoms have appeared; also called *Pap test*, *Pap smear* or *smear test*

paper in science, an article published in a journal, or read at a conference, that describes an investigation, its results and the conclusions drawn

papillae (singular *papilla*) small nipple-shaped projections; **1.** on the tongue, small projections (some types of which contain taste buds) that give the tongue its rough texture; *see also* diagram of gustatory cells (p. 117) **2.** groups of cells at the base of hair follicles

papilloma a non-cancerous wart

para- a prefix meaning near, beside or beyond (e.g. paranasal means near the nose)

paracrine acting locally (e.g. paracrine signalling by a cell affects nearby cells)

paracrine hormone a hormone that acts over a short distance; carried to the target in the extracellular fluid

parallel evolution the independent evolution of similar forms by unrelated organisms; usually occurs because natural selection is adapting them to similar environments; evolution occurs independently in the same direction but away from the ancestral form; the most striking example is the Old World and New World monkeys, which were separated for up to 50 million years but have evolved along very similar lines; also called *convergent evolution*, *convergence* or *parallelism*

parallelism an alternative term for PARALLEL EVOLUTION

paralysis the loss of ability to move part of the body due to disease or damage to muscles or nerves

paranasal sinuses paired spaces in the bones around the nose; *see* sinus

paranoia a type of mental illness in which the patient suffers from feelings of suspicion, persecution and delusion

Paranthropus a genus of extinct hominids known from fossils found in Africa; some authorities split the australopithecines into two genera, *Australopithecus*, which includes the GRACILE forms, and *Paranthropus*, which includes the ROBUST forms; the two species commonly recognised in this genus are *Paranthropus robustus* and *Paranthropus boisei*

Paranthropus aethiopicus a classification used by some authorities for AUSTRALOPITHECUS AETHIOPICUS

Paranthropus boisei a species of robust australopithecine; many authorities refer to this species as *Australopithecus boisei*

Paranthropus crassidens a hominid represented by fossils from Swartkrans in southern Africa; not usually recognised as a separate species but rather classified as belonging to *Australopithecus robustus*

Paranthropus robustus a species of fossil hominid that lived in Africa between 2 and 1 million years BP, first found by Robert Broom at Kromdraai in South Africa in 1938; now usually referred to as AUSTRALOPITHECUS ROBUSTUS, although some authorities are again using the genus PARANTHROPUS for the robust australopithecines

paraplegia paralysis of both legs; caused by injury to, or disease of, the spinal cord; a person suffering from paraplegia is known as a *paraplegic*; *see also* quadriplegia, hemiplegia

parasite an organism that lives on, or in, another living thing and derives food and shelter from it; may be an internal parasite (or *endoparasite*) such as a tapeworm or liver fluke, or may be an external parasite (*ectoparasite*) such as a flea or louse

parasympathetic division one of the two divisions of the autonomic nervous system; *see* nervous system, also table of effects (p. 186)

parathormone an alternative name for PARATHYROID HORMONE

parathyroid gland one of four small endocrine glands embedded in the thyroid gland; secretes parathyroid hormone; *see also* diagram of parathyroid gland, next page

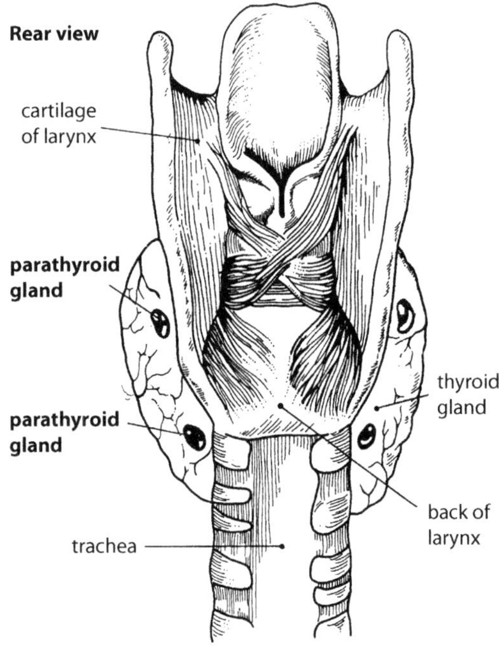

Location of parathyroid glands

parathyroid hormone (PTH) a hormone secreted by the parathyroid gland that regulates calcium and phosphate levels in the blood; stimulates breakdown of bone tissue to cause release of calcium and phosphate into the blood; increases the rate at which kidneys remove calcium from forming urine and therefore increases blood calcium level, but promotes transport of phosphate from blood into urine so that blood phosphate is reduced; with respect to blood calcium, has the opposite effect to calcitonin; also known as *parathormone*; *see also* calcitonin, table of hormones (p. 128)

parental generation in a series of crosses, the male and female of the first cross

parental investment any use of energy or resources by a parent that increases an offspring's chances of survival but reduces the parent's ability to produce more offspring

parietal cell a type of cell located in the gastric glands of the stomach that secretes hydrochloric acid; also called *oxyntic cell*

parietal lobe one of the four lobes of each cerebral hemisphere of the brain; *see* diagram of lobes of the cerebrum (p. 105)

Parkinson's disease a disease of the central nervous system that affects people at about age 60; symptoms are uncontrolled movements of the limbs, rigid face muscles, walking difficulty and generally poor muscle performance; caused by a degeneration of nerve cells in the brain that produce the neurotransmitter DOPAMINE

parotid glands one of the three pairs of SALIVARY GLANDS

partial fracture a break that does not extend right across the bone; *see* fracture

partial monosomy *see* monosomy

partial pressure in a mixture of gases, the pressure exerted by one of those gases

partial trisomy *see* trisomy

partially permeable membrane an alternative name for DIFFERENTIALLY PERMEABLE MEMBRANE

parturition the process of childbirth; it is accompanied by a series of events known as LABOUR

passive immunity IMMUNITY produced by the introduction of antibodies from an outside source

passive process a process that occurs without an input of energy (e.g. diffusion through a cell membrane is passive transport—it does not require any energy from the cell)

passive smoking inhaling smoke as a result of other people smoking; can cause the same health problems as smoking but the risk is less

Patau syndrome a condition caused by having one extra chromosome 13; many embryos with the condition do not survive, and those that do suffer a range of symptoms such as eye and brain defects, cleft palate, hair lip, deformed feet or heart defects; also called *trisomy 13*

patella the small triangular bone that forms the kneecap; *see also* diagram of skeleton (p. 249)

patellar reflex a reflex jerking of the lower limb when the tendon below the knee cap (the patellar tendon) is struck; also called the *knee jerk reflex*

pathogen a disease-causing organism (e.g. many bacteria and viruses are pathogenic in humans)

pathology the scientific study of the origin, nature and course of diseases, or of bodily structures and functions that have been affected by disease; a doctor who specialises in pathology is a *pathologist*

pebble tools stone tools made by crudely chipping flakes off a rounded pebble or small slab of stone; frequently referred to as Oldowan tools, as they were first discovered in Olduvai Gorge; also referred to as *choppers*; *see also* Oldowan tools, diagram (p. 193)

pectoral girdle the bones that make up the shoulder and form an attachment for the arms; the two shoulder blades (scapulae) and the two collarbones (clavicles); also called the *shoulder girdle; see also* diagram of skeleton (p. 249)

pedigree a family tree; shows relationships between members of a family; *see* illustration of a pedigree below

peer review a process where articles intended for publication in scientific journals are checked by independent experts in the field

pellagra a disease caused by a deficiency of the vitamin NICOTINIC ACID (niacin); characterised by swelling and redness of the soft tissues of the mouth and disorders of the nervous system; *see also* table of vitamins (p. 289)

pelvic girdle the bones that form the hips; each side of the girdle is made up

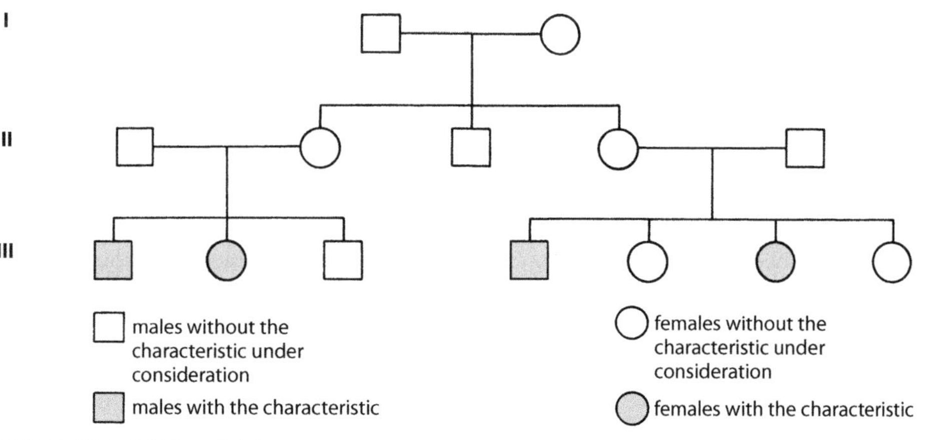

Pedigree

of three embryonic bones (the *ilium*, *ischium* and *pubis*) that are fused together to make one; each side is attached to the sacrum at the rear and they are joined in the front at the pubic symphysis; provides attachment for the legs and supports the weight of the upper body; also called the *hip girdle*; *see also* diagram of skeleton (p. 249)

pelvic inflammatory disease (PID) a bacterial infection of any of the pelvic organs of a female, especially of the ovaries, uterine tubes or uterus; *see also Chlamydia*

pelvis **1.** the basin-like structure formed by the two hip (pelvic) bones, the sacrum and the coccyx; *see also* diagram (p. 262) and diagram of skeleton (p. 249) **2.** the funnel-shaped upper end of the ureter that lies within the kidney; *see also* diagram of kidney (p. 147)

penetrance the proportion of individuals in a population who have a particular allele or a particular genotype

penicillin an antibiotic, originally extracted from the mould *Penicillium*, that is used to treat some bacterial infections

penis the male reproductive organ that is used to transfer sperm to the vagina of the female during sexual intercourse; becomes erect during sexual arousal due to blood filling blood spaces within the body of the penis; also used to pass urine from the body; *see also* diagram of male reproductive system (p. 235)

pentadactyl limb a limb with five fingers or toes; all primates have pentadactyl limbs

pentose sugar a sugar with molecules containing five carbon atoms (e.g. ribose, deoxyribose); *see also* hexose sugar

pepsin an alternative name for GASTRIC PROTEASE

pepsinogen an inactive form of GASTRIC PROTEASE

peptic ulcer an ULCER resulting from oversecretion of stomach acid

peptidase any enzyme that breaks down peptides into amino acids; present in intestinal juice

peptide a compound consisting of two or more amino acids chemically joined by peptide bonds; a *dipeptide* is two amino acids joined together; a *polypeptide* is a chain of amino acids not large enough to be called a protein; *see also* peptide bond

peptide bond a bond joining two amino acids; the amino group of one amino acid ($—NH_2$) is joined to the carboxyl group ($—COOH$) of another; the bonds between the amino acids in all proteins, polypeptides and peptides are formed in this way; *see* diagram of peptide bond below

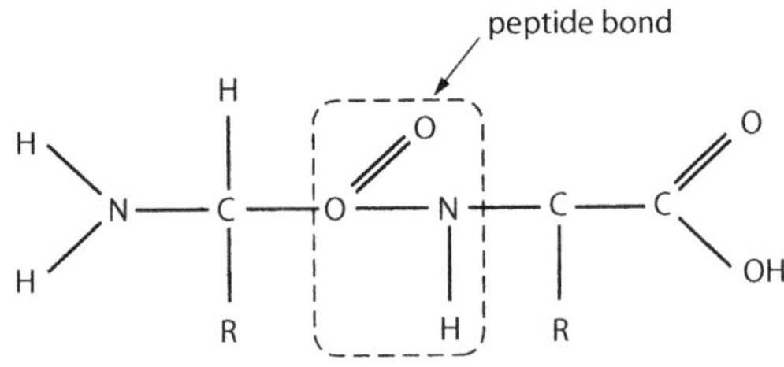

Peptide bond: R represents a group unique to each amino acid

per cent, % the ratio of a number to 100 (e.g. if there are two people in 100 with a particular characteristic, then 2 per cent (2%) of people have the characteristic)

percentage *see* per cent

percentage change the percentage increase or percentage decrease in a value; *see also* per cent

perfusion 1. pumping a liquid into an organ or tissue (e.g. drugs may be perfused into the blood stream) 2. passing fluid from the blood into the extracellular fluid of the tissues

peri- a prefix meaning around (e.g. pericardium—the membrane around the heart)

pericardium a membrane around the heart; holds the heart in position

perichondrium a membrane that covers some types of cartilage

perilymph the fluid between the bony and membranous labyrinths of the ear; *see* labyrinth

period 1. a subdivision of geological time; the eras of geological time are divided into periods and each period may be divided into epochs; *see* table of geological time scale (p. 111) 2. common name for MENSTRUATION

periodical an alternative name for JOURNAL

periosteum the dense, fibrous outer covering of a bone; serves as a means of attachment for ligaments and tendons; contains cells that are responsible for forming new bone during growth and repair; also contains blood vessels, lymph vessels and nerves that pass into the bone

peripheral nervous system that part of the NERVOUS SYSTEM outside the brain and spinal cord; *see* diagram of organisation of nervous system (p. 185)

peripheral resistance resistance to blood flow resulting from friction between the blood and the walls of the blood vessels; depends on the diameter of the blood vessels and the viscosity of the blood; the smaller the diameter of a blood vessel, the greater the resistance to blood flow; arterioles indirectly control blood pressure by changing their diameters and thus changing the peripheral resistance

peripheral thermoreceptor a temperature receptor in the skin

peripheral vascular disease disease of blood vessels in the legs and arms

peristalsis waves of involuntary muscular contraction along a tubular organ, especially the waves of contraction that push food along the alimentary canal; *see also* diagram of peristalsis below

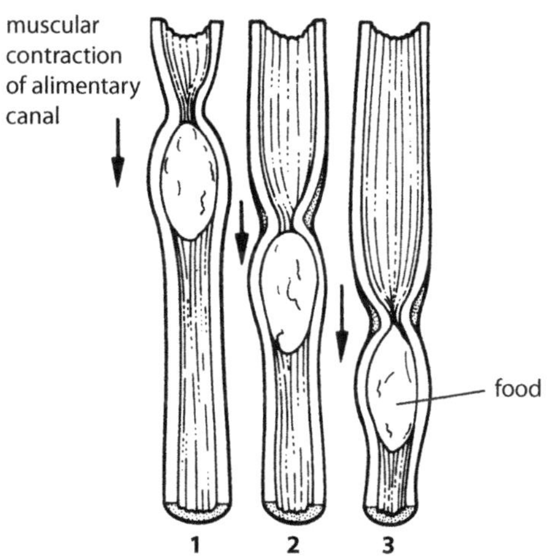

Peristalsis

peritoneal dialysis a treatment that can be performed when the kidneys have ceased to function; the technique involves placing fluid inside the abdominal cavity so that wastes can diffuse from the blood into the fluid, and then removing the fluid again; *see also* dialysis

peritoneum a membrane that lines the ABDOMINAL CAVITY and covers the abdominal organs; secretes fluid that lubricates movement of the organs; inflammation of the peritoneum is *peritonitis*

peritonitis inflammation of the peritoneum, the membrane that lines the abdominal cavity; especially likely to occur after rupture of an inflamed appendix

peritubular capillaries capillaries around the tubules of the kidney NEPHRONS

permanent teeth the second set of TEETH

Permian the last of the 7 periods in the Palaeozoic era of geological time; 299–251 million years ago; the spread of the reptiles occurred during this period and mammal-like reptiles began to evolve; *see* geological time scale (p. 111)

pernicious anaemia insufficient production of red blood cells; *see* anaemia

peroxisomes spherical organelles in the cytoplasm of a cell; similar in size and structure to LYSOSOMES; they hold enzymes that catalyse reactions where hydrogen peroxide (H_2O_2) is a by-product; they always contain the enzyme catalase, which converts the toxic hydrogen peroxide into harmless water and oxygen

persistent generalised lymphadenopathy an illness associated with infection with human immunodeficiency virus (HIV) in which there are swellings in the neck, armpits and groin; other symptoms include diarrhoea, fatigue, fever, weight loss, a continual dry cough, unexplained bleeding and shortness of breath; formerly called *AIDS-related complex*

perspiration an alternative name for SWEAT

pethidine an opioid—a synthetic drug with effects similar to those of opium derivatives; used medically for the relief of severe pain

petri dish a small, flat dish into which a medium is poured for the culture of micro-organisms; consists of two circular dishes, one forming a base, the other, a little wider in diameter, forming a lid; originally made from glass but now made from disposable plastic and supplied in sterile packs; also called a *culture plate* or just *plate*

petrifaction a process by which some fossils are formed; the organic matter in the fossils is replaced by rock-forming minerals

pH a term used to describe how acidic or alkaline a solution is; a measure of the hydrogen ion (H^+) concentration of a solution; the pH scale extends from 1 to 14; values less than 7 represent acidic solutions; values greater than 7 represent basic solutions; a pH of 7 is neutral; *see* pH scale below

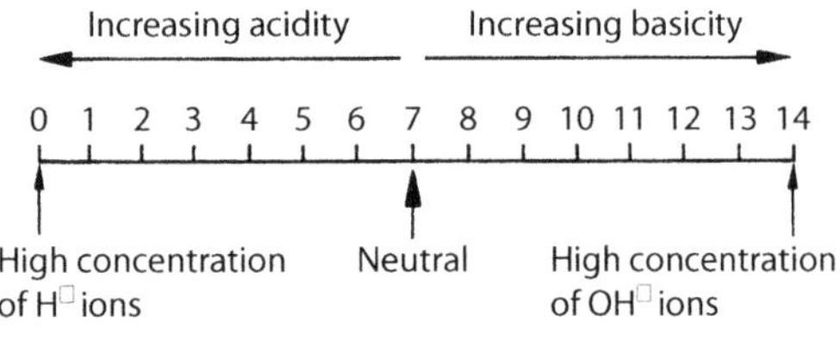

pH scale

phage an abbreviated name for BACTERIOPHAGE

phagocyte a cell that is able to break down foreign materials (e.g. micro-organisms) and cell debris

phagocytosis the engulfing of solid particles by a cell; *see* endocytosis

phalanges (singular *phalanx*) the series of bones that form the fingers and toes; there are three phalanges in each finger and two in the thumb; similarly, the foot has two phalanges in the big toe and three in each of the other toes; *see also* diagram of skeleton (p. 249)

pharmaceutical any drug used for medicinal purposes

pharmacogenetics the study of genetic variations that result in different responses to drugs; often used interchangeably with *pharmacogenomics*, however pharmacogenetics studies single genes and their effects on individual differences in drug metabolising enzymes, while pharmacogenomics studies the influence of the entire genome on drug responses

pharmacogenomics the study of the effect of genetic variations on the way patients respond to chemical substances; it is concerned with the influence of the whole genome on drugs; *see also* PHARMACOGENETICS

pharmacology the scientific study of drugs, including their preparation, uses and effects, in the treatment of disease

pharyngeal arches arches of tissue that develop in the neck region of vertebrate embryos

pharynx the throat; joins the mouth cavity to the oesophagus and larynx; *see* diagram of digestive system (p. 77)

phase contrast microscope a microscope that produces contrast between normally transparent structures within a living cell

phenotype the physical appearance or performance of an individual as determined by genetic constitution; the observable characteristics of an organism; the phenotype is determined by the relationship of the two alleles at a given point in the chromosome, the number of genes involved in determining a characteristic, and frequently environmental influences as well; e.g. for ability to roll the tongue there are two possible alleles—*R* for tongue rolling and *r* for non-rolling; this gives three possible genotypes—*RR*, *Rr* and *rr*; since the allele for rolling is dominant to the allele for non-rolling, there are only two phenotypes—rolling (*RR* and *Rr*) and non-rolling (*rr*)

phenylalanine one of the 20 amino acids that are common in proteins; essential in the human diet; *see also* list of amino acids (p. 11)

phenylketonuria (PKU) an inherited disease caused by a recessive allele on a non-sex chromosome; the body is unable to metabolise the amino acid phenylalanine, which accumulates in the blood, resulting in damage to the growing brain and thus extreme mental deficiency; also a tendency towards epileptic seizures and a failure to produce normal skin pigmentation; babies are routinely tested for PKU and, if found to have the condition, prescribed a diet low in phenylalanine

pheromone a chemical substance released in very small amounts by organisms as a means of communication between members of the species; pheromones influence behaviour, growth or development and many are used to attract a member of the opposite sex for mating; in humans, pheromones may be detected by the VOMERONASAL ORGAN

phlebitis inflammation of a vein, often a vein in the leg

phlegm mucus produced in the trachea and discharged through the mouth by coughing

phobia any obsessive fear or aversion; specific phobias are given particular names (e.g. claustrophobia—a fear of confined spaces, arachnophobia—an excessive fear of spiders)

phoneme the smallest unit of sound in spoken language; phonemes are the basic units of speech; the English language uses about 40 phonemes—the 26 sounds of the alphabet and a few more

phosphate group a chemical unit that consists of one atom of phosphorus combined with four atoms of oxygen; often attached to proteins, carbohydrates, lipids and nucleic acids

phosphocreatine an alternative name for CREATINE PHOSPHATE

phospholipid a lipid that contains phosphorus; instead of having the usual glycerol and three fatty acids found in a fat molecule, one of the fatty acids is replaced by a phosphate group; two parallel rows of phospholipids, called *a phospholipid bilayer*, make up a major part of cell membranes

photoreceptor a receptor sensitive to light; the rods and cones in the retina of the eye are photoreceptors; also called *electromagnetic receptor*; *see also* retina

photosynthesis the process by which green plants, algae and certain bacteria produce carbohydrates and other organic compounds from carbon dioxide and water in the presence of light and chlorophyll

phrenic nerves a pair of spinal nerves that control the diaphragm; involved in the control of breathing

phyletic evolution a pattern of SPECIATION

phylogram a branching, tree-like diagram showing relationships between different genera, families, orders or other taxonomic groups

phylum (plural *phyla*) the major category in the classification of organisms; each kingdom is divided into phyla; phyla are divided into one or more classes (e.g. the phylum Chordata includes sea squirts, lancelets and all vertebrate animals)

physical anthropology an alternative name for *biological anthropology*; *see* anthropology

physiology the study of the way in which living organisms function

physiotherapy the use of physical techniques like massage and exercise to relieve pain, restore muscle strength and increase the range of movement; also known as *physical therapy*

pia mater one of the membranes covering the central nervous system; *see* meninges

piles a common name for HAEMORRHOIDS

pili (singular *pilus*) an alternative name for HAIR

piloerection when the hairs on the surface of the skin stand in a vertical position; contraction of the muscles that cause hair to stand erect produces 'goose bumps'; in mammals other than humans piloerection traps a thicker layer of air next to the skin and helps to reduce heat loss

Piltdown skull the name given to fossil specimens found at a gravel pit near Piltdown in England between 1912 and 1915; once thought to be a 'missing link' between apes and humans, as it confirmed the then popular idea that early humans had large brains and ape-like jaws, exposed as an elaborate hoax in 1953 using fluorine analysis

pineal gland the cone-shaped endocrine gland located in the vertebrate midbrain; in mammals it produces the hormone melatonin at night, which may be linked to body rhythms; in humans, the pineal gland starts to degenerate about age 7, and in the adult it is largely fibrous tissue

pinna (plural *pinnae*) the projecting part of the outer ear consisting of elastic cartilage covered in skin and shaped like the flared end of a trumpet; *see also* diagram of ear (p. 82)

pinocytosis the engulfing of liquid droplets by a cell; *see* endocytosis

pituitary body an alternative name for PITUITARY GLAND

pituitary gland a small endocrine gland lying just below the hypothalamus of the brain and connected to it by a stalk-like structure; also called the *hypophysis;* sometimes called the 'master gland' because its hormones regulate many body activities; divided into an anterior (front) lobe, the *adenohypophysis,* and a posterior (rear) lobe, the *neurohypophysis;* the anterior lobe secretes hormones under the control of regulating factors from the hypothalamus—GROWTH HORMONE, PROLACTIN, ADRENOCORTICOTROPHIC HORMONE, THYROID-STIMULATING HORMONE, FOLLICLE-STIMULATING HORMONE, LUTEINISING HORMONE and MELANOCYTE-STIMULATING HORMONE; the posterior lobe does not make hormones but stores OXYTOCIN and ANTIDIURETIC HORMONE produced by the hypothalamus (of the brain); the pituitary gland is also called the *pituitary body* or just *pituitary*; *see also* diagram of pituitary gland below

pivot joint a SYNOVIAL JOINT that allows rotation

placebo an inactive substance, often in the form of a sugar pill, that is used as a control in experiments to test effectiveness of drugs; some subjects are given the drug and others are given the placebo; use of a placebo is necessary because some subjects show improvement when taking any 'medication', whether it is effective or not

placenta the organ that supplies nutrients to, and removes wastes from, the foetus;

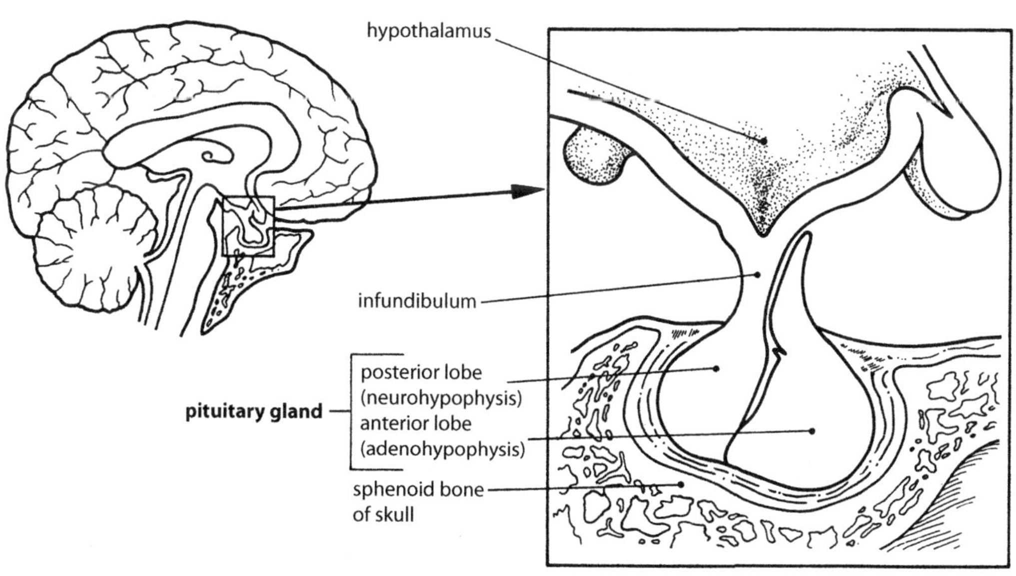

Pituitary gland

connected to the embryo by blood vessels in the umbilical cord; also produces a number of hormones, including oestrogens and progesterone; *see also* diagram of foetal circulation below

placenta praevia a complication of pregnancy where the placenta partly covers the opening of the uterus

plague a serious infectious disease caused by a bacterium that is carried by rats; spreads to humans via flea bites

plain muscle an alternative name for *smooth muscle*; *see* muscular tissue

plantigrade a condition in which the feet of an animal lie flat on the ground

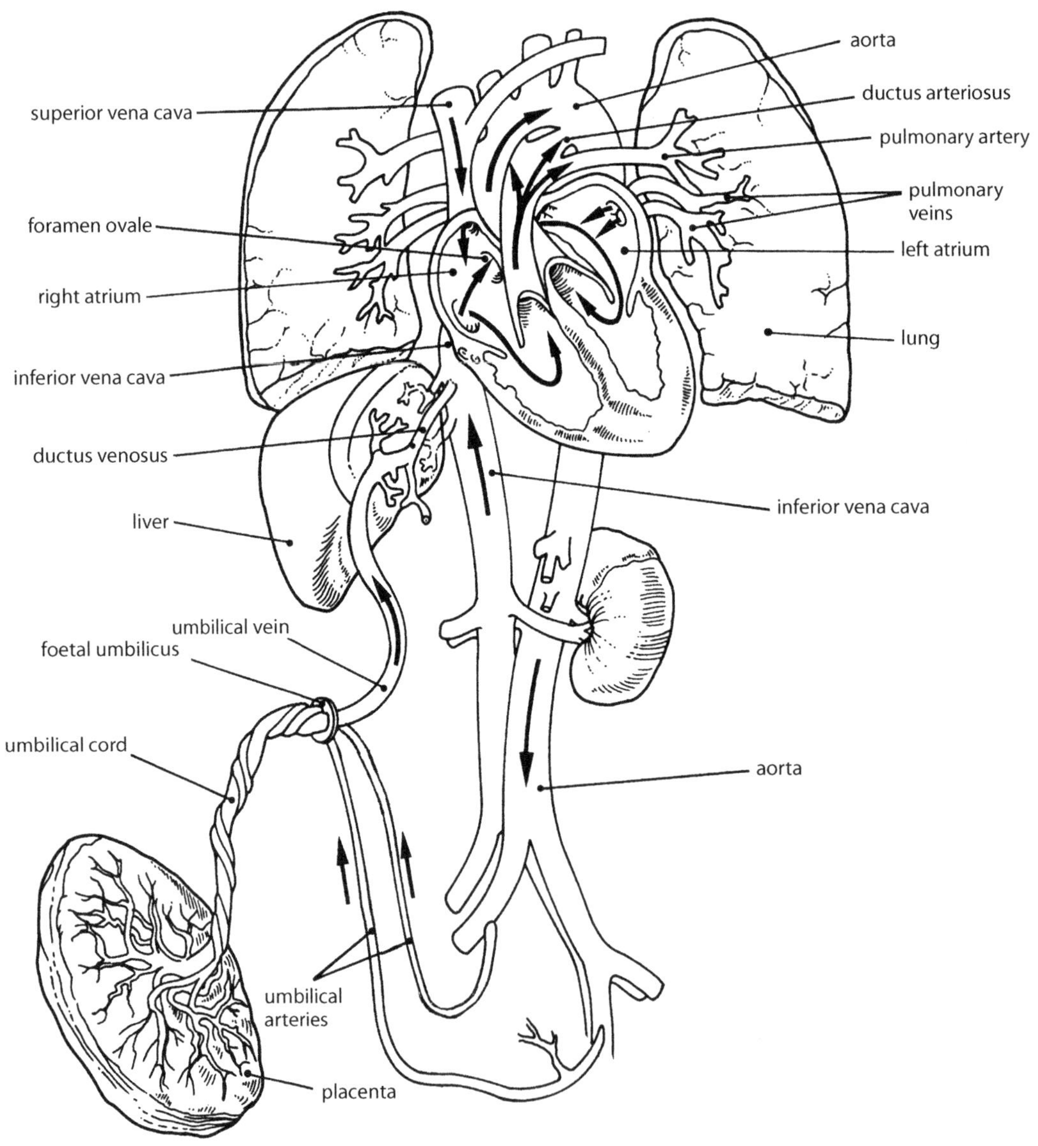

Foetal circulation

when walking; primates and bears have plantigrade gait

plaque **1.** a deposit of fatty material (containing cholesterol) on the inside of the wall of an artery **2.** material that adheres to teeth; *see* dental plaque

plasma the fluid part of the blood; the yellowish liquid left after the formed elements of blood are removed; makes up 55% of the volume of the blood; plasma is 91% water, 7% protein (albumins, globulins and fibrinogen) and 2% of other dissolved substances such as glucose, urea, respiratory gases, hormones, ions (sodium, calcium, chloride and potassium) and many other substances

plasma cell a cell that develops from a B CELL and produces antibodies; *see also* humoral immunity

plasma membrane an alternative name for CELL MEMBRANE

plasma proteins proteins found in the blood plasma; make up about 7% of the plasma; many are also found elsewhere in the body but when in the blood they are called plasma proteins; those in greatest amounts are ALBUMINS, GLOBULINS and FIBRINOGEN

plasmid a circular piece of DNA that is not part of the chromosomes and is able to replicate independently; often used in genetic engineering to carry desired genes into organisms

plate **1.** an alternative name for PETRI DISH **2.** the action of spreading cells over the surface of the medium in a petri dish

platelet one of the formed elements of blood; a fragment of cytoplasm enclosed in a membrane but lacking a nucleus; manufactured in the red bone marrow; when blood vessels are damaged, platelets initiate a series of reactions that result in BLOOD CLOTTING; also called a *thrombocyte*

platyhelminths flat worms; many are parasitic, such as tapeworms and liver flukes

platyrrhine refers to members of a subgroup of the primates called the PLATYRRHINI

Platyrrhini a subgroup of the order primates that is classified according to the positioning of the nostrils (wide apart) and includes all the New World monkeys; members of this group are known as *platyrrhines*; *see* diagram of simplified classification of living primates (p. 219); *see also* New World monkeys, Catarrhini, diagram of platyrrhine monkey (p. 43)

Pleistocene the 6th epoch of the Cainozoic era of geological time; also the earlier of the two epochs in the Quaternary period; dating from 1.8 million to 11 000 years ago; the major event in human evolution during this epoch was the continued development of the genus *Homo*; *see* geological time scale (p. 111)

pleura the membrane on the outside of the lungs and the inside of the chest cavity; *see* pleural cavity

pleural cavity the narrow, fluid-filled space between the outside of the lung and the inside of the chest; a membrane called the *pleura* (or *pleural membrane*) covers the outside of the lungs and lines the inside of the chest cavity; the membrane secretes *pleural fluid*, which completely fills the pleural cavity between the two layers of membrane; the fluid prevents friction between the two layers of the pleural membrane as the lungs move during breathing; *see also* diagram of respiratory system (p. 236)

pleural fluid the fluid that fills the PLEURAL CAVITY

pleural membrane an alternative name for the *pleura*; *see* pleural cavity

plexus a network of blood vessels, lymph vessels or nerves (e.g. the cervical plexus is a network of nerves that supply the skin and muscles of the head, neck and shoulders)

Pliocene the 5th epoch of the Cainozoic era of geological time; also the youngest of the five epochs in the Tertiary period; dating 5.3–1.8 million years ago; the major event in human evolution during this epoch was the origin of the hominids; *see* geological time scale (p. 111)

pluripotent stem cells *see* stem cells

PMT abbreviation for premenstrual tension which is another name for PREMENSTRUAL SYNDROME

pneumonia a severe infection or inflammation of the air sacs in the lungs; the air sacs accumulate fluid and dead white blood cells, reducing the air space; caused by a number of infecting bacteria or viruses

pneumotaxic area a part of the brain involved in the regulation of breathing; *see* respiratory centre

point mutation a change to a single gene; *see* mutation

podocytes cells that make up the inner layer of the glomerular capsule of a kidney nephron; have foot-like extensions that wrap around the capillaries of the glomerulus; spaces between the extensions are called FILTRATION SLITS

polar bodies the three small cells produced from meiosis in the ovary; resulting from the unequal division of cytoplasm during the 1st and 2nd meiotic divisions; one of the four haploid cells gets most of the cytoplasm and becomes the egg; the other three are the polar bodies and have no function; *see also* oocyte, diagram of oogenesis (p. 107)

polarised membrane the membrane of a nerve cell whose inside is at a negative electrical charge compared with its outside; polarisation occurs because sodium ions (Na^+) are actively transported out of the cell; the difference in charge between the inside and outside of the cell membrane is called the *membrane potential*; *see also* depolarisation, sodium–potassium pump

poles the ends of a cell

polio an abbreviation for POLIOMYELITIS

poliomyelitis a disease caused by the polio virus; in the most serious form the virus infects and destroys the cell bodies of motor nerve cells in the central nervous system; causes paralysis of limb muscles, respiratory muscles or even the heart muscle; more common among children and was formerly called *infantile paralysis*; incidence is now rare due to the widespread use of the Sabin polio vaccine

pollutant any product of human activity that harms the environment

pollution contamination of the environment by substances or energy released as a result of human activity; depending on the part of the environment affected, may be classified as *air pollution*, *water pollution* or *soil pollution*

poly- a prefix meaning many (e.g. polysaccharide—a carbohydrate made up of many simple sugars)

polyandry in humans, a form of marriage in which a wife has several husbands; in more general terms, it refers to an adult female animal having several mates; *see also* polygyny

polygenes a group of genes that together determine a characteristic; *see also* polygenic inheritance

polygenic disorder an inherited disease that is controlled by several genes; examples are heart disease and diabetes

polygenic inheritance inheritance of a characteristic that is dependent on many pairs of genes; such characteristics show a wide range of variation (e.g. inheritance of skin colour, eye colour, height); also known as QUANTITATIVE INHERITANCE

polygyny in humans, a form of marriage in which a husband has several wives; in more general terms, it refers to an adult male animal having several mates; *see also* polyandry

polymer a large molecule, consisting of at least five MONOMERS of the same type chemically bonded together; natural polymers include starch, cellulose, glycogen, proteins and DNA; synthetic polymers include plastics and nylon; enzymes that catalyse the formation of polymers from monomers are called *polymerases*

polymerase any enzyme that catalyses the formation of a polymer from monomers

polymerase chain reaction (PCR) a technique used in molecular biology for producing multiple copies of DNA from a sample; used in DNA fingerprinting and in identifying diseases

polymorph the most common type of white blood cell; characterised by granular material in the cytoplasm; important in combating bacterial infections; also called a *neutrophil*

polymorphism when there are two or more different forms of a species

polynucleotide many nucleotides linked together; DNA and RNA are polynucleotides

polyp a tumour on a stalk, usually on a mucous membrane and often in the nose or colon

polypeptide a long chain of amino acids not large enough to be called a protein; *see also* peptide, peptide bond

polyribosome more commonly called a POLYSOME

polysaccharide a CARBOHYDRATE composed of three or more simple sugar molecules chemically combined

polysome a cluster of ribosomes in a cell; may be attached to the endoplasmic reticulum or may remain free in the cytoplasm; also called a *polyribosome*

polyunsaturated fat a FAT that contains a high proportion of polyunsaturated fatty acids

polyunsaturated fatty acid a FATTY ACID molecule that could hold more than one extra hydrogen atom; *see also* fat

pons varolii part of the brain stem; connects the spinal cord with the brain and the parts of the brain with each other; also contains the pneumotaxic and apneustic areas (*see* respiratory centre); also called the *pons*; *see also* diagram of brain (p. 35)

population a group of organisms of the same species living together in a particular place at a particular time; for the human species, the number of people in a given area

population density the number of people per unit area of land (usually per square kilometre)

population doubling time commonly referred to as DOUBLING TIME

population genetics the study of genetics at the population level; includes study of gene frequencies within populations

population growth an increase in the size of a population; occurs when birth rate and immigration rate exceed death rate and emigration rate

population pyramid a graph showing the age distribution of the males and females in a country's population; usually the population is divided into five-year age-groups, with each group shown as a horizontal bar extending on either side of a vertical centre line; males are shown to the left of the centre line and females to the right; also referred to as the *population structure* or an *age profile*; *see* illustration, below

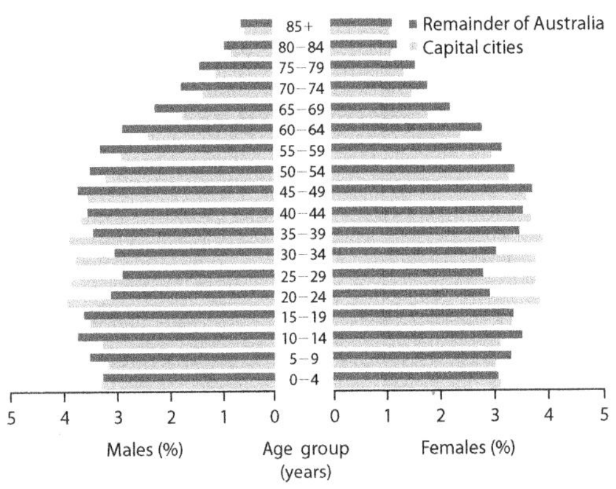

Population pyramid for Australia, 2007

population structure *see* population pyramid

positive feedback feedback that reinforces the stimulus that initiated a change; *see* feedback system

post- a prefix meaning after or beyond (e.g. postcranial—that part of the body beyond the head)

postcranial describes that part of the body or skeleton below the neck

posterior towards the hind end of an animal's body (e.g. the rear limbs are posterior to the front limbs); *see* diagram (p. 13)

posterior root one of the two roots that link a spinal nerve to the spinal cord; *see* root, diagram (p. 255)

posterior tibialis one of the muscles that provides support for the longitudinal arch of the foot

postganglionic neuron a nerve cell in the autonomic nervous system that connects a GANGLION with an EFFECTOR; *see also* preganglionic neuron

Postinor-2 trade name for the MORNING-AFTER PILL

postnatal the term used to describe events following the birth of a child; also called *postpartum* (after parturition)

postpartum an alternative term for POSTNATAL

postsynaptic neuron a nerve cell that carries nerve impulses away from a synapse; *see also* presynaptic neuron

potassium–argon (K/Ar) dating calculation of the absolute age of a rock stratum by measuring the ratio of potassium-40 to argon-40; potassium-40 decays to calcium-40 and argon-40 at a known rate, so the more argon present the older the rock; can only be used on rocks that were once heated to a very high temperature, such as volcanic rocks; may be used to date such rocks that are older than 200 000 years

potential difference the difference in electrical charge between two points; also known as *voltage*

power grip the grasping of an object between the undersides of the fingers and the palm of the hand, as in holding a hammer; *see also* precision grip, illustration below

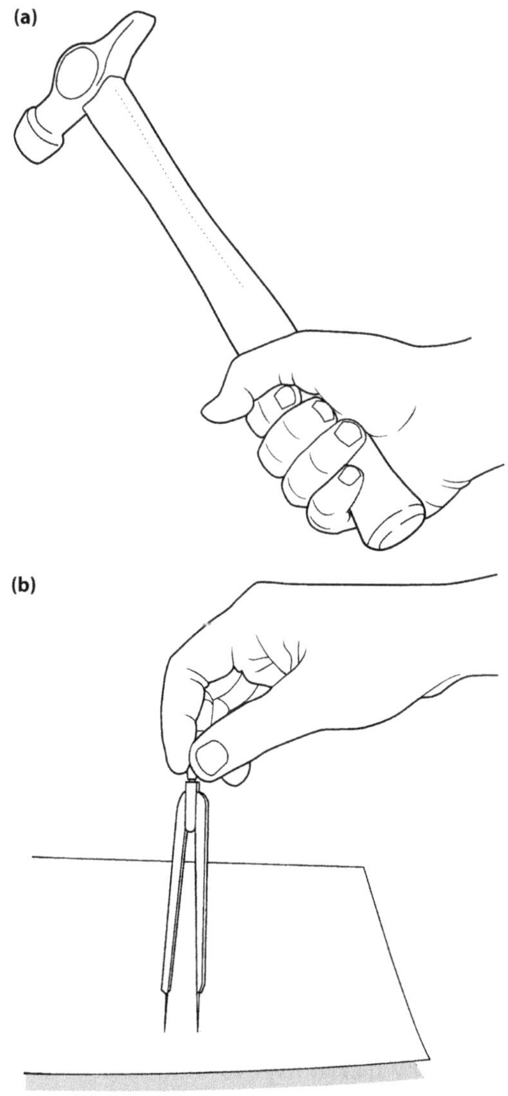

(a) Power grip; (b) precision grip

pre- a prefix meaning before (e.g. prehistory—events that took place before the keeping of written records)

Pre-Cambrian literally 'before Cambrian'; once used for the division of geological time older than the Phanerozoic eon (which includes the Cainozoic, Mesozoic and Palaeozoic eras); now used more informally to refer to the Proterozoic, Archaeozoic (Archean) and Hadean (Azoic) eras; the time in the earth's geological history that preceded the first major appearance of life, although some primitive aquatic algae and marine invertebrates did begin to appear

precapillary sphincter a circular band of smooth muscle that controls blood flow into individual capillaries

precision grip the grasping of an object between thumbtip and fingertip as in holding a pencil when writing; humans have mastered this grip to a far greater extent than any other primate; has allowed the development of tools to reach a very high level; *see also* power grip, diagram of precision grip, left

pre-embryo **1.** a fertilised egg before cell division **2.** cells in the first days after fertilisation when the cells are not differentiated; many of these early cells will not become part of the embryo but will make up parts of the placenta or membranes

preganglionic neuron a nerve cell in the autonomic nervous system that connects the central nervous system with a GANGLION; *see also* postganglionic neuron

pregnancy the sequence of events that begins with fertilisation, includes the implantation of the developing embryo into the uterine lining, and a series of developmental stages that involve both the mother and foetus; normally terminates in the birth of the baby

prehensile adapted to grasp an object; primates have prehensile hands and feet, and some (such as certain New World monkeys) have prehensile tails

prehensility the ability to grasp an object (e.g. fingers wrapping around a hammer handle, or the prehensile tail of a monkey grasping a branch); *see also* diagram of prehensility, below

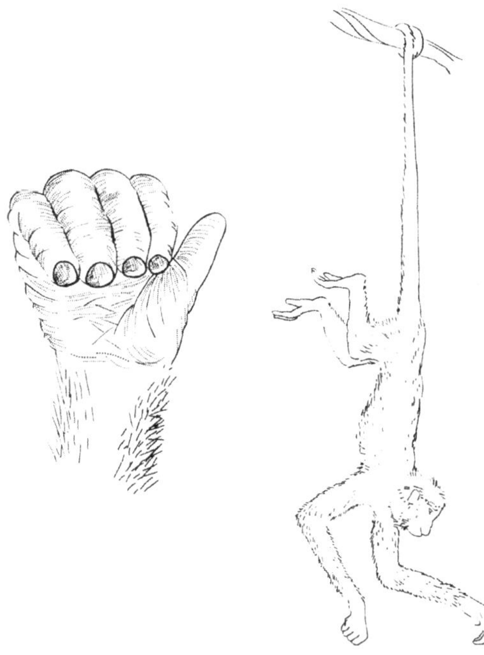

A prehensile hand (left) and tail (right) are able to grip objects

prehistory events before the time of written records; such events are known mainly through archaeological research; often used to refer to study of prehistoric humans

pre-implantation genetic diagnosis (PGD) testing IVF embryos before they are implanted in the uterus; used to decrease the chance of a child being born with a genetic disorder; also called *embryo screening*

premenstrual syndrome (PMS) a condition occurring in some females about a week before menstruation; includes irritability, depression, pain, nervous tension and crying spells; other problems that may occur are fatigue, headaches, bloating, joint pain, swelling and tenderness of the breasts; also known as *premenstrual tension* or *PMT*

premenstruation the last phase of the menstrual cycle; a period of about seven days when the CORPUS LUTEUM degenerates and the lining of the uterus begins to deteriorate

premolar a tooth with two cusps, found between the canines and molars; *see* teeth

prepuce the loose fold of skin covering the slightly enlarged region at the end of the penis in males and the clitoris in females; also called *foreskin*

presbyopia loss of elasticity of the lens of the eye, resulting in inability to focus clearly on close objects; develops with advancing age; corrected by spectacles for close work such as reading

pressoreceptors receptors capable of responding to changes in blood pressure; located in certain arteries and veins; send nerve impulses to regulate heartbeat; often referred to as *baroreceptors* or *pressure receptors*

presynaptic neuron a nerve cell that carries nerve impulses towards a synapse; *see also* postsynaptic neuron

primary germ layers the three embryonic tissues from which all tissues and organs of the body will develop; the outermost layer, the *ectoderm*, gives rise to the outer covering of the body (skin, nails and hair) and also the brain, spinal cord, nerves and sense organs; the middle layer, the *mesoderm*, gives rise to the muscles, connective tissues (including bone and cartilage), blood and blood vessels, and the outer layer of the skin; the inner layer, the *endoderm*, gives rise to the lining of the digestive tract and the glands associated with it, the respiratory tract, and parts of the excretory and reproductive systems

primary motor area an area of the brain's cerebral cortex that sends out nerve impulses to cause muscle contractions

primary motor cortex an alternative name for the PRIMARY MOTOR AREA

primary oocyte a cell in the ovary that divides by meiosis; *see* oocyte

primary response the immune system's response to the first exposure to an ANTIGEN; *see also* secondary response

primary sex organs the organs that produce the gametes; the *gonads*; the testes in males and the ovaries in females; *see also* accessory sex organs

primary spermatocyte a cell that divides by meiosis to produce sperm; *see* spermatocyte

primary structure in proteins, the amino acid sequence of the peptide chains

primate a member of the order PRIMATES

Primates the order of mammals that includes the prosimians, monkeys, apes and humans; have many characteristics that are adaptations to life in trees, even though many modern primates live on the ground—limbs with five digits, an opposable first digit, nails instead of claws, forward-facing eyes with stereoscopic vision, a reduced sense of smell and a large brain; an animal classified in this order is referred to as a *primate*; *see also* classification of primates below

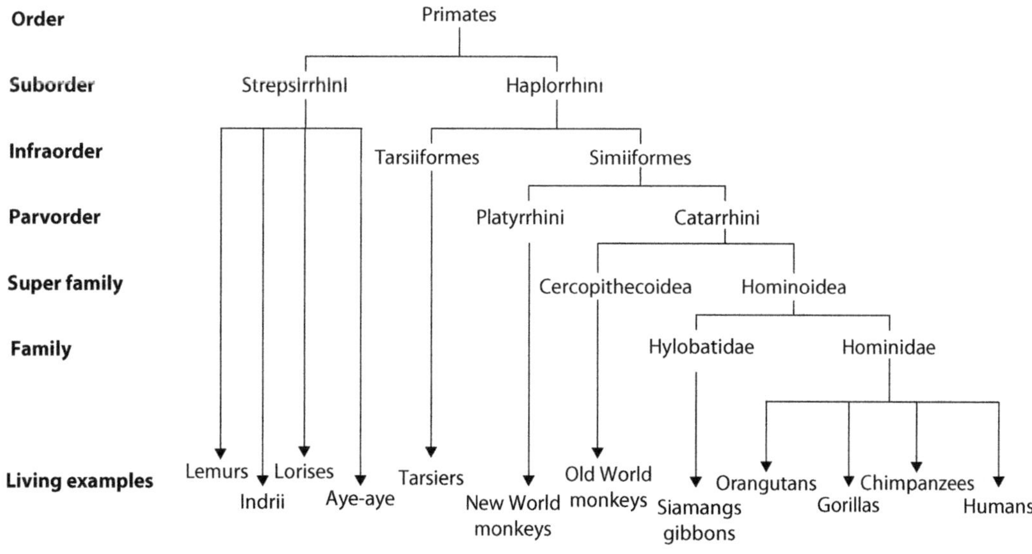

Simplified classification of living primates

primatology the study of the primates (the prosimians, monkeys, apes and humans)

prime mover an alternative name for a muscle acting as an AGONIST

primer a strand of DNA or RNA that serves as a starting point for DNA replication; new nucleotides are added to the primer; a primer is necessary because most of the enzymes that catalyse the replication of DNA cannot begin a new DNA strand from scratch, they can only add to an existing strand of nucleotides

primitive trait in biological terms, a characteristic that has remained unchanged for much of the organism's evolutionary history (e.g. the five digits of the human hand and foot are primitive traits inherited from earlier vertebrate ancestors)

primordial soup a liquid rich in chemicals that was present on the earth before life began; according to a hypothesis proposed by Stanley Miller (an American) and Alexander Oparin (a Russian), exposure to lightning, volcanic eruptions and other environmental influences brought about chemical reactions and gradual changes in the liquid so that eventually complex chemicals with the ability to make exact copies of themselves were formed; the primordial soup thus gave rise to the first life forms

principle of independent assortment *see* independent assortment, Mendel's laws

principle of segregation *see* Mendel's laws

principle of superposition the principle that the upper layers in a sequence of rock strata are younger than those below them; when sedimentary rocks are formed, younger sediments are deposited on older sediments; *see also* diagram, below

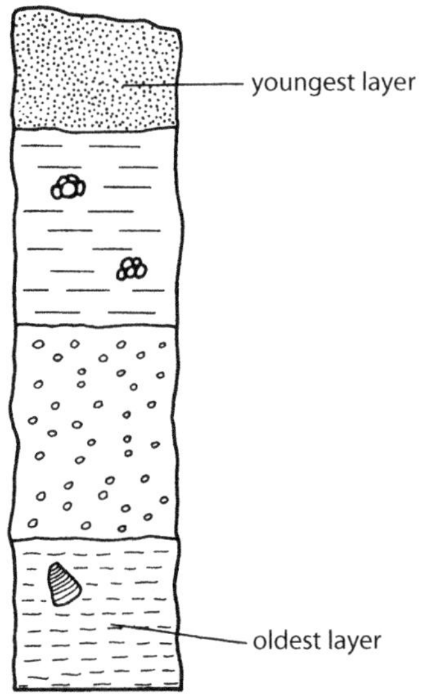

Principle of superposition

prion an infective modified protein; able to reproduce in host cells like a virus but lacking nucleic acid; prions are responsible for *scrapie*, a disease of sheep, and *bovine spongiform encephalitis* or *mad cow disease*; in humans they are known to cause *Creutzfeldt-Jakob disease* and *kuru*; prions cause the nervous system to accumulate myeloid proteins and to degenerate

pro- a prefix that often means first or before (e.g. prophase is the first phase of cell division, and a prokaryote is a type of cell believed to have developed before cells with nuclei)

probability a measure of the likelihood, or chance, that a certain event will occur

problem drug use use of a drug, or drugs, to the detriment of a person's health or economic or social adjustment, or to the detriment of society

proceedings in science, a publication containing the PAPERS presented at a conference

product a substance produced as a result of a chemical REACTION

progeny offspring; descendents

progesterone a female sex hormone produced by the CORPUS LUTEUM of the ovary; helps prepare the uterine lining for a fertilised egg; also secreted by the placenta when pregnancy occurs; prepares mammary glands for milk secretion; *see also* table of hormones (p. 128)

progestin a synthetic form of the hormone PROGESTERONE; used in contraceptives

progestogen an alternative name for PROGESTIN; also spelt *progestagen*

prognathism a protruding jaw; in human evolution, the jaws of *Homo erectus* show prognathism when compared to fully modern *Homo sapiens*; members of some present-day populations also show prognathism; such a jaw shape is described as *prognathic* or *prognathous*; the opposite of ORTHOGNATHISM; *see also* diagram of prognathism below

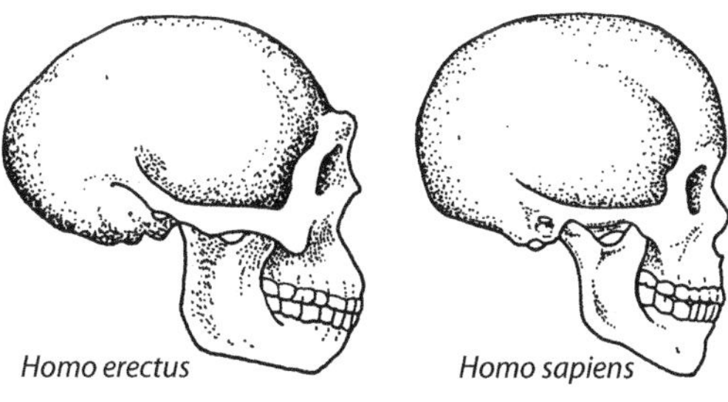

The jaws of Homo erectus *show prognathism when compared with those of* Homo sapiens

prognosis a forecast of the probable course of a disorder including prospects for recovery, time for recovery, increase or decrease in severity of symptoms and possibility of permanent disability

prokaryote an organism in which the genetic material is not separated from the rest of the cell contents by a nuclear membrane; prokaryotic cells also lack membranous organelles; either a bacterium or a blue-green alga; also spelt *procaryote*; *see also* eukaryote

prolactin (PRL) a hormone secreted by the anterior lobe of the pituitary gland that, with other hormones, initiates and maintains milk secretion from the mammary glands; also called *lactogenic hormone* or *luteotropic hormone*; *see* table of hormones (p. 128)

proliferation in cell biology, the division of cells many times so that one, or a few, cells produce large numbers of cells

proline one of the 20 amino acids; found in many proteins; can be synthesised in humans; *see also* list of amino acids (p. 11)

promoter gene a gene that facilitates TRANSCRIPTION from one or more nearby STRUCTURAL GENES; part of an operon; *see also* operator gene

pronation the term used to describe the inward rotation of the forearm so that the palm faces downward or backward; *see also* diagram showing movements at joints (p. 1)

pronuclear stage tubal transfer (PROST) a form of IN-VITRO FERTILISATION

pronuclei the nucleus of the sperm and the nucleus of the egg after fertilisation (after the sperm has penetrated the egg) but before the two nuclei unite

prophase the 1st phase of MITOSIS and of MEIOSIS

proprioceptor a receptor in a joint, tendon or muscle that provides information about the position of body parts and their movements

prosencephalon the front part of the brain

prosimian a term commonly used to describe the more biologically primitive primates; prosimians include lemurs, lorises and bushbabies, and many are nocturnal; they have smaller brains, longer, more pointed snouts and rely more on their sense of smell than monkeys and apes; the term is derived from traditional classifications in which primates were subdivided into two suborders, Prosimii and Anthropoidea

Prosimii in traditional classifications, one of the two suborders of the order Primates; animals in this suborder are called PROSIMIANS

prostaglandins a group of lipids that are produced by almost all cells; secreted into the blood in very small quantities and act on the cells and tissues in the immediate surroundings; are therefore called *local* or *tissue hormones* as distinct from *circulating hormones*, which are produced by endocrine glands and may act on target cells that are remote from the gland; believed to regulate cell metabolism by increasing or decreasing cyclic AMP formation; have many other functions relating to smooth muscle contraction, blood flow, function of blood platelets, transmission of nerve impulses, the immune response and many more; inflammation causes increased production of prostaglandins

prostate gland a single gland, shaped a little like a doughnut, that surrounds the urethra just below the bladder; secretes a thin, milky, alkaline fluid that becomes part of the semen; *see also* diagram of male reproductive system (p. 235)

prostatitis a condition in which the prostate gland becomes swollen, obstructing the flow of urine; often due to bacterial infection

prosthesis an artificial replacement for a part of the body (e.g. an artificial arm or leg)

protactinium dating a method of dating marine sediments using the decay rate of the radioactive elements protactinium and thorium

protease an enzyme that breaks down proteins into smaller molecules; also called a *proteolytic enzyme*; *see also* gastric protease, pancreatic protease

protein an organic compound made up of the elements carbon, hydrogen, oxygen and nitrogen and usually sulphur, consisting of long chains of amino acids (*see also* list of amino acids found in proteins, p. 11); proteins make up much of the structure of cells and nearly all enzymes are proteins; proteins are made in cells at the ribosomes but they are also essential in the diet as a source of amino acids; dietary proteins that contain all 10 of the amino acids that are essential for normal body functioning are called *complete proteins* or *first-class proteins*; proteins that make up the structure of the body may be *fibrous proteins* or *globular proteins*; fibrous proteins have molecules in long strands or flat sheets (e.g. actin, myosin, fibrin, collagen, keratin); globular proteins have molecules in which the chain of amino acids is folded to form a roughly spherical shape (e.g. haemoglobin, immunoglobulins); *see also* amino acid, protein synthesis

protein-energy malnutrition (PEM) malnutrition involving a dietary deficiency of energy and/or protein; kwashiorkor and marasmus are both types of protein-energy malnutrition; *kwashiorkor* occurs in infants and young children; caused by a diet low in protein but rich in carbohydrate; commonly occurs in less affluent countries where young children are fed on a cereal diet after breastfeeding stops; *marasmus* is severe wasting of the body, especially in children, due to insufficient food

protein synthesis the manufacture of proteins in cells; the DNA in the cell nucleus carries a code that determines the order of amino acids in the proteins made by the cell; the code is the sequence of nucleotide bases in the DNA molecule; when protein is to be made, part of the two strands of the DNA molecule separate; using a DNA strand as a template, a messenger RNA molecule forms—a process known as *transcription*; each sequence of three nucleotide bases on the messenger RNA (known as a *codon* or *triplet*) is the code for a particular amino acid; the messenger RNA takes the code from the nucleus to the ribosomes, which 'read' the code; each transfer RNA molecule has a sequence of three nucleotides, called an *anticodon*, that is complementary to a codon in the messenger RNA; by matching codons and anticodons the transfer RNA assembles the amino acids in the correct sequence to form the protein molecule; protein synthesis is also known as *translation* because the sequence of nucleotide bases in the DNA is translated into the sequence of amino acids in the protein molecule; *see also* ribonucleic acid

protein targeting sending newly synthesised protein to the organelle in the cell where that protein is required; sequences in the amino acids of each protein determine the destination of the protein after packaging by the GOLGI APPARATUS of a cell

proteolytic enzyme an enzyme that breaks down proteins into smaller molecules; *see also* protease

Proterozoic one of the major subdivisions of geological time; the 3rd era, 2500–620 million years ago; this era is characterised by fossil stromatoliths (columns of calcium carbonate produced by CYANOBACTERIA); also referred to as an EON in some time scales; *see* geological time scale (p. 111)

prothrombin an inactive protein in the blood plasma that is converted to the enzyme thrombin when the blood clots; *see also* blood clotting

proto-oncogene a gene that may be activated to become an ONCOGENE

proton a positively charged particle in the nucleus of an atom; its mass is approximately the same as that of a NEUTRON; the positive charge of one proton exactly balances the negative charge of one ELECTRON

protoplasm the entire contents of a cell within the cell membrane (i.e. the cytoplasm and the nucleus); *see also* diagram of cell structure (p. 44)

protozoan a single-celled animal

proximal nearer to the point of origin, as in 'proximal convoluted tubule'; *see also* distal

proximal convoluted tubule the first set of convolutions of the kidney tubule, between the glomerular capsule and loop of Henle; as with other parts of the tubule, reabsorbs some substances into the blood and secretes others into the forming urine; *see also* diagram of kidney nephron (p. 183)

proximodistal describes development outwards from the centre of the body; refers to the pattern of motor development in young children; movements of parts of limbs close to the body (e.g. the upper arm) are able to be controlled before movements of parts further away from the body (e.g. the fingers)

psychedelic drug an alternative name for a hallucinogenic drug; *see* hallucinogen

psychiatry the study and practice of the treatment of mental diseases

psychology the study of behaviour and the mental processes that control behaviour

psychosis (plural *psychoses*) a severe mental disorder that makes a person unable to meet the ordinary demands of life

psychosomatic disorder a physical ailment caused entirely, or partly, by emotional factors

psychotherapy a technique used to treat social, behavioural and emotional problems; approaches are varied but all involve a close helping relationship with a therapist

ptyalin a former name for SALIVARY AMYLASE

puberty the time when an individual reaches SEXUAL MATURITY; occurs early in the adolescent period; in males puberty is marked by the first ejaculation of semen with mature sperm, accompanied by other changes such as enlargement of penis and testes, deepening of voice, growth of facial and body hair; in females puberty is marked by the first menstrual bleeding and other changes such as enlargement of the vagina and uterus, growth of body hair and breast development

pubic bone one of two bones that form the front of the PELVIC GIRDLE; also called the *pubis*

pubic lice insects that cause intense itching in the genital area, usually transmitted during sexual intercourse; often referred to as 'crabs'

pubic symphysis an alternative name for SYMPHYSIS PUBIS

pubis an alternative name for the PUBIC BONE

puerperium the period following childbirth during which the reproductive organs of a female gradually return to their non-pregnant state

pulmonary arteries the arteries that carry deoxygenated blood to the lungs; *see also* diagram of major arteries (p. 20)

pulmonary circulation the flow of blood from the right ventricle, through the lungs and back to the left atrium; *see also* systemic circulation, diagram of double circulation (p. 80)

pulmonary trunk the large blood vessel that leaves the right ventricle and divides into the left and right pulmonary arteries; *see also* diagram of major arteries (p. 20)

pulmonary veins blood vessels that carry oxygenated blood from the lungs to the left atrium of the heart; *see also* diagram of heart (p. 121), diagram of major veins (p. 284)

pulp the soft tissue inside a tooth; *see also* diagram of tooth (p. 267)

pulp cavity the cavity inside a tooth; contains nerves and blood vessels and a soft tissue, the pulp; *see also* diagram of tooth (p. 267)

pulse the rhythmic expansion and recoil of the arteries as blood is pushed into them by the heart; the pulse that is felt in arteries near the surface of the body is due to contractions of the left ventricle; used as a measure of heart rate; can be felt at the wrist, the temple and in the neck

punctuated equilibrium a theory of evolution that proposes that relatively short periods of rapid evolutionary change are punctuated by long periods in which there is little evolutionary change (stasis); also called *saltatory evolution*; first proposed by Niles Eldredge and Stephen Jay Gould in 1972; the theory accounts for the fact that the fossil record does not contain a huge range of intermediate forms of organisms; *see* gradualism for an opposing view of evolution

Punnett square a table used to solve genetics problems; all possible combinations of alleles in the gametes of a mating between two individuals are shown and the genotypes resulting from random fertilisation can be worked out; devised by R. C. Punnett, a British geneticist, in the early part of the 20th century (1903–05); *see also* illustration of a Punnett square below

Gametes		Female	
		R	r
Male	R	**RR**	**Rr**
	r	**Rr**	**rr**

Punnett square

pupil the opening in the centre of the iris of the eye, through which light enters the eyeball; the diameter of the pupil is changed by contraction or relaxation of muscles in the iris; *see also* diagram of eye (p. 94)

pure breeding an alternative term for HOMOZYGOUS

purines one of the two types of bases found in DNA and RNA; the molecule has a double ring structure; adenine and guanine are purines; in the two strands of DNA they always pair with pyrimidine bases; *see also* pyrimidines

Purkinje fibre an alternative name for CONDUCTION MYOFIBRE; *see* diagram (p. 200)

pus a thick fluid that accumulates as a result of INFLAMMATION

pyloric sphincter a ring of smooth muscle between the stomach and the duodenum; controls flow of material from the stomach to the duodenum; *see also* diagram of digestive system (p. 77)

pyloric stenosis a blockage between the stomach and the small intestine; caused by an enlargement of the circular muscle fibres at the pyloric sphincter; must be corrected by surgery

pyramidalis muscle a muscle in the front part of the lower abdomen; absent in about 20% of people

pyridoxine one of the B complex of vitamins; water soluble; stored in muscle, liver and brain; good sources are tomatoes, corn, yeast, whole grains and liver; essential for amino acid and protein metabolism; deficiency causes skin disorders; also called *vitamin* B_6

pyrimidines one of the two types of bases found in DNA and RNA; the molecule has a single ring structure; thymine, cytosine and uracil are pyrimidines; in the two strands of DNA, thymine and cytosine always pair with purine bases; uracil and cytosine occur in RNA; *see also* purines

pyruvic acid a substance produced by the ANAEROBIC RESPIRATION of glucose, in which one molecule of glucose is broken down into two molecules of pyruvic acid; if oxygen is available, pyruvic acid is broken down by AEROBIC RESPIRATION, and if no oxygen is available it is converted to lactic acid

Q

quadriceps a common name for the quadriceps femoris group of four muscles that lie at the front of the thigh; attached at the upper end to the pelvis and the femur (thigh bone) and at the lower end, through the kneecap, to the tibia (shin bone); contract to straighten the leg at the knee; *see* diagram below

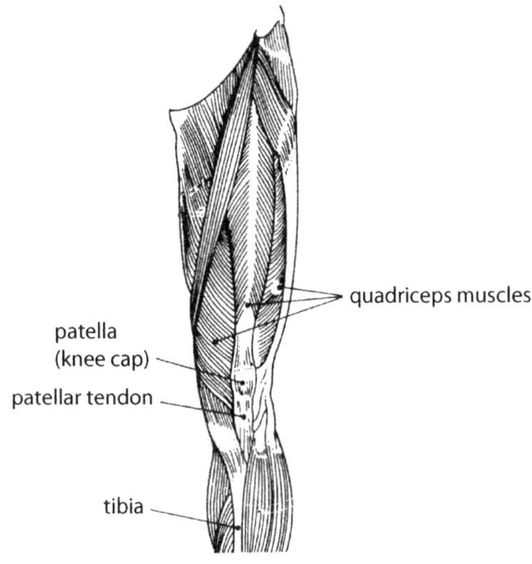

Quadriceps

quadriplegia paralysis of the legs and the arms; caused by damage to the spinal cord in the neck region; a person suffering from quadriplegia is known as a *quadriplegic*; also spelt quadraplegia; *see also* paraplegia, hemiplegia

quadrupedal locomotion the form of movement in which all four limbs are of equal length and make contact with the ground, with the spine roughly parallel to the ground; monkeys are typical quadrupedal primates; also called *quadrupedalism*; *see* diagram below

Quadrupedal locomotion

quadrupedalism an alternative name for QUADRUPEDAL LOCOMOTION

qualitative describes results of an experiment or survey that are descriptions rather than measurements or numbers; *see also* quantitative

qualitative inheritance inheritance of characteristics that are controlled by only one gene or a small number of genes; such characteristics (known as *qualitative characters*) are distinctly different from each other and can be used to group organisms into distinct classes based on their appearance (e.g. Mendel was able to group pea plants into those with yellow and those with green seeds); *see also* quantitative inheritance

quantitative describes results of an experiment or survey that are measurements or are expressed as numbers; *see also* qualitative

quantitative inheritance inheritance of characteristics that are determined by the cumulative action of many genes; such characteristics (known as *quantitative characters*) show gradual, continuous variation between two extremes, as in human skin pigmentation, height and intelligence; also referred to as *polygenic inheritance*

Quaternary the later of the two periods in the Cainozoic era of geological time; spans the past 1.8 million years; subdivided into the Holocene (or Recent) and Pleistocene epochs; the period during which the genus *Homo* evolved; *see also* geological time scale (p. 111)

quaternary structure the fourth level of protein complexity, two or more amino acid chains joined together to form a complex protein

quickening the time during pregnancy when the developing foetus begins to make movements that are clearly felt by the mother; the time when the foetus begins to show signs of life; occurs about halfway through pregnancy

questionnaire a set of questions given to people to obtain information; used in some investigations

R

rabies a serious viral disease that affects humans, dogs, foxes, cattle and other mammals; causes degeneration of the central nervous system and eventual death; the virus is usually transmitted to humans by the bite of an infected animal; does not occur in Australia and quarantine laws are aimed at keeping the country free of such diseases

racism the idea that the distinctive physical characteristics attributed to a race or a particular ethnic group somehow determine the behaviour, morality and intelligence of members of that group; on this basis, racism often manifests itself as an exclusion of some people from social, cultural, sporting, educational or other activities because of their physical characteristics; it also leads to the belief that one's own race is superior to others

radiation **1.** the emission of energy in the form of electromagnetic waves (e.g. light, heat, ultraviolet radiation, X-rays) **2.** the emission of particles such as alpha, beta and gamma rays **3.** heat radiation—the transfer of heat from one object to another without physical contact

radioactive dating DATING rocks or fossils using radioactive elements

radioactive substance a substance with unstable atoms that break down to release radiation (from the nuclei of the atoms) in the form of alpha, beta or gamma particles

radioactivity the property of releasing radiation (from the nucleus of the atom) in the form of alpha, beta or gamma particles; radioactive substances have radioactivity

radiocarbon dating the calculation of the age of a fossil or an artefact by measuring the amount of radioactive carbon (CARBON-14) in the specimen and using the known rate of decay of carbon-14 to nitrogen; also called *carbon-14 dating*; *accelerator mass spectrometry* (*AMS*) *radiocarbon dating* is a technique used to date samples of material as small as 100 micrograms; it involves breaking the sample into its constituent atoms so that the actual number of each form of carbon atom can be counted; *see also* half-life

radioisotope a radioactive ISOTOPE

radiometric dating DATING rocks or fossils using radioactive elements

radius the lower arm bone that meets the wrist on the thumb side; the smaller of the two bones in the forearm; *see also* diagram of skeleton (p. 249)

ramus (plural *rami*) a branch of a nerve, artery, vein or bone (e.g. the ramus of the lower jaw bone—the part that projects up into the jaw socket)

random errors errors that occur in all experiments because measurements cannot be made with absolute precision

random genetic drift occurrence of characteristics in a population as a result of chance rather than natural selection; only occurs in small populations or could be due to random sampling of a larger gene pool, as in the FOUNDER EFFECT; also called *genetic drift* or the *Sewall–Wright effect*

random mating selection of mates by chance; choice of partner is not influenced by phenotype or genotype; humans show random mating for some characteristics (such as blood groups), but for other characteristics (such as skin colour) choice of partner may not be random; also known as *panmixis*; *see also* assortative mating

range **1.** the area of distribution of a species **2.** the area within which an individual normally lives, better known as its *home range*; *see also* territory **3.** in a set of measurements, the difference between the highest and the lowest

rapid eye movement (REM) sleep one of two types of normal SLEEP

rate in scientific measurement, a ratio of how long it takes to do something (e.g. growing at a rate of 2 mm per day)

ratio the relationship between two variables (e.g. the ratio of boys to girls in a class was 3 to 1)

reabsorption a process in which substances are transported from the urine in the kidney tubules into the body fluids; known as *re*absorption because they were once part of the body fluids and are being absorbed again; the process occurs in the tubule of the kidney nephron and may be called *tubular reabsorption*; the body is able to select those substances (and the quantities) that will be reabsorbed, so that the process is also called *selective reabsorption*

reactant a substance that takes part in a chemical REACTION

reaction **1.** a response to a stimulus **2.** a chemical change; during a chemical reaction one or more substances, the *reactants*, are changed into a new substance or substances, the *product/s*

Recent an alternative name for the HOLOCENE epoch of geological time

receptor a structure that detects a stimulus (e.g. the skin contains receptors for touch, heat, cold, pain and pressure); *see also* sense organ

receptor neuron an alternative name for SENSORY NEURON

receptor protein a protein in the cell membrane or cytoplasm that binds to a signalling molecule (a LIGAND)

recessive in genetics, the term used to describe a form of a gene (an allele) that is masked by alternative forms (dominant alleles); the characteristic determined by the *recessive allele* is called the *recessive characteristic*; it only shows in the absence of a dominant allele (usually when two recessive alleles are present); e.g. the inability to roll the tongue is a recessive characteristic determined by a recessive allele (symbol *r*); a person will be unable to roll the tongue only if he or she has two recessive alleles (*rr*); if a dominant allele (*R*) is present (*Rr* or *RR*), the person will be a tongue roller; *see also* dominance

recessive characteristic a characteristic that is determined by a recessive allele; the characteristic only appears if there is no dominant allele to mask its effect; *see* recessive

recognition site in recombinant DNA technology, the point at which an enzyme cuts a strand of DNA at a specific sequence of nucleotides

recombinant DNA synthetic DNA; made using genetic engineering techniques in which genes from one source are inserted into a DNA molecule from a different source; the procedures involved are known as *recombinant DNA technology*

recombination **1.** a change in the order of alleles along a chromosome; occurs as a result of CROSSING OVER and results in the formation of a new linkage group; *see also* gene linkage **2.** may sometimes be used to refer to new combinations of alleles as a result of INDEPENDENT ASSORTMENT; also called *genetic recombination*

recovery oxygen an alternative term for OXYGEN DEBT

recruitment increasing muscle tone (or force) by activating more motor units

rectum the last part of the large intestine; between the colon and the anus; stores faeces before defecation; *see also* diagram of digestive system (p. 77)

red blood cell one of the formed elements of the blood; contains the pigment haemoglobin, which is red and gives the cell its colour; transports oxygen, which combines with the haemoglobin; also transports some carbon dioxide; mature red blood cells have no nucleus and are biconcave in shape; produced in the red bone marrow and lives for about 120 days; also called *red corpuscle* or *erythrocyte*; *see also* diagram of red blood cell, right

red bone marrow soft tissue in some bones that produces blood cells; *see* marrow

red cell concentrate blood from which most of the plasma and platelets have been removed

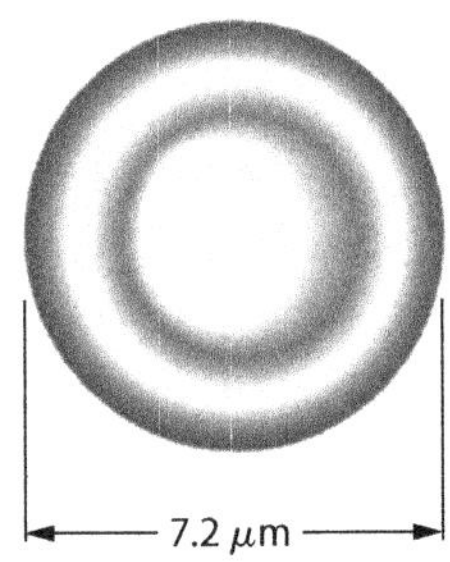

Red blood cells

redox reactions an abbreviation for *reduction–oxidation reactions*; *see* oxidation–reduction reactions

reducing sugar a sugar that is able to reduce an oxidising agent; all monosaccharides and maltose (but not sucrose) are reducing sugars; presence of reducing sugars can be detected using Benedict's test or Fehling's test

reduction the addition of electrons and hydrogen atoms to a molecule; *see* oxidation–reduction reactions

reduction division an alternative name for MEIOSIS

referred pain pain that is felt in an area different from the source of the pain (e.g. pain of a heart attack is felt in the skin of the chest and down the left arm)

reflex a rapid, automatic response to a change in the external or internal environment; tries to restore homeostasis; reflexes that are concerned with contraction of the skeletal muscles (such as withdrawing a limb from a painful stimulus) are called *somatic reflexes*; those causing secretion by glands (e.g. salivation) or contraction of heart muscle or smooth muscle are called *visceral reflexes*; reflexes handled entirely by the spinal cord are called *spinal reflexes*; reflexes that are inbuilt and do not have to be learned are *innate reflexes*; those that are learned are *acquired reflexes*

reflex arc the basic pathway travelled by nerve impulses from receptor to effector; the pathway travelled by nerve impulses during a simple reflex; the basic components of a reflex arc are the RECEPTOR, a SENSORY NEURON, a control centre, a MOTOR NEURON and an EFFECTOR; *see also* diagram below

refraction the bending of light rays as they pass from one medium to another; light rays are refracted as they pass through the cornea and lens of the eye; *see also* diagram of refraction next page

refractory period a short period following a stimulus during which a nerve cell or a muscle fibre cannot respond to a second stimulus

regenerative medicine medical treatment in which STEM CELLS are used to produce new tissues or organs

regulating factor a substance that controls the amount of a hormone circulating in the blood; secreted by the hypothalamus and forming an important link between the nervous and endocrine systems; a regulating factor that stimulates release of a hormone into the blood is called a *releasing factor* (e.g. thyrotropin releasing factor stimulates release of thyroid-stimulating hormone from the anterior lobe of the pituitary gland); a regulating factor that slows down release of a hormone is called an *inhibiting factor* (e.g. during certain phases of the menstrual cycle prolactin inhibiting factor inhibits the release of prolactin from the anterior lobe of the pituitary gland)

regulatory gene *see* gene

relative age the age of a fossil or artefact relative to another fossil or artefact (i.e. whether older or younger); *see also* absolute age, dating

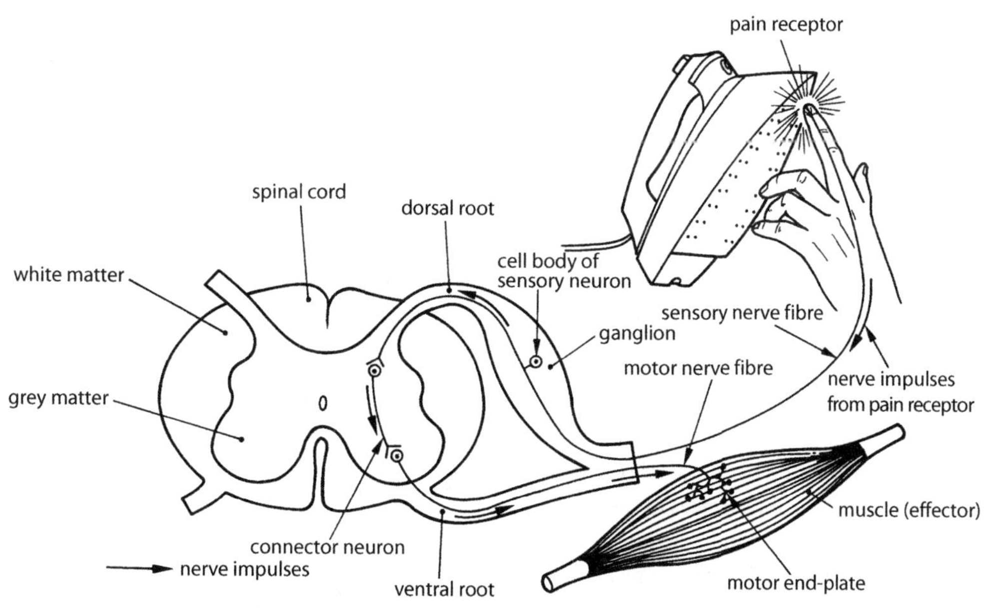

Reflex arc

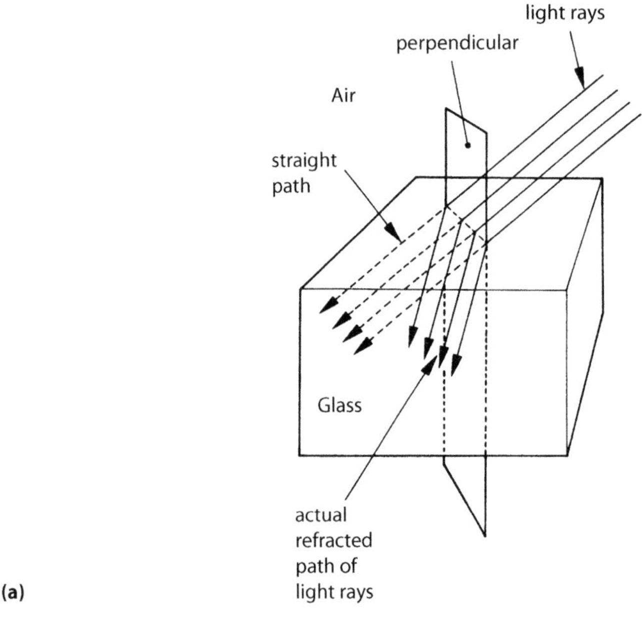

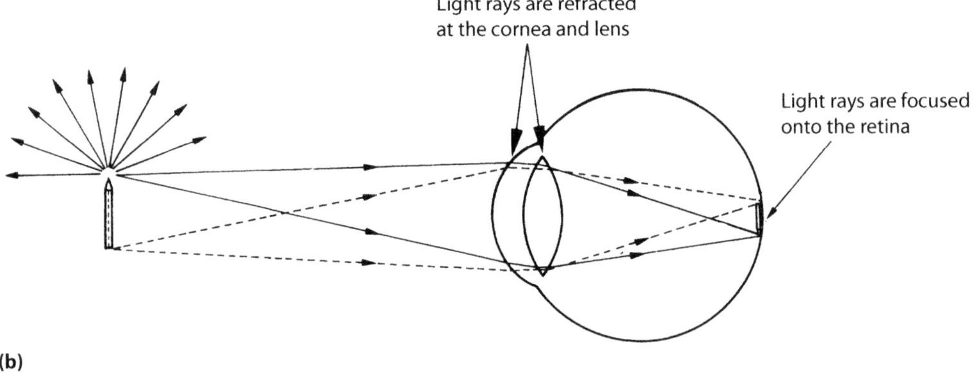

Refraction (a) of light passing through glass; (b) of light entering the eye

relative dating determining whether one object is older or younger than another; *see* dating

relaxin a hormone secreted by the ovaries, especially by the corpus luteum during pregnancy, and by the placenta; relaxes the joint between the front surfaces of the hip bones, and helps soften and relax the cervix of the uterus before childbirth; some evidence suggests that it plays a role in increasing the movement of sperm in the oviducts

relay neuron alternative name for a CONNECTOR NEURON

releasing factor a hormone whose purpose is to control the release of another hormone

reliability the extent to which the same result can be achieved when an investigation or measurement is repeated

REM sleep an abbreviation for *rapid eye movement sleep*; *see* sleep

remodelling the replacement of old bone tissue with new bone tissue, a process that goes on throughout adult life; cells called *osteoclasts* absorb the old bone tissue while *osteoblasts* deposit the calcium of new bone; there is a delicate balance between bone removal and bone production

renal arteries the arteries that carry blood to the kidneys; *see also* diagram of major arteries (p. 20)

renal calculus a KIDNEY STONE

renal column a part of the internal structure of the KIDNEY

renal corpuscle part of a kidney nephron; the glomerulus together with its surrounding capsule (Bowman's capsule); *see also* diagram of renal corpuscle (p. 112) and of kidney nephron (p. 183)

renal cortex the outer, reddish-coloured region of the KIDNEY; *see also* diagram of kidney (p. 147)

renal failure the inability of the kidneys to function normally; glomerular filtration stops or is reduced; may occur suddenly due to kidney stones, damage to kidney tubules by toxins or reduced output from the heart; may occur gradually due to bacterial infection causing inflammation of the kidneys or as a reaction to bacterial toxins

renal medulla the central, reddish-brown portion of the KIDNEY; *see also* diagram of kidney (p. 147)

renal papilla (plural *papillae*) the tip of a RENAL PYRAMID; part of the internal structure of the KIDNEY

renal pelvis the cavity of the KIDNEY that collects urine before it passes to the ureter

renal pyramid a triangular structure within the RENAL MEDULLA; part of the internal structure of the KIDNEY

renal tubule the tubule part of a NEPHRON; *see* illustration (p. 183)

renal veins the veins that carry blood away from the kidneys; *see also* diagram of major veins (p. 284)

renin an enzyme secreted into the blood by certain kidney cells when blood pressure is low or in response to stimulation by sympathetic nerves to the kidney; causes *angiotensinogen*, a protein in blood plasma, to be converted to *angiotensin*; angiotensin stimulates the adrenal cortex to secrete more aldosterone, which in turn promotes reabsorption of water and sodium ions in the kidney tubules; angiotensin is also a powerful vasoconstrictor

rennin an enzyme in the stomach secretions of children; coagulates milk protein (casein), which is then broken down by pepsin; important in infants because their main food source is milk

repetition (of an experiment or investigation) repeating the procedure many times to make sure the results are reliable

replication **1.** the process by which new virus particles are produced in a host cell **2.** the process by which a DNA molecule forms an exact copy of itself; the double helix of the DNA unwinds and the two halves split into single strands; matching bases join up with each of the single strands and are joined by sugars and phosphates to form two new molecules of DNA that are identical to the original molecule; thus each new double helix consists of one old strand and one newly-formed strand—this is known as *semiconservative replication* **3.** having a number of identical experiments running together or performing the experiment on a large number of subjects at the same time

report in science, a formal account of an investigation; includes a description of the procedure followed, the results, an analysis of the results and conclusions drawn from the results

reproduction the process by which animals or plants produce one or more organisms similar to themselves; may be sexual or asexual

reproductive isolating mechanism any structural, behavioural or biochemical feature that prevents the individuals of one species from successfully breeding with individuals from another species; *see also* reproductive isolation

reproductive isolation the prevention of breeding between two populations; the two gene pools remain distinct and there is no gene flow between them; such isolation may be due to behavioural, cultural or physiological differences, or to geographical barriers; *see* geographical isolation

reproductive system the organs associated with the production of sex cells, mating and fertilisation, development of the embryo and birth of the baby; *see* diagrams of male and female reproductive systems opposite

residual volume the volume of air left in the lungs after a person has forced out as much air as possible; *see* lung volumes, diagram (p. 155)

resolution the minimum distance between two points that can be distinguished separately with a microscope; that is, if the points were any closer together they would be seen as one; with a light microscope the resolution is about half the wavelength of light; only by using electromagnetic radiation with a smaller wavelength, such as in the electron microscope, can resolution be improved

resolving power the ability of a microscope to distinguish as separate points two points that are very close to each other; *see also* resolution

respiration **1.** one inspiration plus one expiration **2.** the exchange of gases between organisms and their environment, also known as *external respiration* **3.** the chemical reactions of cellular respiration, also called *internal respiration*; *see also* anaerobic respiration, aerobic respiration

respiratory centre the part of the brain that regulates breathing rate; consists of three groups of nerve cells; one group, located in the medulla oblongata, controls the basic rhythm of breathing and has separate groups of nerve cells for inspiration and expiration; a second, the *apneustic area* within the pons, is concerned with prolonging inspiration and inhibiting expiration; a third, the *pneumotaxic area* in the pons, sends inhibiting impulses to the inspiratory area so that inspiration ceases before the lungs become too full of air, thus limiting the period of inspiration; when the pneumotaxic area is active it overrides the apneustic area

respiratory distress syndrome the collapse of the lung in a newborn infant; *see* surfactant

respiratory failure a condition in which the respiratory system is unable to supply sufficient oxygen for normal body functions, or is unable to get rid of enough carbon dioxide to allow normal functions to continue; may be caused by disease of the

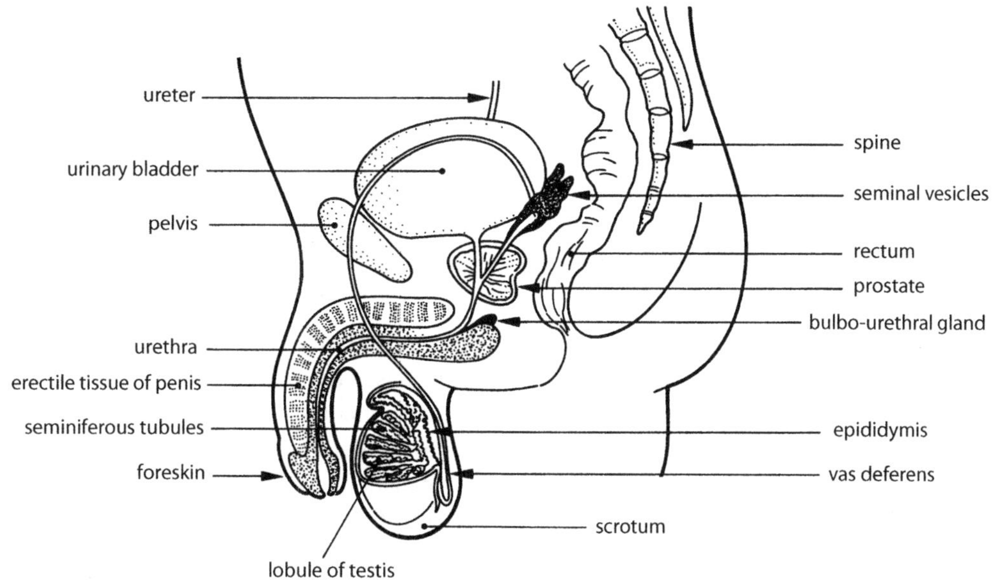

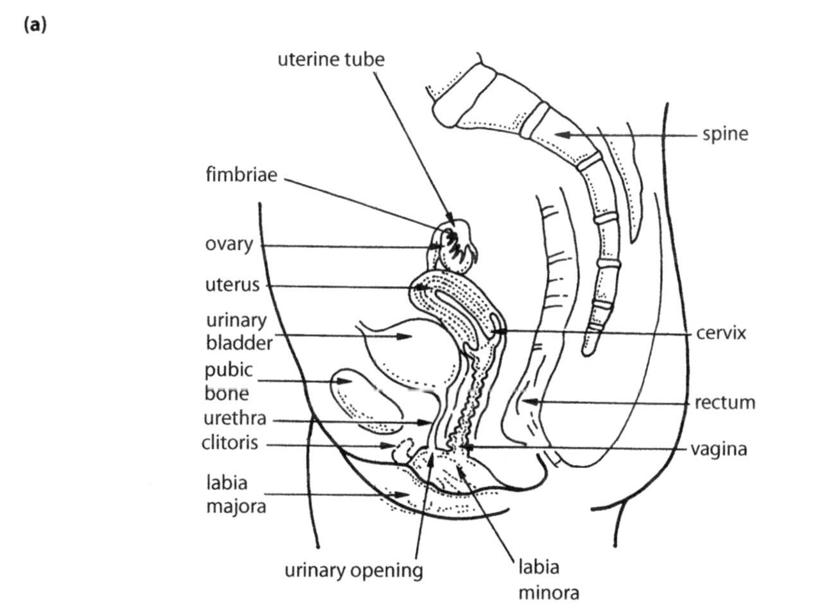

(a) Male reproductive system; (b) female reproductive system

lung, stroke, carbon monoxide poisoning or drugs that affect the respiratory centre in the brain

respiratory minute volume the volume of air taken into the lungs in one minute; *see* lung volumes

respiratory system the organs involved in exchange of gases; the nose, pharynx, larynx, trachea, bronchi and lungs; *see* diagram below

respirometer an alternative name for SPIROMETER

responding variable an alternative name for *dependent variable*; *see* variable

response any change or action that occurs as a result of a stimulus (e.g. when one touches something hot the stimulus of heat or pain causes the response of withdrawing the hand)

resting potential an alternative name for MEMBRANE POTENTIAL

restriction endonuclease alternative name for a RESTRICTION ENZYME

restriction enzyme an enzyme that cuts a DNA molecule at any site where a specific sequence of bases occurs; so named because in bacteria restriction enzymes cut up DNA molecules carried into the cell by viruses—thus *restricting* entry of foreign DNA; used in genetic engineering to isolate genes by cutting them from their chromosomes; also known as *restriction endonucleases*

resuscitation the process of reviving an unconscious person; *expired air resuscitation* (*EAR*) is mouth-to-mouth (or mouth-to-nose) resuscitation; attempting to revive a patient by breathing expired air into the patient's lungs; *external cardiac compression* involves attempting to restore heartbeat by pushing rhythmically on a patient's

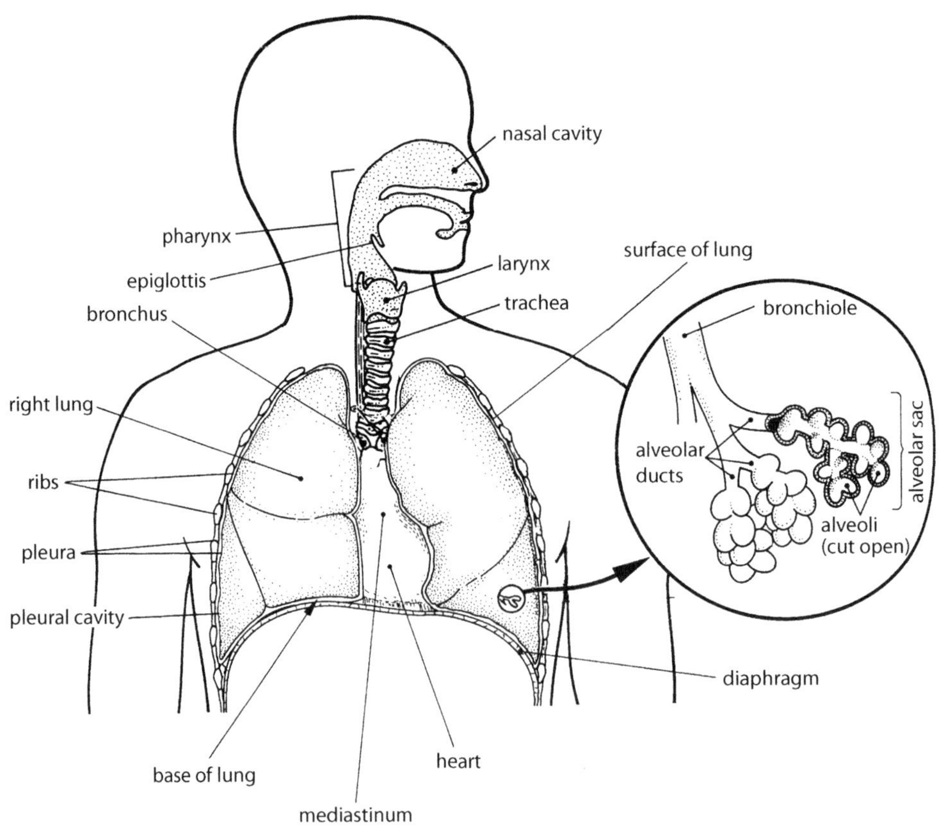

Respiratory system

breastbone; *cardiopulmonary resuscitation (CPR)* is a technique used to restore heartbeat and breathing to a dying person; a combination of expired air resuscitation and external cardiac compression

reticular activating system an alternative name for RETICULAR FORMATION

reticular formation a region of grey matter containing a network of nerve fibres located in the medulla of the brain; thought to be essential in maintaining consciousness and alertness by acting as an arousal or activating system; also called the *reticular activating system*

retina the innermost layer of the eyeball; contains light-sensitive nerve endings, some of which are rod-shaped and others cone-shaped; *rods* (or *rod cells*) are sensitive to dim light; they provide discrimination between shades of light and dark and enable us to see movement and shapes but are not sensitive to colour; rods are absent from the small depression (the fovea) in the centre of the retina and increase in number towards the sides of the retina; *cones* (or *cone cells*) are sensitive to bright light and to colour; they are most densely concentrated in the fovea; *see also* diagram of eye (p. 94)

retinol the chemical name for VITAMIN A

retrovirus a VIRUS that contains RNA instead of the usual DNA

reverse transcription making a double-stranded DNA molecule using single-stranded RNA as a template

Rh blood group a blood group based on the presence or absence of certain antigens (known as *Rh factor*) on the surface of the red blood cells; individuals who have the antigens are called Rh+ and those who do not are Rh–; if an Rh– person is given Rh+ blood, the body starts to make anti-Rh antibodies, which will remain in the blood; a second transfusion of Rh+ blood will result in a severe reaction; Rh is an abbreviation for *rhesus*—this blood group system was first worked out using the blood of the *Rhesus* monkey

Rh factor antigens involved in the RH BLOOD GROUP

Rhesus factor an alternative name for *Rh factor*; *see* Rh blood group

Rhesus incompatibility a condition in which a pregnant female and her foetus have incompatible Rhesus (Rh) blood groups; the mother's blood group is Rh– and the foetus is Rh+; the incompatibility can lead to ERYTHROBLASTOSIS

rheumatism any painful condition of the bones, joints, ligaments, tendons or muscles

rheumatoid arthritis a severe form of ARTHRITIS that affects the synovial membrane at a joint

rhodopsin a purple pigment in the rods of the RETINA of the eye; broken down by low light intensities, and the energy released triggers off a nerve impulse; vitamin A is essential in the diet for normal formation of rhodopsin; also called *visual purple*

rhythm method a method of birth control that relies on abstinence from sexual intercourse for the period just prior to, during, and just following ovulation

rib cage the bony structure that surrounds the chest cavity; consists of the ribs, sternum and thoracic vertebrae

riboflavin one of the B complex of vitamins; water soluble; not stored in large amounts in the body; good sources are liver, beef, lamb, eggs, yeast, whole grains and green vegetables; a component of certain enzymes involved in carbohydrate and protein metabolism; deficiency can lead to cracking of the skin (especially at the corner of the mouth), blurred vision and cataracts; also known as *vitamin B_2*

ribonuclease an enzyme that breaks down ribonucleic acid (RNA); *see* nuclease

ribonucleic acid (RNA) a nucleic acid consisting of one strand of nucleotides; each nucleotide consists of a sugar (ribose), a phosphate group and one of four bases (containing nitrogen); the four bases in RNA molecules are *adenine*, *uracil*, *guanine* and *cytosine* (note that DNA has the base thymine instead of uracil); RNA is involved, with DNA, in the production of proteins; there are three types—*messenger RNA* (*mRNA*) transfers the code for making a protein from the DNA in the nucleus to the ribosomes in the cytoplasm; *ribosomal RNA* (*rRNA*) is part of the ribosomes and forms the site of protein synthesis; *transfer RNA* (*tRNA*) brings amino acids to the ribosomes to be linked together in the correct sequence for the protein; *see also* protein synthesis

ribose a 5-carbon (pentose) sugar that is part of the RNA molecule; also a part of co-enzymes like NAD and FAD and nucleotides like AMP, ADP and ATP

ribosome a structure in the cytoplasm of a cell that makes proteins; ribosomes are normally attached to the endoplasmic reticulum but may be suspended in the cytoplasm of a cell; they often cluster together as polysomes; *see also* diagram of cell structure (p. 44), and the definition of protein synthesis

ribs flattened, curved bones that enclose the chest cavity; in humans there are 12 pairs; all are attached to vertebrae at the rear; the 1st to the 7th ribs, known as *true ribs*, are attached at the front to the breastbone (sternum) by strips of cartilage; the five lower pairs of ribs, which are not directly attached to the breastbone, are called *false ribs*; the cartilages on the ends of the 8th, 9th and 10th ribs attach to each other and then to the cartilage of the 7th rib; the 11th and 12th ribs are not even indirectly attached to the sternum; because they are unattached at the front they are called *floating ribs*; *see also* diagram (p. 63)

rickets a disease of young children in which the bones remain soft; caused by deficiency of vitamin D or of calcium in the diet, or possibly insufficient exposure to sunlight (as vitamin D can be made in the skin under the influence of ultraviolet light); *see also* table of vitamins (p. 289)

rickettsias a group of micro-organisms; usually classified as bacteria but smaller than bacteria; all are parasitic, alternating between arthropod hosts (such as ticks, lice and fleas) and mammals; can be transmitted to humans by the bite of an infected arthropod; several types cause serious diseases such as typhus, Q-fever and Rocky Mountain spotted fever

right lymphatic duct a vessel that drains lymph from the upper right side of the body and empties into the right subclavian vein, a vein at the base of the neck; *see also* diagram of lymphatic system (p. 156)

rigor mortis the state of partial contraction of muscles following death; caused by a lack of ATP, which results in the filaments of the muscles becoming attached to each other, thus preventing relaxation

Ringer's solution a solution of salts similar to, and the same concentration as, tissue fluids; used to keep cells alive outside the body (in vitro)

ringworm a fungal infection of the skin; symptoms are circular areas of red swellings on the skin

RNA an abbreviation for RIBONUCLEIC ACID

RNA polymerase an enzyme that is active during transcription of RNA from DNA; the double strands of DNA temporarily separate and the RNA polymerase moves along the DNA template, joining nucleotides in the order specified by the DNA; it is active in the synthesis of messenger RNA, transfer RNA and ribosomal RNA; also called *RNA transcriptase*

RNA transcriptase an alternative name for RNA POLYMERASE

robust heavy in form; used to refer to the body build of primates, especially the more recent ancestors of humans such as the australopithecines (e.g. *Australopithecus robustus* and *Australopithecus boisei* are robust forms in contrast with the more slender, gracile forms of *Australopithecus afarensis* and *Australopithecus africanus*); *see also* gracile

rod a receptor in the RETINA of the eye sensitive to dim light

root **1.** the part of a tooth that fits into a socket in the jawbone; *see also* diagrams of tooth (p. 267) **2.** one of two points of attachment of a spinal nerve to the spinal cord; the *ventral* (or *anterior*) *root* is located towards the front of the body; it conducts nerve impulses away from the spinal cord; that is, it contains axons of motor nerve cells; the *dorsal* (or *posterior*) *root* is located towards the back of the body; it contains nerve fibres that conduct nerve impulses into the spinal cord; that is, it contains axons of sensory nerve cells; the *dorsal root ganglion* (sometimes called a *spinal ganglion*) is a group of nerve cell bodies located in the dorsal root of the spinal nerve; it contains the cell bodies of the sensory nerve cells; *see also* diagram of reflex arc, spinal cord (p. 231)

Ross River virus an infection caused by a mosquito-borne virus; symptoms include fever, a rash and joint pain; also called *Ross River fever*

rotation movement of a bone around its long axis; *see also* diagram showing movements at joints (p. 1)

rough endoplasmic reticulum ENDOPLASMIC RETICULUM with ribosomes attached to it

roughage indigestible material in food that stimulates the movements of the walls of the alimentary canal; mostly cellulose from the cell walls of plant foods

round window a small, membrane-covered opening between the middle and inner ear; *see* oval window and diagram of the ear (p. 82)

roundworm a type of non-segmented worm; many are parasitic and live in the intestines

rRNA an abbreviation for *ribosomal RNA*; *see* ribonucleic acid

rubella a mild viral infection caused by the *Rubella* virus; highly contagious and, if contracted by a pregnant female, can have serious consequences for her unborn child; the child may be born deaf, blind or with heart malformations; the major period of risk to the baby is during the first 3 months of pregnancy; a vaccine is available, and in Australia it is usually given in combination with the measles and mumps vaccines; also called *German measles*

rugae (singular *ruga*) large folds in the mucous membrane lining an empty, hollow organ; occur in the stomach, vagina and urinary bladder

S

S phase the period in interphase of the CELL CYCLE when DNA is duplicated

SA node an abbreviation for *sinoatrial node*; *see* pacemaker

Sabin vaccine a vaccine for immunisation against poliomyelitis; contains active but weakened polio virus; is taken by mouth and is the usual vaccine used today; *see also* Salk vaccine

saccule a chamber of the membranous labyrinth of the inner ear; contains a receptor organ that gives information about the position of the head; also called the *sacculus*; *see also* diagram of structures of the inner ear (p. 140)

sacculus an alternative name for the SACCULE in the inner ear

sacrum a single bone formed by five vertebrae fused together; joined to the hip bones of the pelvic girdle; *see* vertebral column

saddle joint a type of SYNOVIAL JOINT

sagittal a vertical plane that divides an animal or organ into right and left halves; in a *midsagittal plane* the two halves are roughly equal and mirror images of each other (e.g. midsagittal sections are often drawn of organs like the brain to show the internal structure); in a *parasagittal plane* the left and right portions are unequal

sagittal crest an erect ridge of bone along the midline of the skull to which the jaw muscles are attached; male gorillas have a pronounced sagittal crest; also prominent in some australopithecines

saliva a clear digestive juice secreted by glands that open into the mouth; contains water, mucus, salts and the enzyme salivary amylase; saliva lubricates, dissolves and begins the chemical breakdown of food

salivary amylase an enzyme in saliva that breaks down some of the starch in food to the disaccharide maltose; formerly known as *ptyalin*

salivary glands three pairs of glands that secrete saliva and have ducts that open into the mouth cavity; the *parotid glands* are located below and in front of the ears between the skin and the jaw muscle; the *submandibular glands* are below the base of the tongue at the rear of the mouth; the *sublingual glands* are in front of the submandibular glands

Salk vaccine the first vaccine developed for immunisation against poliomyelitis; contains killed polio virus and is given by injection; now superseded by the SABIN VACCINE

Salmonella a group of rod-shaped bacteria found in the alimentary canal of animals; many species cause disease in humans (e.g. *Salmonella typhimurium* causes gastroenteritis (food poisoning), *Salmonella typhi* causes typhoid fever)

saltatory conduction the conduction of a nerve impulse along a myelinated nerve fibre; the impulse seems to jump from one NODE OF RANVIER to the next, so that speed of conduction is much faster than along an unmyelinated fibre

saltatory evolution an alternative term for PUNCTUATED EQUILIBRIUM

sarcolemma the thin, transparent membrane surrounding a muscle cell or muscle fibre; *see* diagram of structure of skeletal muscle (p. 176)

sarcoma a cancer of connective tissue (e.g. cancer of the bone, bone marrow or cartilage)

sarcomere the contractile unit of skeletal muscle; the skeletal muscle fibres are composed of cylindrical structures called myofibrils; each myofibril is divided into units called sarcomeres; between each sarcomere is a band of dense material called a *Z line* (or *Z membrane*); within the sarcomere are actin and myosin filaments; a darker, denser area, called the *A band*, represents the thicker myosin filaments; between the A bands are lighter areas, called *I bands*, that are composed only of the thinner actin filaments; actin filaments are attached to the Z lines and project into the sarcomere on each side of the line; the *H zone* is an area in the centre of the A band where the actin and myosin filaments do not overlap; when the myofibril contracts, the H zone decreases in length and it is thought that the actin and myosin filaments slide over one another; *see also* sliding filament model, muscle fibres, diagram of structure of sarcomere right

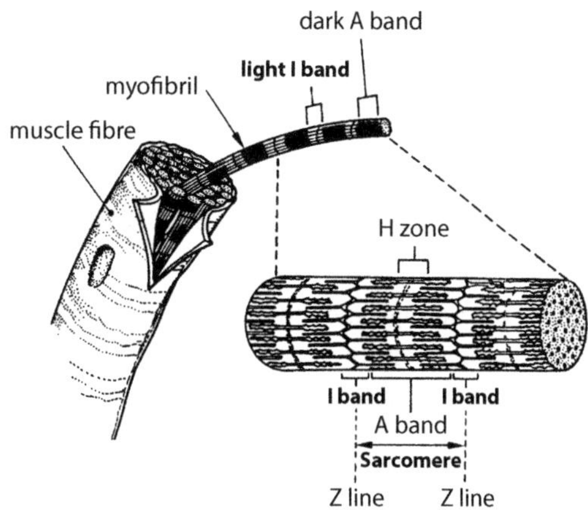

Sarcomere

sarcopenia degenerative loss of skeletal muscle mass and strength, associated with ageing

sarcoplasm the cytoplasm of striated muscle fibres

SARS *see* severe acute respiratory syndrome

saturated fat FAT with a high proportion of saturated fatty acid molecules

saturated fatty acid a FATTY ACID molecule with the maximum number of hydrogen atoms

scabies a mite that causes intense itching, often in the genital area; frequently transmitted during sexual intercourse; often referred to as 'the itch'

scaffold in tissue engineering, a template for tissue growth; collagen and fibrin are natural materials that have been used to construct scaffolds

scala media a membranous channel in the COCHLEA of the ear

scala tympani the lower channel in the COCHLEA of the ear

scala vestibuli the upper channel in the COCHLEA of the ear

scanning electron microscope an ELECTRON MICROSCOPE that produces a three-dimensional image of the surface of an object

scapula the flat, triangular bone at the rear of the rib cage; the shoulder blade; with the clavicle, forms the socket into which the upper arm bone (humerus) fits; provides attachment for many of the muscles that move the arms; *see also* diagram of skeleton (p. 249)

scarlet fever a bacterial infection that occurs more commonly in children; symptoms are a red rash, sore throat and fever

schistosomiasis a disease of humans common in tropical areas; caused by infection with the blood fluke *Schistosoma*, the larvae of which enter the body by burrowing through the skin from contaminated water; the blood fluke lives in the abdominal veins and causes weakness, lethargy and anaemia

schizophrenia a severe mental disorder; symptoms may include hallucinations, delusions and bizarre thought patterns

Schwann cell a cell that wraps around a nerve fibre forming the MYELIN SHEATH

sciatica inflammation of the sciatic nerve, a nerve that runs down the back of the thigh and the inside of the leg; pain occurs in those areas

science a process of inquiry involving observation and experimentation; also, the knowledge gained by that process

scientific method a systematic approach to the solution of problems; usually involves proposing a hypothesis based on observations, and then testing the hypothesis using suitable experiments

scientific model a simplified representation of a complex idea or process; may be a physical model, a flowchart, a diagram or a simplified explanation of a process (e.g. the lock-and-key model that explains the way enzymes work)

scientific name a name made of two words—the genus and the species to which an organism belongs; the words are Latin or Greek, or made to sound like Latin or Greek; each species of plant, animal or micro-organism has its own scientific name (e.g. the scientific name of the human species is *Homo sapiens*)

scientific report *see* report

sclera a white coat of fibrous tissue that forms the outer coat of the eyeball (except at the front of the eye); gives the eyeball its shape and protects the interior; *see also* diagram of eye (p. 94)

scoliosis an abnormal curvature of the vertebral column; sideways bending of the spine, usually in the chest region

scrotal sacs an alternative name for SCROTUM

scrotum a skin-covered pouch in which the testes are held outside the abdominal cavity so that the sperm can develop at a temperature slightly lower than normal body temperature; also called the *scrotal sacs*

scurvy a disorder caused by vitamin C deficiency; characterised by weakness and internal bleeding around the bones and teeth; symptoms can be reversed by increasing consumption of vitamin C (ascorbic acid); *see also* table of vitamins (p. 289)

sebaceous gland an oil-secreting gland in the inner layer of the skin (the dermis); not present in the palms of the hands or soles of the feet; associated with hair follicles, except for on the lips, eyelid, labia and penis; secretes *sebum* into the neck of the hair follicle; sebum, an oily substance that is a mixture of fats, proteins, cholesterol and salts, helps to waterproof the skin and keep it soft and pliable; also called *oil gland*

sebum an oily substance secreted by glands in the skin; *see* sebaceous gland

second filial generation (F_2) a term used in Mendelian genetics to refer to the second generation produced from a mating of two individuals; *see also* first filial generation

second meiotic division the second of two cell divisions that take place during meiosis; during the second division the chromatids separate

second messenger a molecule inside a cell that is produced when a chemical messenger (a hormone or a neurotransmitter) binds to the outside of the cell membrane; the second messenger brings about responses inside the cell to the message from outside the cell (the first messenger)

secondary group a formal, impersonal social group devoted to a particular function or goal; *see* social group

secondary oocyte a cell resulting from the first meiotic division of a primary OOCYTE; *see also* gametogenesis

secondary response the response to a second or subsequent exposure to an ANTIGEN; the secondary response is quicker and more intense than the PRIMARY RESPONSE

secondary sexual characteristic a characteristic associated with an individual's sex but not involved in production of sex cells; develops at puberty under the influence of sex hormones; in males includes deep voice, facial hair, hair on chest, muscle development and narrow pelvis; in females includes breasts, deposition of fat under the skin giving rounded body shape, broad pelvis

secondary spermatocyte a cell resulting from the first meiotic division of a primary spermatocyte; *see* spermatocyte, gametogenesis

secondary structure the second level of complexity in the structure of a protein; a coiled or folded shape of the amino acid chain that makes up the protein

secondary tumour a cancerous growth that results from the spreading of a primary tumour; *see also* metastasis

secretin a hormone secreted by the lining of the small intestine; decreases secretion of gastric juice; reduces movements of the alimentary canal; stimulates secretion of bile, intestinal juice and pancreatic juice

secretion **1.** the process of producing and releasing a fluid from a cell; usually refers to a useful fluid and not a waste product (e.g. a hormone produced by cells of an endocrine gland) **2.** the fluid that is released from the cell (e.g. a hormone produced by an endocrine gland may be referred to as the secretion of that gland)

section **1.** a thin slice through a tissue, organ or other structure that is mounted on a glass slide for microscopic examination **2.** a diagram showing the appearance of a structure when sliced through; a *longitudinal section* is a slice through the long axis of a structure; a *transverse section* (or *cross-section*) is a slice through the short axis of a structure

sedative a drug that has a calming and relaxing effect; may also induce sleep (e.g. minor tranquillisers like *Valium*, barbiturates)

sedentary describes a lifestyle that includes little physical activity

sedentism living in one place, rather than having a nomadic lifestyle

segmentation of the small intestine, localised contractions of the circular muscle of the wall of the small intestine so that the diameter is narrowed and the contents are sloshed back and forth

segregation principle alternative name for Mendel's law of segregation; *see* Mendel's laws

selection (in evolution) where certain characteristics have a greater chance of being passed on to later generations

selection pressure refers to natural selection; the pressure exerted by environmental factors in causing the death (or failure to reproduce) of organisms with characteristics that are not suited to the environment, and in favouring the reproductive success of those that are suited to the environment

selective agent any factor that causes the death of organisms with certain characteristics but has no effect on individuals without those characteristics

selective reabsorption in the kidney tubule, REABSORPTION of some substances and not others

selectively permeable membrane an alternative term for DIFFERENTIALLY PERMEABLE MEMBRANE

self the body's own substances, which do not act as antigens (i.e. do not provoke an immune response); foreign substances (those that are not part of an individual's usual body chemistry) are described as *non-self*—they stimulate formation of antibodies; sometimes the distinction between self and non-self breaks down and antibodies are produced that attack the body—this is an *auto-immune disease*

self-antigen any large molecule produced in a person's own body; does not cause an immune response in that person

semen the liquid, containing sperm, that is ejaculated from the male penis; nourishes and aids the transport of sperm; made up of secretions of the seminal vesicles, prostate gland and bulbo-urethral glands; also known as *seminal fluid*

semen analysis an examination of a specimen of semen to determine the quality and quantity of sperm; also referred to as a *sperm count*

semi- a prefix meaning half (e.g. semicircular canals—canals in the ear shaped like a half circle)

semicircular canals three bony channels projecting from the vestibule of the inner ear; each canal is approximately at right angles to the other two; inside each are membranous canals (known as *semicircular ducts*) filled with fluid; a swelling at one end of each canal, the *ampulla* (plural *ampullae*), contains hair (receptor) cells that are stimulated by movement of fluid in the canal when the head moves; *see also* diagram of ear (p. 82)

semicircular ducts *see* semicircular canals

semiconservative replication REPLICATION of DNA; occurs in such a way that each new DNA molecule consists of one old strand and one new strand

semilunar valves valves at the junction between the ventricles and the arteries that take blood out of the heart; prevent backflow of blood into the ventricles after ventricular contraction; *see also* diagram of heart (p. 121)

seminal vesicles a pair of pouch-like organs located on either side of the vas deferens and just behind the urinary bladder of the male; secrete a thick fluid that is rich in sugars and makes up about 60% of the volume of semen; also called *vesicula seminalis*; *see also* semen, diagram of male reproductive system (p. 235)

seminiferous tubules tightly coiled ducts, located in each TESTIS

semipermeable membrane an alternative term for DIFFERENTIALLY PERMEABLE MEMBRANE

senescence the process of growing old; the gradual decline in the performance of all the organ systems as a person ages

senility a loss of mental or physical ability that may occur with the ageing process

sensation conscious awareness of a stimulus

sense organ a concentration of nerve endings specialised to detect stimuli (e.g. the eye or the ear); may also be called a *receptor*

sensory area a part of the cerebral cortex of the brain that interprets impulses from receptors

sensory division an alternative name for the *afferent division* of the peripheral nervous system; *see* nervous system

sensory neuron a nerve cell that carries messages from receptors to the brain or spinal cord; also called *receptor neuron*; *see also* diagram (p. 187), diagram of reflex arc (p. 231)

septicaemia the presence of toxins or disease-causing bacteria in the blood; causes fever; also called *blood poisoning*

septum any dividing wall or partition (e.g. of the heart—the wall between the two sides of the heart; *see also* diagram of heart (p. 121); of the nose—the tissue between the two nostrils)

serine one of the 20 amino acids that are common in proteins; not essential in the human diet; *see also* list of amino acids (p. 11)

seroconversion illness an illness developed by some people after infection with HIV; occurs within two weeks of exposure to the virus; symptoms are lack of energy, fever, headache, dry cough, and pain in muscles and joints; recovery occurs within 3 to 14 days

serosa an alternative name for SEROUS MEMBRANE

serotonin a neurotransmitter that occurs in the brain stem; thought to be involved in regulation of body temperature, sensory perception, induction of sleep and regulation of mood

serous membrane a membrane that lines a body cavity and covers the organs that lie within the cavity; consists of layers of loose connective tissue with a layer of squamous epithelium on the outside; secretes fluid to lubricate movement of the organs; the pleura, peritoneum and pericardium are serous membranes; also called *serosa*; the outer layer of squamous (flattened) epithelial cells is called *mesothelium*

Sertoli cells the supporting cells of the walls of the seminiferous tubules of males; surround and nourish the developing sperm

serum **1.** the liquid that is left after blood clots; blood plasma without any clotting proteins **2.** a liquid containing antibodies or antitoxins that can be injected to combat a disease; also called *antiserum* (e.g. tetanus antiserum can be injected if injury results in the possible entry of tetanus bacteria)

set point in a feedback system, the level at which a variable is to be maintained (e.g. the set point for human body temperature is about 37°C; if body temperature rises above or falls below this, the set point mechanisms come into operation to return temperature to the correct level)

severe acute respiratory syndrome (SARS) a respiratory disease first reported in China in 2003; commonly called *SARS*

Sewall–Wright effect an alternative name for RANDOM GENETIC DRIFT

sex chromosomes a pair of chromosomes, found in all mammals and many other animals, that determine the sex of the individual; may be an *X*-chromosome or a *Y*-chromosome; female mammals have two *X*-chromosomes; males have one *X*-chromosome and one *Y*-chromosome; humans have 44 non-sex chromosomes (autosomes) and two sex chromosomes

sex hormones the hormones that determine sexual characteristics and which are involved in the production of sperm and eggs, (e.g. oestrogen and testosterone)

sex-influenced characteristic a characteristic in which an allele is dominant in one sex and recessive in the other (e.g. baldness)

sex linkage the control of certain characteristics by genes in the sex chromosomes; the occurrence of *sex-linked characteristics* differs for males and females because females have two *X*-chromosomes, and males have an *X*-chromosome and a *Y*-chromosome; *X-linkage* is where a characteristic is controlled by genes in the *X*-chromosome; *X-linked characteristics* (may also be loosely called *sex-linked characteristics*), which are determined by recessive alleles, are much more common in males than in females (e.g. red-green colour-blindness, haemophilia and Duchenne muscular dystrophy are inherited conditions that are *X*-linked); *Y-linkage* is where a characteristic is controlled by a gene in the *Y*-chromosome; *Y-linked characteristics* are determined by alleles on the *Y*-chromosome (e.g. the genes for male characteristics are *Y*-linked)

sex-linked characteristics characteristics controlled by genes located in the sex chromosomes; *see* sex linkage

sex-linked genes genes that are carried in the sex chromosomes and therefore show different patterns of inheritance between males and females (e.g. the gene for red-green colour-blindness in humans is sex-linked and is carried on the *X*-chromosome)

sex selection attempting to produce a baby of a particular sex

sexual dimorphism refers to any difference in body size and shape between adult males and females of a species; primate species show sexual dimorphism in body size with adult males being, on average, larger than adult females; some species, like gorillas and orang-utans, show significant differences in size between males and females; the saggital crest that occurs only on the skull of male gorillas is another example of sexual dimorphism; differences between the sexes may also occur in coat colour, size of canine teeth and many other characteristics

sexual intercourse the insertion of the erect male penis into the female vagina; the act in which sperm are transferred from male to female; also called *coitus* or *copulation*

sexual maturity the age when a person is able to reproduce

sexual reproduction the type of reproduction in which new individuals are produced by the union of sex cells (GAMETES), usually a sperm and an egg

sexual response cycle the physiological changes that occur in both sexes during sexual intercourse; divided into four phases—excitement, plateau, orgasm and resolution

sexual selection a form of natural selection in which characteristics likely to give an advantage in reproduction are selected, rather than those that give a survival advantage; usually the male has characteristics that enable him to compete favourably for mates against other males; sexual selection explains the presence of secondary sexual characteristics (e.g. the tail of a peacock, the antlers of a deer and the mane of a lion)

sexually transmitted disease (STD) a disease transmitted by direct sexual contact, usually during sexual intercourse (e.g. syphilis, gonorrhoea, genital herpes, AIDS); also called *venereal disease* (*VD*)

sheath of Schwann an alternative name for *neurilemma*; *see* myelin sheath

shin splints soreness or pain along the shin bone (tibia); probably due to inflammation of the membrane around the bone (periostium); often occurs after a period of vigorous activity following prolonged inactivity

shingles infection of the spinal nerves by a virus called *Herpes zoster*; the virus may remain dormant for many years but when activated it causes pain along the pathways of the affected sensory nerves; groups of vesicles develop in the skin overlying the nerve pathways; thought to be caused by the same virus that causes chickenpox

shivering the involuntary contraction of a muscle; generates heat when body temperature begins to fall

shock a condition in which there is insufficient blood supply, and therefore oxygen supply, to the tissues; results from severe illness or injury; symptoms include faintness, pale skin, weak pulse, low blood pressure and shallow breathing

short sight a common name for MYOPIA

short tandem repeats repeating sequences of bases found at different positions in the human GENOME; used in DNA profiling—the number of repeats in two different samples of DNA are compared; *see also* DNA fingerprinting

shoulder girdle an alternative name for PECTORAL GIRDLE

shovel-shaped incisors incisor teeth with very concave inner surfaces; a characteristic shared by many members of Asian populations; also referred to as *shovelling*

shunt in medicine, a hole or vessel that allows liquid to flow from one part of the body to another; may occur as a developmental defect or may be placed artificially to solve a medical problem

siamang the common name for the small, slender, long-armed, tree-dwelling ape belonging to the species *Symphalangus syndactylus*; closely related to the gibbons, but slightly heavier in build and with a shaggier coat; found in South-East Asia

Siamese twins *see* conjoined twins

sibling rivalry competition between brothers and/or sisters for affection from parents or for some other form of recognition

siblings brothers and sisters; offspring of the same parents; commonly abbreviated to *sibs*

sibs an abbreviation for SIBLINGS

sickle-cell allele the allele that causes production of abnormal haemoglobin; a person with one such allele has sickle-cell trait; if two such alleles are present, the individual has sickle-cell anaemia; *see also* anaemia

sickle-cell anaemia an inherited disease causing early death in persons inheriting two alleles for sickle-cell anaemia; *see* anaemia

sickle-cell trait an inherited condition resulting in abnormal haemoglobin that causes the red blood cells to become sickle-shaped when oxygen concentration is low; *see* anaemia

SIDS an abbreviation for SUDDEN INFANT DEATH SYNDROME

sign a term used when describing a disease; any objective evidence of a disease, such as an open wound, swelling or fever; *see also* symptom

Silurian one of the 7 periods in the Palaeozoic era of geological time; between the Ordovician and Devonian periods; 444–416 million years ago; the first fish with jaws evolved during this period, and plants began to colonise the land; *see also* geological time scale (p. 111)

simple epithelium epithelium consisting of a single layer of cells; *see* epithelial tissue

simple fracture a FRACTURE where the broken ends of a bone do not break through the skin

simple sugar an alternative name for *monosaccharide*; *see* carbohydrate

single-crystal-laser-fusion dating dating of volcanic rock using a laser beam to melt a single crystal of the mineral feldspar so that argon gas is released; the quantity of gas released can be measured and, since argon accumulates at a known rate, the age of the rock can be determined; used to determine the age of rocks formed from volcanic ash, such as those at Hadar in Ethiopia, and thus the fossils contained in those rocks

single-gene disorder an inherited disease caused by an allele of a single gene (e.g. Huntington's disease and cystic fibrosis); also called a *monogenic disorder*

single-species hypothesis a model of early hominid evolution developed in the 1960s that stated that there was only one species of hominid in existence at any point in time; later fossil discoveries have led to the rejection of this model, and today the term *hominid* is used to refer to orang-utans, chimpanzees, gorillas and humans, and their extinct ancestors

sinoatrial node the PACEMAKER of the heart

sinus a space or cavity; in human anatomy it is most commonly used to describe **1.** *lymphatic sinuses*—irregular channels in lymph nodes **2.** *paranasal sinuses*—paired spaces in the bones around the nose; these sinuses are lined by a mucous membrane covered with ciliated epithelium; secretions of the membrane drain into the nasal cavity; inflammation of the paranasal sinuses is called *sinusitis*

sinusitis inflammation of mucous membranes in spaces in the bones of the face; *see also* sinus

skeletal muscle muscle that is joined to the bones and is under voluntary control; made up of thousands of long cylindrical cells or fibres; also called *voluntary muscle*, *striated muscle* or *striped muscle*; *see also* muscle fibres, muscular tissue, diagram of structure of skeletal muscle (p. 176)

skeletal system the bones and cartilage that provide structural support for the body

skeleton the body's bony framework; maintains the shape of the body and supports the soft tissues; can be moved at joints by contraction of skeletal muscles; *see* diagram below

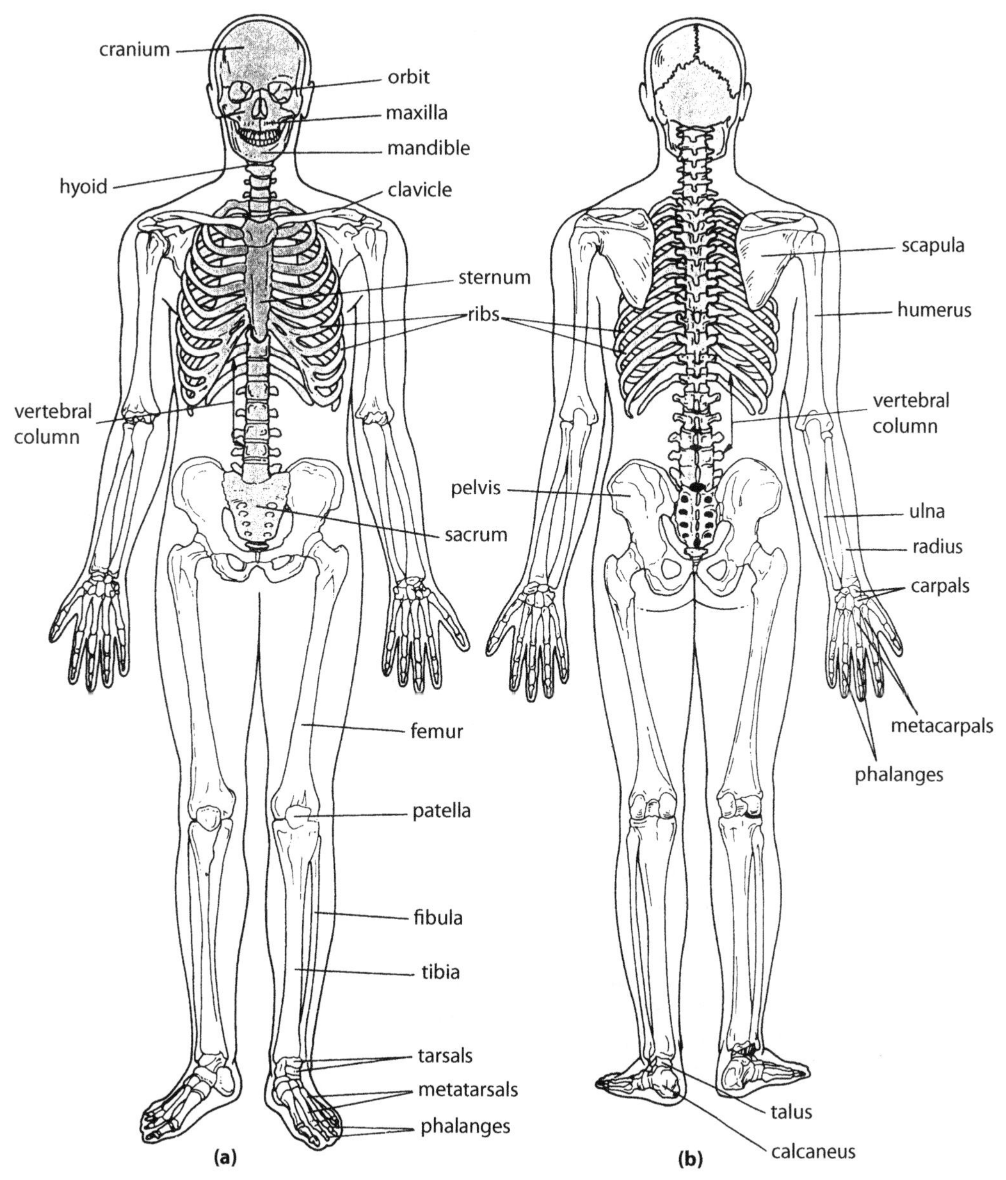

Skeleton

skin an organ covering the surface of the body (classified as an organ because it contains several tissues); functions include temperature regulation, prevention of water loss through evaporation, protection against entry of micro-organisms, perception of stimuli, excretion, production of vitamin D; consists of an outer, thinner layer of epithelial tissue called the *epidermis*, an inner, thicker layer of connective tissue called the *dermis* and a *subcutaneous layer* that attaches the skin to underlying structures; *see also* epidermis, diagram of skin below

skull the bony covering of the head; consists of the cranium (which encloses and protects the brain) and the facial bones; *see* diagram of skeleton previous page

skull cap the upper part of the cranium; *see also* diagram of skull cap, top right

Skull cap from a Homo erectus *fossil*

sleep a state of partial consciousness from which a person can be aroused by stimuli such as noise, touch, heat or cold; sleep is of two types—*non-rapid eye movement* (*NREM*) *sleep* (or *slow-wave sleep*) is characterised by low-frequency electrical waves in the brain, a regular and deep breathing cycle, slow heart rate and lowered blood pressure; about every 90 minutes during a normal night's sleep, electrical waves in the brain speed up for periods of 5–20 minutes and the eyes move rapidly; this is *rapid eye movement* (*REM*)

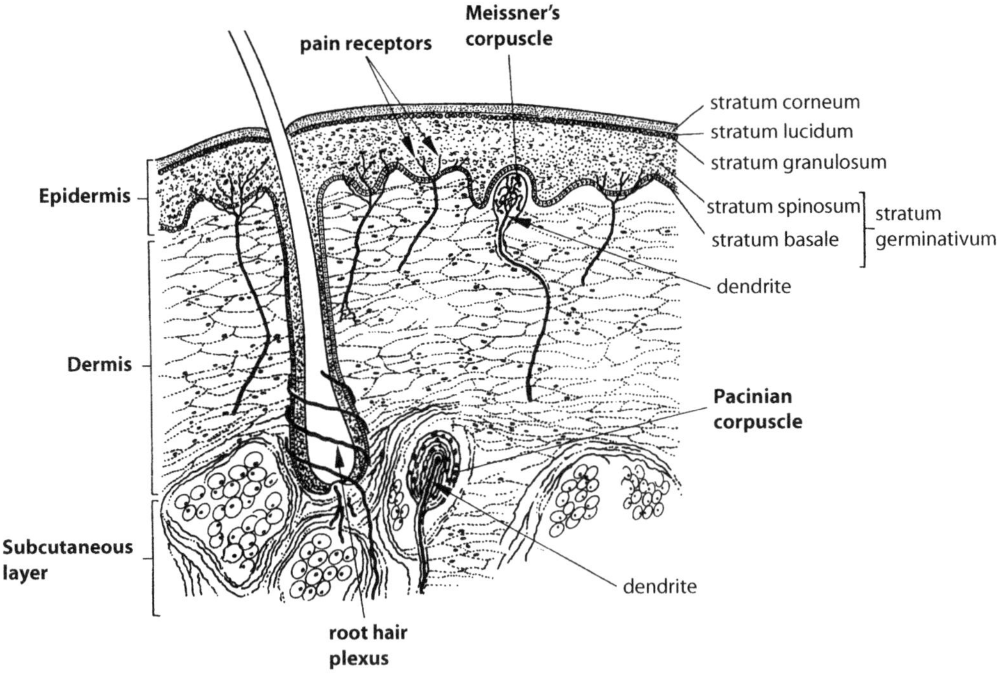

Section through the skin

sleep (or *paradoxical sleep*); during REM sleep, breathing and heart rate are irregular, blood pressure may rise or fall and most dreaming occurs

sleep apnoea a disorder in which breathing ceases momentarily during sleep

slide a rectangular piece of glass on which a specimen is placed for viewing with a light microscope; more correctly called a *microscope slide*

sliding filament model the most commonly accepted explanation for skeletal muscle contraction; it is believed that when a skeletal muscle fibre contracts, the actin and myosin filaments slide past one another, pulling the Z lines towards each other and shortening the sarcomere; also called the *sliding filament hypothesis* or *sliding filament theory*; *see* diagram of structure of sarcomere (p. 241)

slightly movable joint a JOINT with only limited movement between the bones

slime layer a jelly-like outer covering on some bacteria

slipped disc occurs in the cartilage between vertebrae; the fibrous outer cartilage sheath ruptures, allowing the softer inner cartilage to bulge out; more correctly called a *herniated* or *ruptured disc*

slow-twitch fibres muscle fibres that contract gradually; they have little power but are resistant to fatigue; used when endurance is needed

small intestine the part of the alimentary canal between the stomach and large intestine; a very long tube (about 6 metres) divided into three parts—the duodenum, the jejunum and the ileum; responsible for most digestion and absorption; *see also* diagram of digestive system (p. 77)

smallpox a highly infectious viral disease; it was eradicated in 1977 by a world-wide vaccination program

smear test an alternative name for PAPANICOLAOU TEST

smooth endoplasmic reticulum ENDOPLASMIC RETICULUM with no ribosomes attached

smooth muscle muscle that is not under conscious control; *see* muscular tissue

sneezing the involuntary expulsion of air from the lungs; caused by irritation of the nose

social group a group of individuals acting together in a joint effort to satisfy common needs, and who share a sense of common identity; a *primary group* consists of people who know one another intimately, as in a family or a circle of close friends; a *secondary group* is a formal, impersonal group devoted to a particular function or goal, such as a department in a bureaucracy; the term social group also applies to animals other than humans; *see also* mother–infant group, one-male group, multimale group

socialisation the process whereby children learn to behave in a manner that conforms to the culture to which they belong

sociobiology the study of social behaviour; concerned with evolutionary explanations of social behaviour, focusing on the role of natural selection; a controversial area of study because it looks for genetic links to human behaviour

sociocultural barriers barriers to interbreeding that are due to social and cultural factors

sociology the study of the way human society is organised and of the way it functions; now often considered to be a part of cultural anthropology

sodium chloride common salt; an important constituent of body fluids

sodium–potassium pump a mechanism in the membrane of a nerve cell that transports sodium ions out of the cell and potassium ions into the cell; *see* membrane potential

soleus the flat calf muscle lying beneath the gastrocnemius muscle; joined to the lower leg bones (tibia and fibula) at one end and to the heel bone (calcaneus) at the other; contracts to bend the foot up at the ankle; *see also* diagram of muscular system (p. 178), and diagram of calcaneal tendon (p. 38)

solute in a solution, the substance that is dissolved; the solute is the substance that is in the lesser quantity in a solution (e.g. in a sugar solution the sugar is the solute); *see also* solvent

solution a mixture of one substance dissolved in another, usually a solid or gas dissolved in a liquid; a combination of solute and solvent

solvent in a solution, the substance in which other material is dissolved; the solute is the substance that is in the greater quantity in a solution (e.g. when sugar is dissolved in water, the water is the solvent); *see also* solute

somatic cell any cell that makes up the body of an individual, except for the reproductive cells; any cell that is not a gamete

somatic mutation a change occurring in a gene in a body cell; *see* mutation

somatic nervous system part of the efferent division of the peripheral nervous system; *see* nervous system

somatic reflex a REFLEX involving contraction of skeletal muscles

somatic sensory neuron a nerve cell that carries messages from receptors in the skin, muscles, bones and joints into the brain and spinal cord

somatogenetic adaptability an alternative term for ACCLIMATISATION

somatotropin somatotropic hormone (STH); an alternative name for GROWTH HORMONE

somatotype body type; the three basic somatotypes are ectomorph, endomorph and mesomorph; an *ectomorph* has a thin angular frame, thin legs and arms, narrow shoulders and hips and little body fat; an *endomorph* has a rounded body shape, large head, large stomach, heavy build and a lot of fat; a *mesomorph* is muscular with a large head, broad shoulders, relatively narrow hips and little body fat; *see also* illustration below

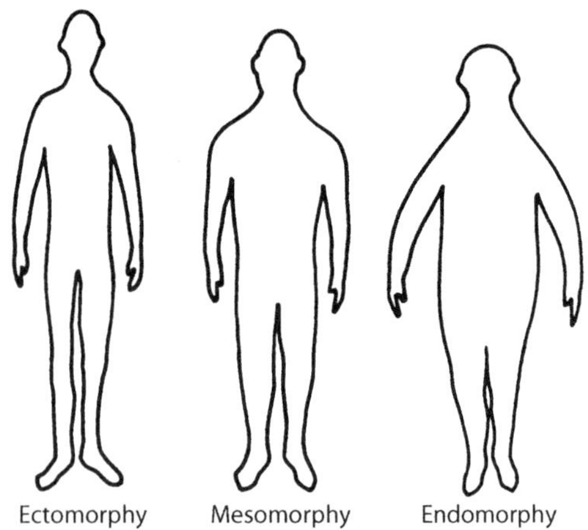

Somatotypes

spasm an abnormal, sudden and involuntary contraction of a muscle for a short duration

special creation the belief that each species (or genus) was individually created by God

speciation the process of producing a new species; two patterns of speciation are revealed by the fossil record; *anagenesis* (or *phyletic evolution*) is when a single species changes so much over time that it is justifiable to rename it as a new species; *cladogenesis* (or *branching evolution*) is when one or more new species develops from a species that continues to exist; cladogenesis is the more common pattern and may occur when populations of a species become reproductively isolated and natural selection causes them to evolve along different lines

species the basic unit of biological classification; members of a species show consistent differences from other organisms and are usually capable of interbreeding and producing fertile offspring; humans all belong to the species that has the scientific name *Homo sapiens*; similar species are grouped together in the same genus

species selection a hypothesis stating that species that live the longest, and develop into the greatest number of new species, determine the direction of major evolutionary trends; the hypothesis is controversial and is not accepted by all evolutionary biologists

specific defence an immune response that is directed towards a particular ANTIGEN; *see also* non-specific defence

specific immunity resistance to a particular disease; *see* immunity

sperm an abbreviation for SPERMATOZOA

sperm count an alternative name for SEMEN ANALYSIS

sperm duct a common name for VAS DEFERENS

sperm mother cells an alternative name for SPERMATOGONIA

sperm sorting separating sperm with an *X*-chromosome from those with a *Y*-chromosome; a technique used to choose the sex of a baby

spermatids immature spermatozoa (sperm); *see* spermatocyte

spermatocyte a *primary spermatocyte* is a diploid cell (with 46 chromosomes) that will divide by meiosis to produce sperm; the first meiotic division produces *secondary spermatocytes*, which are haploid (with 23 chromosomes), but each of the chromosomes consists of a pair of chromatids; the second meiotic division separates the chromatids, producing haploid cells called *spermatids* or immature sperm; *see* diagram of spermatogenesis (p. 107)

spermatogenesis the production of sperm; *see* gametogenesis

spermatogonia (singular *spermatogonium*) immature cells that line the seminiferous tubules; develop into primary spermatocytes that then divide by meiosis to become sperm; also called *sperm mother cells*

spermatozoa (singular *spermatozoon*) the male gametes; each consists of a head containing a nucleus with the haploid number of chromosomes (23), a midpiece packed with mitochondria and a flagella or tail for locomotion; also called *sperm*; *see* diagram, on next page top; *see also* diagram of spermatogenesis (p. 107)

spermicide a substance that kills spermatozoa

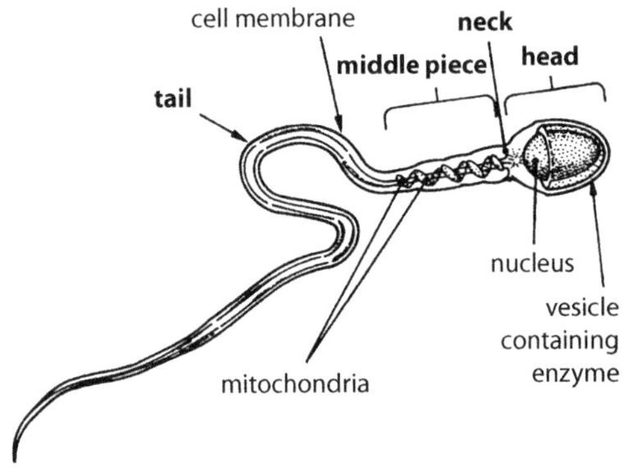

A spermatozoon

sphincter a circular muscle around a body opening or around a tube; controls the size of the opening or tube (e.g. the pyloric sphincter between the stomach and small intestine, the anal sphincter around the anus)

sphygmomanometer an instrument used for measuring BLOOD PRESSURE in the arteries

spina bifida one of a group of conditions called *neural tube defects*; occurs when some vertebrae do not form a complete bony arch around the spinal cord; in severe cases the spinal cord may protrude through the skin; effects vary depending on the location of the affected vertebrae and the extent of exposure of the spinal cord; symptoms can range from minor problems such as leg weakness, or poor sensation, to lack of bowel or bladder control, lack of movement in the legs and misshapen feet; increased intake of the vitamin folic acid before and during pregnancy can reduce the chances of a baby being born with a neural tube defect

spinal column an alternative name for VERTEBRAL COLUMN

spinal cord the nerve cord that extends from the brain to about waist level; enclosed in the vertebrae; 31 pairs of *spinal nerves* arise from the spinal cord and each is joined to the cord by dorsal and ventral roots; the main function of the cord is to carry nerve impulses between the parts of the body and the brain, and to carry out reflexes; *see also* reflex, diagram of reflex arc (p. 231), diagram of spinal cord, opposite

spinal ganglion an alternative name for DORSAL ROOT GANGLION; *see* root

spinal nerve one of the 31 pairs of nerves that arise from the SPINAL CORD

spinal reflex a REFLEX involving only the spinal cord

spinal reflex arc the pathway travelled by the nerve impulses in a simple reflex involving the spinal cord; *see also* reflex arc

spindle a framework of fine fibres (micro-tubules) that forms in a cell during prophase of mitosis and meiosis; chromosomes attach themselves to the spindle and spindle fibres contract to pull the chromosomes apart; *see* diagrams of the phases of mitosis (p. 172)

spine an alternative name for VERTEBRAL COLUMN

spiral fracture a break of a bone in which the two parts are twisted apart at the break; *see* fracture

spiral organ an alternative name for ORGAN OF CORTI

spirillum a bacterial cell that is twisted, spiral or corkscrew in shape; *see* bacteria

spirochaete a bacterial cell in the form of a tight spiral; *see* bacteria

spirometer an apparatus used to measure the volume of air that can be moved into and out of the lungs; also called a *respirometer*

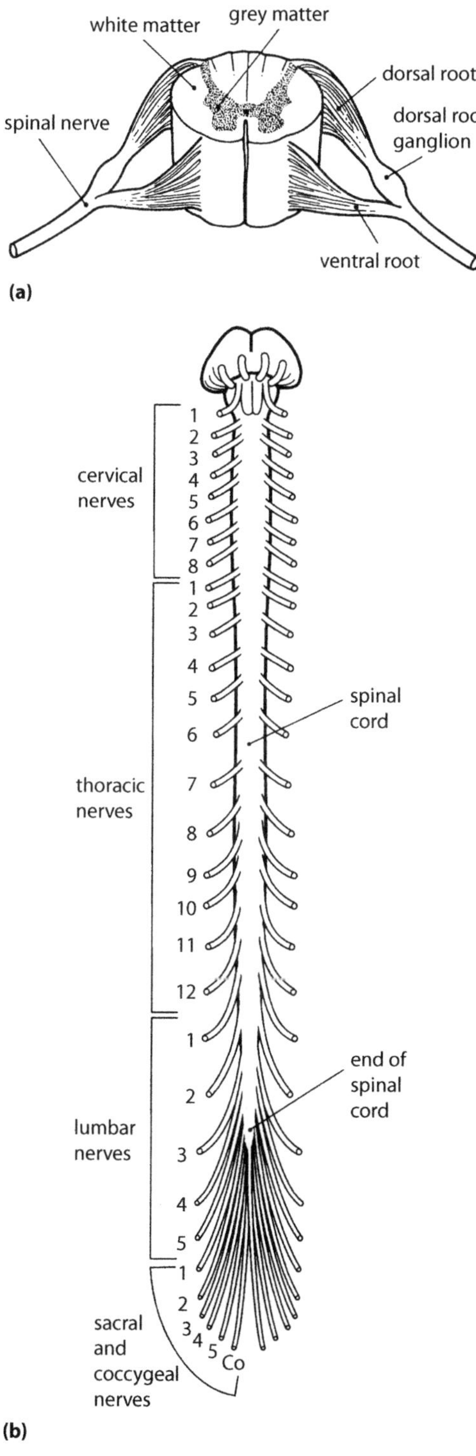

(a) A part of the spinal cord showing a pair of spinal nerves; (b) with spinal nerves

spleen a mass of lymphoid tissue between the stomach and the diaphragm; produces B lymphocytes; stores and releases blood; the cells of the spleen engulf and destroy bacteria and worn out or damaged red blood cells and platelets; if the spleen is surgically removed due to disease or injury, the functions are taken over by other organs, particularly the bone marrow

spongy bone an alternative name for cancellous bone; *see* bone

spontaneous abortion an alternative name for MISCARRIAGE; *see also* abortion

spontaneous generation the idea that life can arise from non-living matter; a common belief up to the 19th century but now discredited

sprain when a ligament is torn or separated from its bony attachment; twisting or wrenching of the joint stresses ligaments beyond their usual capacity

squamous epithelium epithelium consisting of a layer of flattened cells; *see* epithelial tissue

staggered cut the cut produced when a restriction enzyme creates fragments of DNA with unpaired nucleotides that overhang at the break in the strands; the result is known as *sticky ends*

stain a chemical used to colour parts of a tissue that have a similar chemical composition, usually used to enhance viewing with a microscope (e.g. methylene blue can be used to colour the nuclei of cells dark blue and the cytoplasm light blue, Leishman's stain colours the nuclei of white blood cells purple); the use of stains in preparing material for microscopic examination is called *staining*

stapes one of the bones in the middle ear; *see* auditory ossicles

Staphylococcus (plural *staphylococci*) a genus of bacteria with spherical cells that live on the skin or in the body; some types produce toxins that can cause disease if taken in with food; *Staphylococcus aureus* produces pimples, boils and other skin conditions

starch a POLYSACCHARIDE made up of many glucose molecules; a component of many plant foods

start codon the initial three bases in messenger RNA which begin the translation of the code into a sequence of amino acids

stasis static equilibrium; used in discussion of evolution to refer to long periods of time in which there is little or no evolutionary change; *see also* punctuated equilibrium

statins drugs that lower the level of low-density lipoprotein in the blood; prescribed to lower blood cholesterol

steady state HOMEOSTASIS; maintaining a constant internal environment that is the optimum for normal cell functions; balancing inputs and outputs to maintain the relatively constant internal environment

steady state control system a negative FEEDBACK SYSTEM that helps to maintain homeostasis

stem cells the general term for any cells that have not yet undergone DIFFERENTIATION (e.g. stem cells in bone marrow are the source of all blood cells and platelets); *totipotent stem cells* can differentiate into any of the cell types in the human body and also into the cells that make up the embryonic membranes; *pluripotent stem cells* can differentiate into any type of body cell but not into cells of the embryonic membranes; *multipotent stem cells* can differentiate into some but not all types of body cell; *embryonic stem cells* are pluripotent and come from the INNER CELL MASS of the early embryo; *adult stem cells* are multipotent and are found in many tissues; stem cells are often derived from umbilical cord blood and are then called *cord blood stem cells*

stereoscopic vision the ability to see objects in three dimensions; *see* binocular vision

sterile **1.** free from living micro-organisms (e.g. a sterile petri dish) **2.** unable to conceive or produce offspring

sterility the condition of being STERILE; an alternative name for INFERTILITY

sternum a flat bone at the front of the chest to which many of the ribs are attached; the breastbone; *see also* diagram of skeleton (p. 249)

steroids a group of complex lipids; includes vitamin D, cholesterol and sex hormones such as testosterone, oestrogen and progesterone

stethoscope a device used to listen to sounds that occur inside the body (such as heartbeat and breathing)

sticky ends the overhanging ends produced by a STAGGERED CUT of a sequence of nucleotide bases; sometimes called *cohesive ends*

stimulant any drug that tends to increase the activity of the central nervous system and thus behavioural activity (e.g. caffeine, nicotine, cocaine)

stimulus (plural *stimuli*) any change, internal or external, that causes a response in an organism; can also be defined as anything that will trigger off a nerve impulse

stimulus–response model a model that describes the sequence of events and the structures involved in a simple response to a stimulus; the stimulus is detected by a receptor, then a message (hormonal or nervous) is sent to a modulator, which in turn sends a message to an EFFECTOR, which actually carries out the response; *see also* diagram of *reflex arc* (p. 231)

stirrup one of the bones in the middle ear; *see* auditory ossicles

stomach the part of the alimentary canal between the oesophagus and small intestine; mixes and partly digests the food; the mixture of partly digested food and secretions, known as *chyme*, passes into the small intestine through the pyloric sphincter; *see also* diagram of digestive system (p. 77)

stop codon an alternative name for TERMINATION CODON

strabismus a disorder of the eye muscles where the muscles of each eye do not move together so that movements of the two eyes are not co-ordinated when the person looks at an object; may be referred to as a *squint* or as *crossed eyes*

straight cut the cut produced when a RESTRICTION ENZYME makes a clean break across the two strands of DNA so that the ends (called *blunt ends*) terminate in a base pair

strain **1.** a condition that results from overstretching a muscle or tendon **2.** a group of organisms within a species that has particular distinguishing characteristics (e.g. a *strain* of a bacterium that is resistant to antibiotics)

stratified epithelium epithelium consisting of several layers of cells; *see* epithelial tissue

stratigraphy the study of a sequence of rock layers as a means of relative dating; lower strata (and the fossils or artefacts they contain) are usually older than the strata above them; *see also* principle of superposition

stratum basale a part of the stratum germinativum of the EPIDERMIS of the skin

stratum corneum the outer layer of the EPIDERMIS of the skin; consists of dead cells in which the cytoplasm has been replaced by keratin; the outermost cells of this layer are continually worn away

stratum germinativum the deepest layer of the EPIDERMIS of the skin; where mitosis occurs

stratum granulosum a layer of the EPIDERMIS of the skin, above the stratum germinativum; a layer of flattened cells in which granules develop, the nucleus disintegrates and the cells die

stratum lucidum a layer of the EPIDERMIS of the skin prominent only in areas of thick skin

stratum spinosum the part of the stratum germinativum of the EPIDERMIS of the skin where cells become more flattened in shape

strepsirrhine a primate with a moist nose, such as a loris or lemur; in some classifications included in a suborder of primates called Strepsirrhini, which separates the lemurs and lorises from tarsiers, monkeys, apes and humans; the latter are included in the suborder Haplorrhini; *see also* haplorrhine

streptococci spherical bacteria that occur in pairs or chains

streptomycin an antibiotic; kills certain bacteria by inhibiting protein synthesis

stress any circumstance that threatens a person's well-being; also defined as any stimulus that causes an imbalance in the internal environment of the body; may be external (such as heat, cold or noise) or internal (such as pain, high blood pressure or disturbing thoughts)

stretch receptor a receptor in a muscle that records the degree of stretch of the muscle fibres; more specifically, often taken to mean receptors in the walls of the lungs, bronchi and bronchioles that send impulses to the respiratory centre to prevent overinflation of the lungs

striated muscle an alternative name for SKELETAL MUSCLE

striding gait walking in such a way that the hip and knee are fully extended; that is, the leg is straight for part of each step; often called a *free striding gait*; the mode of locomotion used by hominids; *see* illustration of striding gait below

striped muscle an alternative name for SKELETAL MUSCLE

stroke a condition in which some brain tissue is destroyed because it does not receive enough oxygen; may be due to a ruptured blood vessel in the brain or a blockage in a blood vessel (due to a blood clot or the narrowing of the vessel due to deposits of plaque); effects vary depending on the location of the brain tissue destroyed; strokes are known to doctors as *cerebrovascular accidents*; a mini-stroke, or TRANSIENT ISCHAEMIC ATTACK produces symptoms similar to a stroke but they do not last as long

stroke volume the volume of blood pumped from the left ventricle of the heart during one contraction; average is about 70 mL in a resting adult; *see also* cardiac output

stroma the tissue that forms the framework of an organ, rather than its functional parts (e.g. the mass of connective tissue within the ovary)

structural gene a gene that has the code for the amino acid sequence in a protein; *see also* operator gene, promoter gene

sub- a prefix meaning under or below (e.g. a suborder is a level of classification below order)

subarachnoid space a space between two of the layers of MENINGES; contains CEREBROSPINAL FLUID

subclavian artery an artery that supplies blood to an arm; *see also* diagram of major arteries (p. 20)

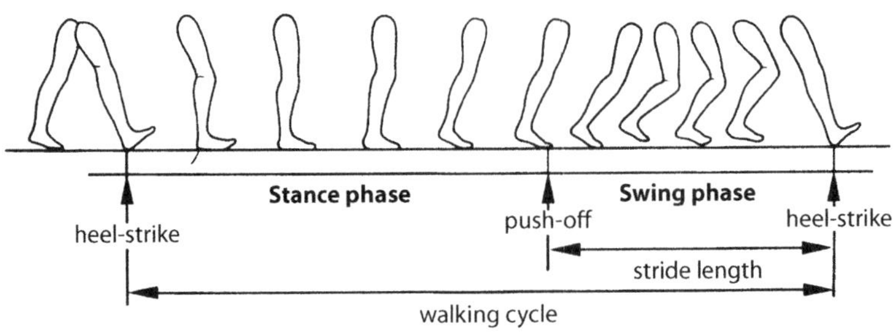

Striding gait

subclavian vein a vein that carries blood away from an arm; *see also* diagram of major veins (p. 284)

subculture in biology, a subculture is prepared by transferring some organisms from an existing CULTURE (definition 2) into a new growth medium

subcutaneous layer a layer of tissue that attaches the skin to underlying structures; *see* diagram of skin structure (p. 250)

sublingual glands one of the three pairs of SALIVARY GLANDS

submandibular glands one of the three pairs of SALIVARY GLANDS

submucosa a layer of connective tissue that is located below a mucous membrane, e.g. below the mucous membrane (mucosa) that lines the alimentary canal

suborder a category of classification between order and family; a major subdivision of an order (e.g. most biologists now divide the order Primates into the suborders Strepsirrhini and Haplorrhini; humans, apes, Old World monkeys and tarsiers are members of the suborder Haplorrhini)

subphylum a category of classification between phylum and class; a major subdivision of a phylum; humans, and all other animals with backbones, belong to the subphylum Vertebrata

subspecies a level of classification below species; a group of organisms that shows differences from other populations but is still considered to be part of the same species (e.g. the major human races may be considered as subspecies of *Homo sapiens*)

substrate a substance with which an enzyme reacts during a chemical change; the substrate combines with the enzyme at the enzyme's active site to form an enzyme–substrate complex; the substrate is then changed to become the product and the enzyme is released intact; *see also* enzyme; diagram of Lock and key model of enzyme action (p. 154)

substrate–enzyme complex an alternative term for ENZYME–SUBSTRATE COMPLEX

subthreshold stimulus a stimulus that is not strong enough to cause a response; *see* threshold

subunit vaccine a vaccine that contains only a fragment of the disease-causing micro-organism (e.g. hepatitis B vaccine)

succus entericus a general name for all the digestive juices secreted by the lining of the small intestine

sucrose $C_{12}H_{22}O_{11}$ a disaccharide CARBOHYDRATE consisting of a molecule of glucose and a molecule of fructose chemically combined; may be called *cane sugar* when obtained from sugar cane or *beet sugar* when obtained from sugar beet; also called *table sugar*

sudden infant death syndrome (SIDS) the sudden and unexplained death of a baby that appeared to be normal and healthy; often referred to as 'cot death'

sudoriferous gland an alternative name for SWEAT GLAND

sulcus (plural *sulci*) a groove, such as the posterior median sulcus of the spinal cord; the term is particularly used to describe the shallow grooves between the convolutions (folds) of the cerebrum of the brain; *see* diagrams, next page

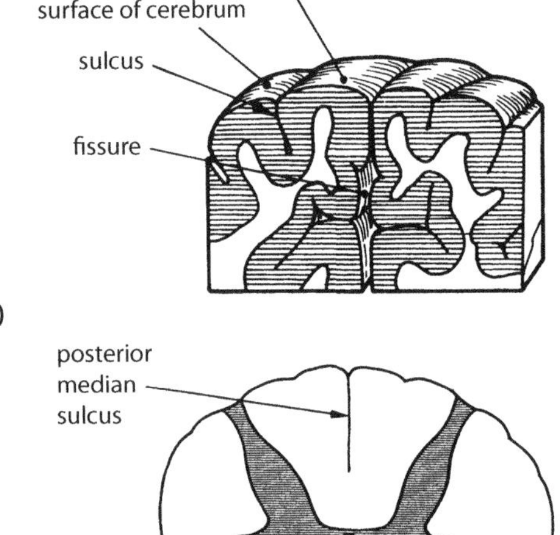

Sulcus of (a) the cerebrum; (b) the spinal cord

summation an accumulation of effects; especially used in relation to nervous or muscular activity (e.g. a nervous response may be brought about by the summation of a number of nerve impulses)

sunstroke an alternative name for HEATSTROKE

super- a prefix meaning above (e.g. superfamily—a level of classification above family)

superfamily a category of classification above the level of family but below the order level; humans, along with the apes, are members of the superfamily Hominoidea

superior describes that which is higher, or above another, in place or position (e.g. superior vena cava); *see also* inferior, diagram (p. 13)

superior vena cava one of the two veins returning blood to the heart after circulation through the body; *see* vena cava

supernatant the liquid that floats above a sediment or precipitate

superposition describes the upper layers in a sequence of rock strata, which are normally younger than those below them; when sedimentary rocks are formed, younger sediments are deposited on older sediments; *see also* principle of superposition, stratigraphy

supination the term used to describe the outward rotation of the forearm so that the palm faces upward or forward; *see also* diagram showing movements at joints (p. 1)

suppository a solid capsule that is inserted into the rectum, vagina or urethra to deliver drugs

suppressor T cell a type of T lymphocyte; *see* T cells

supraorbital torus an alternative name for BROW RIDGE

suprarenal glands an alternative name for ADRENAL GLANDS

surface area to volume ratio the relationship between the surface area of an organism, or a cell, and its volume; for an organism, this ratio is important in heat retention or heat loss (e.g. the Inuit (Eskimos), with their short, stocky build, have a low surface area to volume ratio in order to reduce heat loss, whereas Sudanese people from the Sahara Desert are very tall and slim, with a high surface area to volume ratio to promote heat loss); for a cell, a large ratio is important in allowing materials to pass across the cell membrane

surfactant a phospholipid secreted by special cells in the alveoli of the lungs; forms a thin covering inside the alveoli and reduces surface tension so that when the lungs deflate the sides of the alveoli do not stick together; the amount of surfactant in the lungs of a foetus is not sufficient to prevent collapse of the alveoli until about the 30th week of pregnancy, so that premature babies are likely to experience breathing difficulties; where there is insufficient surfactant to prevent lung collapse in infants, the condition is called *respiratory distress syndrome* or *hyaline membrane disease*

surrogacy when a woman agrees to bear a child for a couple because the female partner is unable to become pregnant; such a woman is then known as a *surrogate mother*

survey collecting information from a group of people using interviews or QUESTIONNAIRES

suspensory climbing and hanging the ability to raise the arms above the head and hang on to branches, and to climb in this position; the apes are suspensory climbers and hangers

suspensory ligament 1. of the lens of the eye, one of a number of ligaments that hold the lens in place; *see* diagram of eye (p. 94) **2.** of the breast, one of the strands of connective tissue between the lobules of the breast that supports the breast **3.** of the ovary, a ligament that helps to hold the ovary in place; extends from the ovary to the abdominal wall

suture bones small additional bones found in the cranium; occur more commonly in some populations than in others; *see also* diagram of suture bones, top right

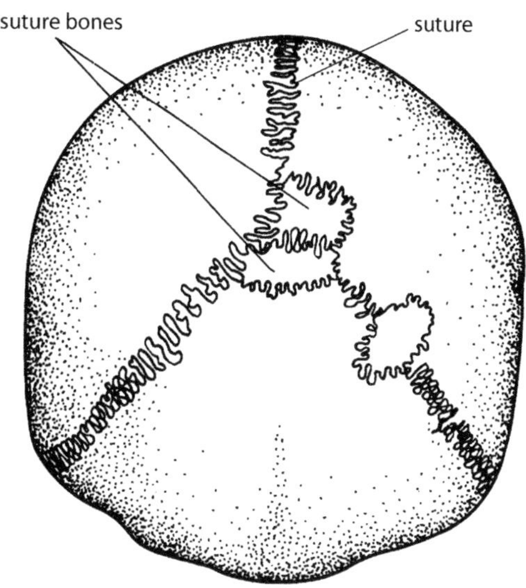

Suture bones

sutures 1. immovable joints that hold the bones of the skull together; between the bones is a thin layer of dense connective tissue; the irregular nature of the sutures increases their strength; *see also* joint, diagram of the sutures of the skull, above **2.** fine threads (stitches) used to join tissue during surgery

sweat the liquid produced by the sweat glands in the skin; consists of water with dissolved salts (mainly sodium chloride), uric acid, urea, lactic acid, amino acids, ascorbic acid and ammonia; main function is to cool the body by evaporation from the skin; also helps to excrete wastes; also called *perspiration*

sweat glands glands in the skin that produce sweat; also called *sudoriferous glands*; there are two types—*apocrine sweat glands* occur mainly in the armpits, the pubic region and around the nipple of the breast; they do not begin to function until puberty; *eccrine sweat glands* are distributed in most parts of the skin; they function throughout life and the sweat produced is more watery than that of apocrine sweat glands

symbiosis an association between two organisms, of different species, in which at least one of the organisms derives benefit from the association; forms of symbiosis include *commensalism*, where one species benefits and the other is unaffected, *mutualism*, where both species derive benefit, and *parasitism*, where one species benefits and the other is harmed; *see also* parasite

sympathetic division one of the two divisions of the autonomic nervous system; *see* nervous system; also table of effects (p. 186)

symphysis a joint in which the connecting material between the two bones is a broad, flat disc of fibrocartilage (e.g. between the body of the vertebrae); *see also* symphysis pubis

symphysis pubis the cartilaginous joint that joins the two hip bones at the front of the body; also called the *pubic symphysis*; *see also* diagram of symphysis pubis, below

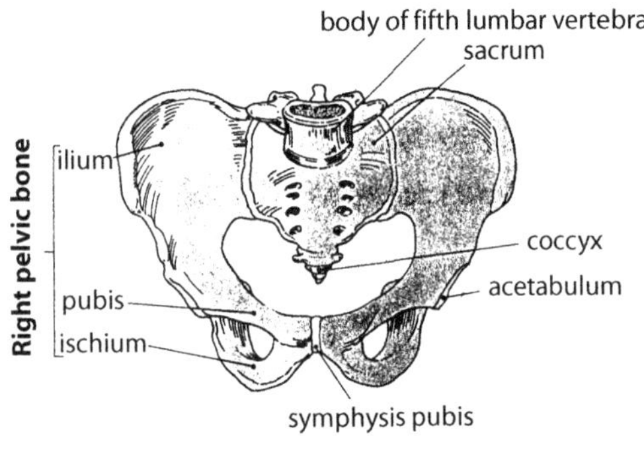

Pelvic girdle showing the symphysis pubis

symptom a change from the normal functioning of the body that indicates the presence of a disease or disorder; *see also* sign

symptothermal method used in birth control; a method of determining the time of OVULATION using both body temperature and changes in the mucus of the CERVIX

synapse the junction between the branches of adjacent nerve cells; in most cases actually a small gap between the branches of adjacent nerve cells, the nerve impulse being carried across the gap by molecules known as NEUROTRANSMITTERS; the gap between the branches of one nerve cell and the next, or between a nerve cell and a muscle fibre, is called a *synaptic cleft*; the expanded ends of the branches of a nerve cell are *synaptic knobs* (or *synaptic end bulbs*); *see also* diagram of synapse opposite

synapsis the pairing of like (homologous) chromosomes that occurs during prophase of the 1st division of meiosis

synaptic cleft the gap between adjacent nerve cells or between a nerve cell and a muscle fibre; *see* synapse

synaptic end bulb an alternative name for *synaptic knob*; *see* synapse

synaptic knob the expanded end of a branch of a nerve cell; *see* synapse

syndrome a characteristic group of signs and symptoms that occur together to indicate a particular disease or abnormal condition

synergist a muscle that acts indirectly in steadying a joint during a particular movement; the main muscles involved in a desired movement may also cause undesired actions, such as rotating a bone when a joint is bent—synergists would oppose the undesired rotation; also called *synergistic muscles*; *fixators* are synergists that act to immobilise a joint or a bone

synergistic effect **1.** in muscles, where two muscles working together produce a much stronger force than either on its own **2.** when two or more drugs used together have a more powerful effect than each individual drug

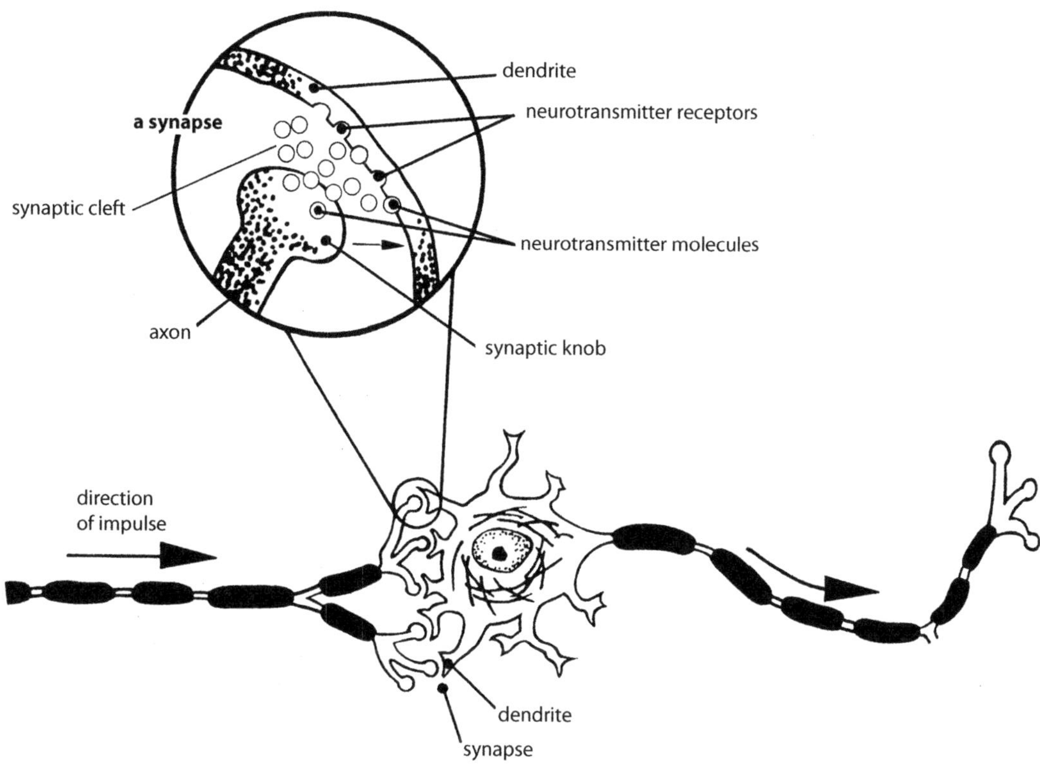

Transmission of nerve impulses across a synapse

synonym 1. a word or phrase that means essentially the same as another **2.** in biology, two or more names for the same classification group (e.g. *Pithecanthropus erectus* and *Homo erectus*)

synovial cavity the space between articulating bones in a SYNOVIAL JOINT

synovial fluid the fluid that fills the cavity of a synovial joint; secreted by the synovial membrane; lubricates the joint and provides nourishment for the cells of the cartilage that covers the bones (articular cartilage); *see also* diagram of structure of synovial joint, next page

synovial joint a JOINT in which there is a space between the adjacent bones called the *joint* (or *synovial*) *cavity*; a capsule surrounds the joint and encloses the cavity, which is filled with synovial fluid; the joint is freely movable, the amount of movement possible being limited by ligaments, muscles, tendons and adjoining bones; *see* diagram of structure of synovial joint, next page; in *gliding joints*, the surfaces of the bones are usually flat, allowing a side-to-side or back-and-forth movement, e.g. the joints between the carpal bones of the hand, the tarsal bones of the foot, the breastbone and collarbone (sternum and clavicle), the shoulder blade and collarbone (scapula and clavicle); *hinge joints* allow movement in one plane only, such as at the knee and elbow; formed when the convex surface of one bone fits into the concave surface of another; *pivot joints* allow rotation about the longitudinal axis of a bone, such as the movement of the head when the first vertebra rotates about the second; *saddle joints* have articulating surfaces that are saddle-shaped (convex in one direction and concave in the other); the bones fit together to allow both sideways and back-and-forth movement, such as where the thumb joins the palm of the hand; *ball-and-socket joints* are the most movable of all synovial joints; formed when the spherical head of one bone fits into a cup-like cavity of another (e.g. the shoulder and hip joints)

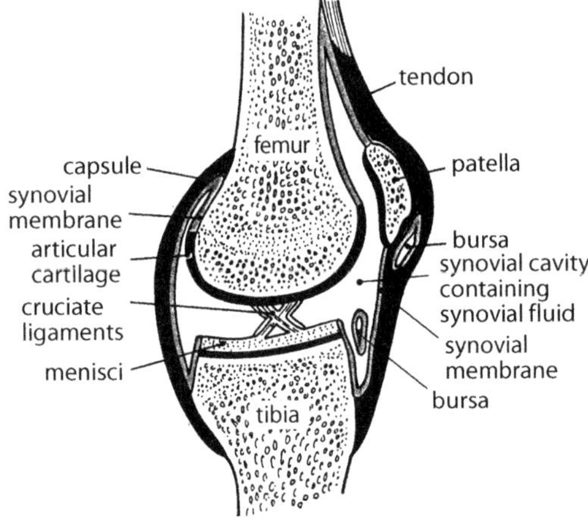

Structure of the knee joint

synovial membrane the inner layer of the capsule around a synovial joint; secretes synovial fluid; composed of loose connective tissue with elastic fibres and some fat storage tissue

synthesis in biochemistry, the bonding together of small molecules to make larger ones (e.g. joining glucose molecules to make glycogen or joining amino acids to make protein); also called *anabolism*

syphilis one of the sexually transmitted diseases; caused by the bacterium *Treponema pallidum*, which is spread by sexual contact; often referred to as 'the pox'

system a group of associated organs that work together for a common purpose (e.g. circulatory system, digestive system, endocrine system, excretory system, immune system, muscular system, nervous system, respiratory system, skeletal system)

systematic errors errors in measurement such that the values measured are consistently too high or too low; could be caused by faulty measuring equipment or by faulty experimental design

systemic describes something that affects the whole body (e.g. systemic vascular disease)

systemic circulation the flow of blood from the left ventricle, through the capillaries of the body and back to the right atrium; the blood flow through the whole of the body except the lungs; *see also* pulmonary circulation, diagram of double circulation (p. 80)

systole that part of the heart cycle when the heart muscle is contracting; *see* cardiac cycle

systolic blood pressure the pressure of blood on the arterial walls while the ventricles are contracting; *see* blood pressure

T

T cells LYMPHOCYTES responsible for cellular immunity; derived from cells in the bone marrow that migrate to the thymus, where they become T cells and then become embedded in lymphoid tissue; there are thousands of types of T cells, each able to respond to a specific antigen; exposure to an antigen makes the particular responding type of T cell differentiate into a killer T cell, helper T cell, suppressor T cell or memory T cell, each of which is involved in cell-mediated immunity; *killer T cells* secrete substances that destroy antigens, enhance the activity of macrophages and inhibit replication of viruses; *helper T cells* enhance antibody production by B cells; *suppressor T cells* help to 'turn off' the immune response after an infection has been controlled; *memory T cells* retain the ability to recognise the original invading antigen so that a subsequent invasion can be dealt with so rapidly that symptoms of the disease may not occur

T lymphocyte an alternative name for T CELL

table the orderly arrangement of DATA, usually in columns and rows, in a rectangular format

talus one of the tarsal (ankle) bones; the only ankle bone to articulate with the lower leg bones; receives the entire weight of the body, which it transmits to the other ankle bones; *see also* diagram of skeleton (p. 249)

tapeworm a ribbon-like parasitic worm that infects the intestinal tract of humans and other verebrates; the body is divided into many egg-producing segments; eggs pass out with the FAECES; *see also* platyhelminths

Taq polymerase a heat-stable DNA POLYMERASE originating from the heat-loving BACTERIUM *Thermus aquaticus*

target cells cells whose activity is affected by a particular hormone; *see* target organ

target organ an organ whose activity is affected by a particular hormone; contains target cells with receptors that must combine with the hormone molecule before the hormone will produce its effect (e.g. the target organ for the hormone glucagon is the liver)

tarsals the 7 bones that make up the ankle and heel; the largest is the calcaneus (heel bone); collectively these bones are called the *tarsus*; *see also* diagram of skeleton (p. 249)

tarsier a nocturnal primate found today in Indonesia; frequently referred to as a prosimian but has several features that are similar to those of monkeys and apes; unlike other prosimians, tarsiers do not have a moist nose, the orbit of the eye is completely closed with bone at the back, and the skull is more rounded and the snout less projecting; traditionally classified in the superfamily Tarsiioidea, although more recent taxonomic groupings place the tarsiers in an infraorder of their own; *see* diagram of simplified classification of living primates (p. 219)

tarsus the bones of the ankle and heel; *see also* tarsals

taste buds groups of taste receptors (gustatory cells) on the tongue, the soft palate and in the throat; there are about 2000 of them, mostly on the tongue; *see also* gustatory cells, diagram (p. 117)

taste pore an opening in a taste bud; *see* gustatory cells

taxon a level in the hierarchical system of classification of living things; phylum, class, order, family, genus, species and subspecies are taxons

taxonomic group a category of classification based on shared characteristics; the groups are kingdom, phylum, class, order, family, genus and species; species is the group with the greatest number of shared characteristics

taxonomy the study of the classification and scientific naming of living things

Tay-Sachs disease (TSD) a genetic disorder caused by a missing ENZYME that results in fatty substances accumulating in the nervous system; results in disability and death; occurs most frequently in individuals of Jewish descent from Eastern Europe (the so-called *Ashkenazi* Jewish population)

TCA cycle an abbreviation for tricarboxylic acid cycle; *see* Krebs cycle

tear gland a gland that secretes tears onto the surface of the conjunctiva of the eye; there is one above each eye, about the size and shape of an almond; 6–12 ducts carry the tears to the inside of the upper eyelid; also called the *lachrymal* (or *lacrimal*) *gland*

tears the watery secretions of the tear glands, which moisten, lubricate and clean the eyeball; contain mostly water with some mucus and an enzyme, LYSOZYME, that kills bacteria

tectorial membrane a membrane in the ORGAN OF CORTI in the inner ear

teeth (singular *tooth*) structures found in the mouth and used for biting and chewing food; in some primates also used for defence and grooming; types of teeth are incisors, canines, premolars and molars; *incisors* are chisel-shaped teeth specialised for cutting and biting; the front teeth; humans have 4 incisors at the front of each jaw; *canines* are sharp, pointed teeth located between the incisors and premolars; humans have 2 canines in each jaw; mammals normally use these teeth for puncturing and defence, but humans have small canine teeth that function like incisors; *premolars* (or *bicuspids*) are teeth with two cusps, found between the canines and molars; humans have 4 premolars in each jaw; *molars* are teeth with broad crowns and rounded cusps for crushing or grinding food; humans have 3 molars on each side of each jaw; the first set of teeth are the *deciduous teeth* (or *milk teeth*), which are shed and replaced by permanent teeth; the full set of deciduous teeth consists of 4 incisors, 2 canines and 4 premolars in each jaw; all deciduous teeth are lost by about 12 years of age and replaced by *permanent teeth*; in humans these consist of 4 incisors, 2 canines, 4 premolars and 6 molars in each jaw; the last pair of molars (third molars, commonly called the *wisdom teeth*) are the last to erupt, appearing at about 18–20 years of age; *see* diagrams of teeth, opposite

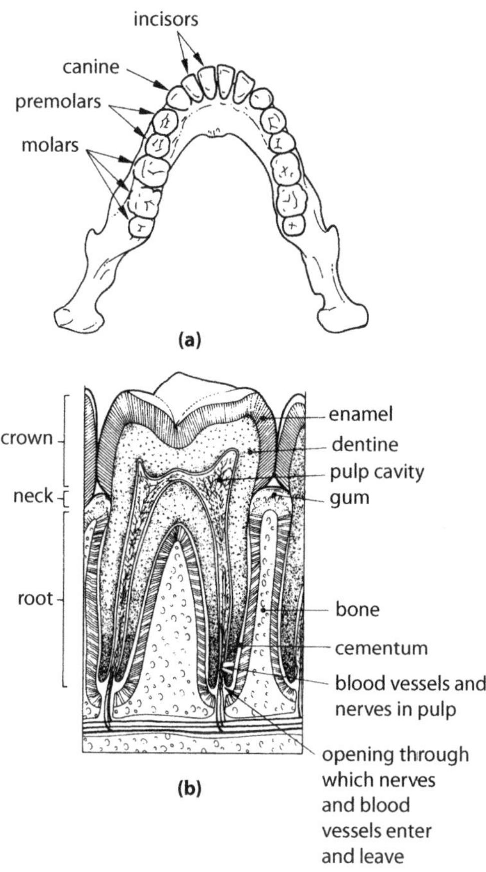

(a) Permanent teeth in the lower jaw; (b) tooth structure

telencephalon one of five vesicles that form the brain by the 5th week of embryonic development; in the adult the telencephalon becomes the CEREBRUM; also called *end brain*

telomeres the sections of DNA occurring at either end of a CHROMOSOME; involved in the replication and stability of the chromosome; act as caps to stop the STICKY ENDS of chromosomes from randomly clumping together

telophase the final phase of MITOSIS, and of each of the two divisions of MEIOSIS

template in general terms, a pattern from which exact copies can be made; in biology, a molecule that serves as a mould or pattern for making identical molecules (e.g. in replication of DNA, the two strands of the DNA molecule separate and each serves as a template for the construction of new strands)

template strand one of the strands of DNA (the other is called the CODING STRAND); the template strand is used to form the sequence of bases in messenger RNA (*see* ribonucleic acid)

temporal distribution the period of time over which a species existed on the earth; *see also* distribution

temporal lobe one of the four lobes of each cerebral hemisphere; contains centres for speech, hearing and smell; *see* diagram of lobes of cerebrum (p. 105)

tendinitis the inflammation of the tendon sheath surrounding certain joints; may lead to the affected sheath becoming swollen and result in severe pain when movement occurs at the affected joint; also called *tenosynovitis*

tendon the fibrous connective tissue that attaches a muscle to a bone; the fibres of the tissue are arranged parallel to one another, giving great strength

tenosynovitis an alternative name for TENDONITIS

teratogen an agent or factor that causes physical defects in a developing foetus (e.g certain medicinal drugs, marijuana, LSD); also known as *teratogenic agents*; *see also* thalidomide

teratogenic agent an agent that causes physical defects in a developing foetus; *see* teratogen

termination codon a sequence of three bases in a messenger RNA molecule that does not code for an amino acid; during the translation phase of PROTEIN SYNTHESIS the termination codon stops the synthesis of the polypeptide chain; also called a *stop codon* or a *nonsense codon*

terrestrial **1.** applied to organisms in general, living on land rather than in water **2.** applied to primates, living on the ground rather than in trees

territoriality a form of behaviour in which an individual defends its living or feeding space against any intruders; *see* territory

territory the specific area or habitat occupied by an individual, a pair of individuals or a group belonging to the same species; the area is defended by the resident members of the species against intrusion by non-resident members of the same species; if the area is not defended, it is called a *home range*

Tertiary the earlier of two periods in the Cainozoic era of geological time; 65–1.8 million years ago; the period during which the origin and evolution of the primates took place; in some classifications this period is divided into two, the Neogene and Palaeogene, with the Neogene period containing the OLIGOCENE, MIOCENE and PLIOCENE epochs, and the Palaeogene period containing the PALAEOCENE and EOCENE epochs; *see* geological time scale (p. 111)

tertiary structure the three-dimensional structure of a protein molecule or nucleic acid; largely determined by the sequence of AMINO ACIDS of which it is composed (the PRIMARY STRUCTURE)

test cross the mating of an individual of unknown genotype for a characteristic with an individual homozygous recessive for that characteristic; the ratio of phenotypes in the offspring of the test cross indicates the genotype of the unknown individual

test-tube baby a baby born as a result of IN-VITRO FERTILISATION

Testape a paper tape impregnated with chemicals that change from yellow to green in the presence of glucose; used to test samples of food, body fluids (such as urine) or other material for presence of glucose; test material must be in liquid form

testis (plural *testes*) the male sex organ; produces sperm and male hormones (androgens) such as testosterone; inside each testis are tightly coiled ducts, the *seminiferous tubules*, where sperm are produced; *see also* gametogenesis, diagram of male reproductive system (p. 235)

testosterone the male sex hormone secreted by endocrine cells within a mature testis; causes development and maintenance of male sex organs such as the penis, scrotum, sperm duct and epididymis; stimulates sperm production and development of male secondary sexual characteristics

tetanus **1.** a prolonged contraction of a muscle; produced by a series of rapid stimuli so that muscle fibres are stimulated to contract before they have relaxed from the previous contraction **2.** a disease caused by toxins from the bacterium *Clostridium tetani*; the bacterium usually enters the body through a deep wound; results in muscle spasms; treated by giving tetanus antiserum; tetanus is uncommon in Australia because most children are immunised in infancy, although booster injections must be given at regular intervals

tetrad a general term meaning a group of four; specifically, **1.** the 4 HAPLOID CELLS that are produced as a result of MEIOSIS **2.** the group of four CHROMATIDS that occurs when two HOMOLOGOUS CHROMOSOMES pair off during the first phase of meiosis; while they are in this arrangement, CROSSING OVER can occur between chromatids

thalamus the two masses of grey matter at the base of the brain; the principal relay station for sensory messages to the cerebral cortex; also interprets some sensory impulses (such as temperature, touch, pressure and pain); *see also* diagram of brain (p. 35)

thalassaemia an inherited form of ANAEMIA

thalidomide a medicinal drug that was once freely available before it was found to cause limb malformations in the developing foetus of a mother who used the drug; a TERATOGEN frequently resulting in the child having severely malformed arms and/or legs

theory a scientific theory is an idea that is generally accepted because it has a lot of evidence to support it; a HYPOTHESIS becomes a theory after it has been successfully tested many times by experiments or observations (e.g. the theory of evolution through natural selection, the atomic theory)

therapeutic cloning *see* cloning

thermography a procedure in which a temperature-sensitive photograph is used to detect areas of slightly raised temperature; was used in the diagnosis of breast cancer but has been superseded by other methods

thermoluminescence a method of absolute dating of fossils and artefacts; natural radiation in soil excites electrons in the crystalline structure of buried objects; the longer an object remains buried, the greater is the accumulation of excited electrons; the number of excited electrons is measured by heating the material, causing those electrons to revert to their normal state and give out light; the amount of light emitted is a measure of how long excited electrons have been accumulating; *see also* electron spin resonance

thermoreceptors receptors sensitive to temperature change; located in the skin and hypothalamus

thermoregulation the regulation of body temperature; the balance of heat gain and heat loss in order to maintain a constant internal body temperature, independent from the environmental temperature; *see also* homeostasis

thiamine one of the B complex of vitamins; water soluble; rapidly destroyed by heat and not stored in the body; involved in metabolism of carbohydrates and many amino acids; good food sources are whole-grain cereals, green vegetables, milk, meat, seafood and poultry; deficiency leads to extreme lethargy and beri-beri; also called *vitamin* B_1

Third World country a term that was once used to refer to countries that were economically underdeveloped or developing; *see* developing country

thirst centre located in the HYPOTHALAMUS and responsive to a number of different signs of DEHYDRATION, including a drop in blood pressure, secretion of ANTIDIURETIC HORMONE, and input from OSMORECEPTORS

thoracic cavity the chest cavity; the cavity between the neck and the diaphragm; contains the lungs and the heart; *see also* diagram of respiratory system (p. 236)

thoracic duct a vessel that drains lymph from the left side of the head, neck and chest, the left arm and the whole body below the chest; empties into the left subclavian vein, a vein at the base of the neck; also called the *left lymphatic duct*; *see also* diagram of lymphatic system (p. 156)

thoracic vertebrae the chest vertebrae; *see* vertebral column

thorax the chest; contains the heart and lungs and is separated from the abdomen by the diaphragm

threadworms small threadlike worms that infest the human intestine and rectum; found particularly in children; includes the pinworm *Enterobius vermicularis*

3TC an alternative name for the drug EPIVIR, used in treating HIV

threonine one of the 20 amino acids that are common in proteins; essential in the human diet; *see also* list of amino acids (p. 11)

threshold the lower limit of stimulus strength that results in a response; below the threshold the stimulus is too weak to cause a response in the tissue being stimulated (e.g. a nerve, sense organ or muscle fibre); a *threshold stimulus* is a stimulus that is just strong enough to cause a nerve impulse in a nerve cell; a *subthreshold stimulus* is one that is not strong enough to cause a nerve impulse

threshold stimulus *see* threshold

thrombin an enzyme formed from prothrombin; *see* blood clotting

thrombocyte an alternative name for PLATELET

thromboplastin an enzyme involved in BLOOD CLOTTING

thrombosis the formation of a blood clot inside a blood vessel; the clot, known as a *thrombus*, may dissolve naturally but if not, tissues may be damaged because the oxygen supply is reduced; such a clot in a blood vessel supplying the heart muscle leads to a heart attack, a *coronary thrombosis*; in the brain it causes a stroke, a *cerebral thrombosis*

thrombus a blood clot formed inside a blood vessel; *see* thrombosis

thrush a CONTAGIOUS DISEASE caused by a yeast-like fungus, *Candida albicans*; commonly occurs as *oral thrush* in young children, characterised by small whitish eruptions on the mouth, throat, and tongue; may also infect the VAGINA, where white patches appear and the surrounding skin becomes red and sore

thymine one of the four bases that form part of the DNA molecule; *see* deoxyribonucleic acid

thymosins a group of hormones secreted by the THYMUS that stimulate the IMMUNE SYSTEM by helping T CELLS to mature

thymus lymphoid tissue near the base of the neck, behind the breastbone and between the lungs; reaches maximum size at puberty, after which much of it is replaced by connective tissue; plays a role in the immune mechanisms of the body by producing T cells and producing a number of enzymes that cause T cells to mature; *see also* diagram of thymus, opposite

thyroid cartilage a cartilage that forms part of the LARYNX; forms the 'Adam's apple' at the front of the neck; *see also* diagram of thyroid gland, opposite

thyroid gland an ENDOCRINE GLAND located in the neck, just below the LARYNX; has two lobes, one on either side of the TRACHEA; some of the hormones it produces are THYROXINE, TRIIODOTHYRONINE and CALCITONIN; embedded in the thyroid gland are the four PARATHYROID GLANDS, which function independently of the thyroid; *see also* hormones by name, table of hormones (p. 128), diagram of thyroid gland opposite

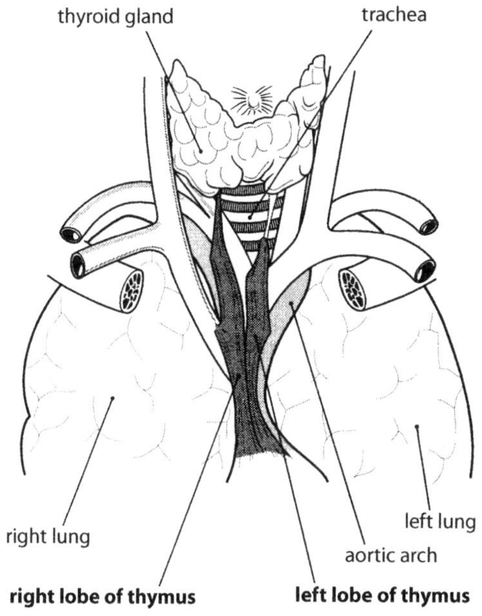

Thymus gland

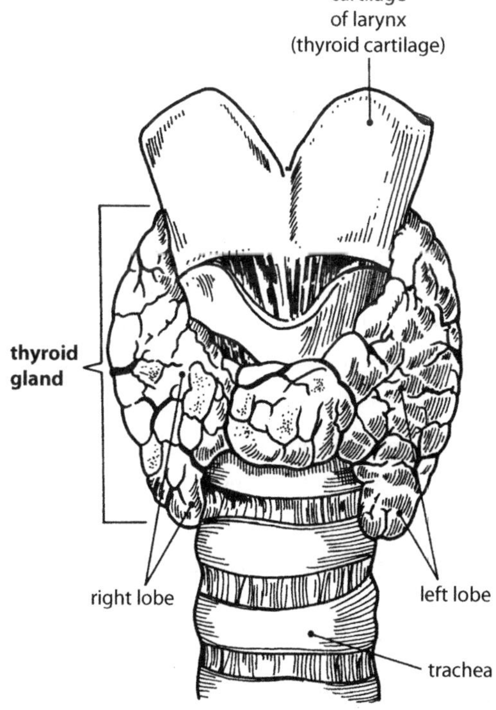

Thyroid gland

thyroid-stimulating hormone (TSH) a hormone secreted by the anterior lobe of the pituitary; stimulates production and release of hormones from the thyroid gland; secretion is controlled by a releasing factor from the hypothalamus; also called *thyrotropin* or *thyrotropic hormone*; *see* table of hormones (p. 128)

thyrotropin *see* thyroid-stimulating hormone

thyroxine a hormone secreted by the thyroid gland that regulates metabolism, growth and development; thyroxine molecules contain iodine; *see* table of hormones (p. 128)

tibia the shin bone; the larger of the two bones of the lower leg; articulates at the upper end with the femur (thigh bone) and at the lower end with the talus bone of the ankle; *see also* diagram of skeleton (p. 249)

tibia stress syndrome an alternative term for SHIN SPLINTS

tic an involuntary, spasmodic twitching of a muscle that is normally under voluntary control (e.g. twitching of muscles of the face such as the eyelid); usually caused by psychological factors

tick a small bloodsucking ECTOPARASITE that can transmit a variety of diseases from one animal to another; attaches to warm-blooded vertebrates to feed

tidal volume the volume of air breathed in and out in one breath; *see* lung volumes

tinea a fungal infection of the skin, often occurring between the toes; commonly called *athlete's foot*

tinnitus a sound such as ringing or buzzing that is heard in one or both ears or in the head, without an external cause; may be caused by an ear infection or blockage, certain drugs, head injury, and some diseases; commonly referred to as 'ringing in the ears'

tissue a large group of cells that are similar in structure and function; the four main types are EPITHELIAL TISSUE, CONNECTIVE TISSUE, MUSCULAR TISSUE and NERVOUS TISSUE; groups of tissues make up the organs of the body

tissue culture **1.** a technique for growing cells outside the body in a special nutrient solution **2.** the group of cells that is grown outside the body

tissue engineering the rebuilding of damaged tissue; uses the disciplines of biology, medicine and engineering to restore, maintain or enhance tissue and organ function; usually refers to methods of repairing or replacing structural tissues such as bone, cartilage, blood vessels and bladder tissue; sometimes referred to as *regenerative medicine*

tissue fluid fluid found in the spaces between the cells; makes up about one-third of all body fluid; acts as a connecting link between blood and cells; derived from the blood, and similar in composition to blood plasma, except that it does not contain as much protein because protein is not easily filtered through the capillary walls; when flowing through lymphatic vessels it is called *lymph*; also known as *interstitial fluid* or *intercellular fluid*

tissue hormone a HORMONE that acts on cells in the area where it is produced

tissue respiration an alternative term for CELLULAR RESPIRATION

tocopherol an alternative name for VITAMIN E

toilet claw a specialised claw (or NAIL) on the feet of certain primates that is used for personal grooming; the toilet claw of STREPSIRRHINES is on the second toe

tolerance a progressive reduction in a person's response to a drug; higher and higher doses of the drug are required to achieve the desired effect

tolerance limits the upper and lower limits between which a value must lie to allow normal functioning

tomography making images of a particular plane within the body; detail is shown in that plane while structures in other planes are blurred; if the images are produced by X-rays, the X-ray tube is rotated around the body; movement of the X-ray tube is co-ordinated by computer and the method is known as *computed tomography* (*CT scan*) or *computed axial tomography* (*CAT scan*); *see also* magnetic resonance imaging

tone partial contraction of skeletal muscles; *see* muscle tone

tonsils masses of lymphoid tissue in the mucous membrane of the throat; protect against invasion of micro-organisms and foreign substances; produce lymphocytes and antibodies; a pair of palatine tonsils is at the back of the throat and becomes enlarged when suffering a sore throat; a pair of lingual tonsils is at the base of the tongue; a single pharyngeal tonsil at the back of the throat is at peak development during childhood and when enlarged is called an *adenoid*

tonus partial contraction of skeletal muscles; *see* muscle tone

tool a device or implement, especially one held in the hand, that is used to carry out mechanical functions (e.g. a flake of stone used to cut the skin from an animal)

tool use model a hypothesis of hominin origins that stated that bipedalism, large brains and small canines all evolved simultaneously during hominid evolution as a consequence of increased reliance on tool use; now rejected because current evidence suggests that bipedalism predates tool use

tooth decay an alternative name for CARIES

topical preparation a DRUG applied to the surface of a particular area of the body or body part; includes ointments, creams, lotions, solutions, gels, oils, foams and powders

torn cartilage a torn articular disc (MENISCUS) in the knee; may occur when there is a sudden change in direction when the weight is on the limb; results in severe pain and swelling of the knee; *see* diagram of knee joint (p. 264)

torso the body excluding the head, neck and limbs; also called the *trunk*

totipotent stem cells *see* stem cells

toxaemia a condition in which the blood contains poisonous products; may be bacterial toxins, without the bacteria themselves present; *toxaemia of pregnancy* is a serious condition but the cause is not known; symptoms include puffy hands and feet due to accumulated fluid, raised blood pressure and protein excreted in the urine; if uncontrolled, toxaemia of pregnancy leads to convulsions or fits, a stage known as *eclampsia*

toxic shock syndrome (TSS) a serious and sometimes fatal disease; most commonly occurs in females who use tampons but can occur in males and children; caused by toxins that are produced by certain strains of the bacterium *Staphylococcus aureus*; infection by the bacteria can occur through cuts, burns, boils or other breaks in the skin; tampons do not cause the condition but for some reason they increase the risk of contracting it; symptoms include high temperature, sore throat, headache and muscular pains—symptoms very similar to influenza

toxin any poisonous substance produced by bacteria, or by other plants or animals; the toxin produced by the bacterium *Clostridium tetani* causes tetanus; diphtheria, and the type of food poisoning known as botulism, are caused by toxins; toxins are also produced by spiders, snakes, blue-ringed octopuses, sea wasps and some toadstools

toxoid an inactivated toxin; toxoids of tetanus and diphtheria are used for immunisation

toxoplasmosis an INFECTION caused by a single-celled PARASITE named *Toxoplasma gondii*; healthy adults usually do not suffer ill effects from infection, but infection of a FOETUS or a newborn baby usually results in damage to the brain and eyes, and can be fatal; individuals with a weakened IMMUNE SYSTEM are also susceptible to infection

trabeculae (singular *trabecula*) thin, criss-crossing, bony plates that make up spongy bone; in some bones, spaces between the trabeculae are filled with red bone marrow

tracer a substance, usually radioactive, that is taken into the body so that its distribution may be determined or so that the steps in complex chemical reactions can be followed; also known as a *tracer element*

trachea the windpipe; the air tube extending from the larynx to the bronchi; the wall is supported by incomplete rings of cartilage; it is lined with a mucous membrane; the mucus traps solid particles and beating of cilia on the lining cells moves mucus and particles up to the throat; *see also* diagram of respiratory system (p. 236)

tracheal cartilage bands of supporting cartilage within the wall of the trachea; there are about 16–20; each is shaped like a C with the opening of the C at the rear, facing the oesophagus; the oesophagus expands into the opening when swallowing

trachoma a serious and contagious disease of the eyes; affects the conjunctiva and the membranes lining the eyelids; reduces vision and eventually leads to blindness; caused by *Chlamydia trachomitis*, a bacterium with some of the characteristics of viruses

tract a general name for a pathway or channel (e.g. the digestive tract); when used on its own, the term usually refers to a bundle of nerve fibres in the central nervous system; *ascending tracts* are sensory nerve fibres that carry impulses towards the brain; *descending tracts* are motor nerve fibres that carry impulses away from the brain

trait an inherited feature of an organism; may also be called a *character* or *characteristic*; in some cases characteristic may mean the type of feature (e.g. eye colour) and trait may mean the form of the characteristic (e.g. blue eyes); pronounced 'tray'

tranquilliser a drug used to relieve anxiety; *major tranquillisers* are prescription drugs used for the treatment of severe mental illness; *minor tranquillisers* are used to treat anxiety and other less serious problems; minor tranquillisers belong to a group of chemicals known as benzodiazepines, which include Librium, Valium, Serepax and Mogadon; may be spelt *tranquilizer*

trans- a prefix meaning across or over (e.g. a transverse fracture—a break across a bone)

trans fat a type of FAT found in many processed foods that contributes to CARDIOVASCULAR DISEASE; created by adding hydrogen to UNSATURATED FATS; used by food manufacturers to improve the stability of vegetable oils or to convert oils into solid fats for use in foods like cakes and pastries; also produced naturally by cows and sheep, so beef, lamb and dairy products contain small amounts of trans fat; health authorities advise people to consume as little trans fat as possible

transamination the transfer of an amino group (NH_2) from one amino acid to form another; occurs in the liver; important in the formation of new amino acids from others that are in excess; only non-essential amino acids (those that do not have to be consumed in the diet) can be formed in this way

transcription the formation of messenger RNA from DNA in the nucleus by matching the sequence of nucleotides in the DNA; *see* protein synthesis

transfer RNA (tRNA) a small RNA molecule that transfers amino acids from the cytoplasm of a cell to the ribosomes, where they are joined to make proteins; each type of amino acid has a specific tRNA molecule; each tRNA has three bases that must complement a sequence of three bases on the messenger RNA (mRNA) molecule; by complementing the code on the mRNA molecules, the tRNA molecules assemble the amino acids in the correct sequence for the particular protein; *see also* protein synthesis

transfusion the transfer of blood, or some of the components of blood (such as red blood cells, plasma or serum), into the circulation of a person; also the transfer of bone marrow into the bloodstream

transgenic describes a cell or an organism that has had DNA from another SPECIES introduced into it artificially (i.e. by GENETIC ENGINEERING)

transient ischaemic attack (TIA) a sudden occurrence that produces symptoms similar to a STROKE that go away within a short period of time; also called a *mini-stroke*; caused by a temporary shortage of blood and oxygen in the brain, frequently due to a narrowing, or blockage, of the CAROTID ARTERIES; also spelt *transient ischemic attack*

translation an alternative name for PROTEIN SYNTHESIS

translocation **1.** in chromosomes, a type of mutation in which they break and exchange pieces **2.** part of the translation process of PROTEIN SYNTHESIS—the MRNA is shifted by one CODON in relation to the RIBOSOME; translocation is essential to the process of protein synthesis and occurs each time a new part of the polypeptide chain is formed, thus enabling it to increase in length

transmissible disease a disease passed from one person to another by infection with micro-organisms (e.g. chickenpox, influenza, AIDS); also called *infectious disease* or *communicable disease*

transmission electron microscope an ELECTRON MICROSCOPE in which the electrons pass through the specimen

transplantation the transferring of an organ or tissue from one part of the body to another (e.g. a skin graft) or from a donor to another person (e.g. a corneal graft or heart or kidney transplant)

transverse arch the arch of the foot running from side to side; *see* arches

transverse fracture a break across a bone; *see* fracture

transverse section a SECTION through the short axis of a tissue, organ or other structure

trapezius the broad, flat, triangular muscle at the rear of the neck and shoulder; joined at one end to the skull and the neck and chest (thoracic) vertebrae and at the other to the collarbone (clavicle) and the shoulder blade (scapula); because of its triangular shape, the movement brought about by the muscle depends on the actual fibres that contract; some movements include lifting the shoulders, squaring the shoulders and moving the head

trauma an injury, either a physical wound or a psychological disorder; caused by an external agent or force, such as a physical blow or an emotional shock

tree ring dating an alternative name for DENDROCHRONOLOGY

tri- a prefix meaning three (e.g. trisomy—having three of a particular chromosome)

triacylglycerol an alternative name for *triglyceride*; *see* fat

trial and error a method of solving problems based on practical EXPERIMENT and experience, by making repeated trials or tests and improving procedures until the correct result is discovered; experimenting until a solution is found; contrasts with an approach that uses insight and THEORY

Triassic the earliest of the three periods in the Mesozoic era of geological time; lasted from 250–205 million years ago; the first dinosaurs appeared during this time, along with the egg-laying mammals; *see* geological time scale (p. 111)

tribe **1.** a level of classification that occurs between subfamily and GENUS; usually contains several genera (e.g. the tribe HOMININI includes modern humans and our extinct ancestors) **2.** any group of people united by ties of descent from a common ancestor, common customs and traditions, or adherence to the same leaders; they commonly identify with one another, share a territory and work together at joint enterprises; in modern anthropology the term *ethnic group* is often preferred

tricarboxylic acid (TCA) cycle an alternative name for the KREBS CYCLE

triceps a common name for the muscle making up the back of the upper arm; full name triceps brachii; one end is attached to the shoulder blade (scapula) and the upper arm bone (humerus), the other to the ulna in the forearm; contracts to straighten the arm at the elbow; antagonistic to the biceps brachii; see diagrams of muscular system (pp. 178, 179)

trichomoniasis an infection caused by the protozoon *Trichomonas vaginalis*, resulting in inflammation of the mucous membranes of the vagina in females and the urethra in males; frequently transmitted during sexual intercourse

tricuspid valve an ATRIOVENTRICULAR VALVE on the right side of the heart, between right atrium and right ventricle

triiodothyronine a hormone secreted by the THYROID GLAND; contains iodine and is the most powerful of the thyroid hormones; affects many body processes, including the regulation of body temperature, growth and heart rate; used in treating HYPOTHYROIDISM

triglyceride a type of FAT with a molecule containing three fatty acids

triplet a sequence of three nucleotides in an RNA molecule that is the code for a particular amino acid; *see* protein synthesis

trisomy a condition in which an individual has three of a particular chromosome, instead of the normal two (e.g. in Down syndrome—trisomy-21—the affected individual has three of chromosome number 21); *partial trisomy* occurs when only part of an extra chromosome attaches to another

trisomy-21 a genetic disorder resulting from an extra copy of chromosome number 21; affected individuals have three of chromosome 21 instead of the normal two; characterised by short stature, flat facial profile, large tongue, broad skull, slanting eyes and mental retardation; more common in children born to older mothers; can be diagnosed during foetal development by procedures such as amniocentesis and chorionic villus sampling; formerly known as *Down syndrome*, *Down's syndrome* or *Mongolism*

trivial name the species name or the subspecies name in the binomial system of nomenclature

tRNA an abbreviation for TRANSFER RNA; *see* ribonucleic acid

true ribs the first 7 pairs of RIBS

trunk the body excluding the head, neck and limbs; also called the *torso*

trypsin an alternative name for PANCREATIC PROTEASE, a digestive enzyme that breaks down proteins

trypsinogen the inactive form of PANCREATIC PROTEASE

tryptophan one of the 20 amino acids that are common in proteins; essential in the human diet; *see also* list of amino acids (p. 11)

tubal embryo stage transfer (TEST) a form of IN-VITRO FERTILISATION

tubal ligation female sterilisation; the removal of a small piece of, or the clamping of, each uterine tube; prevents sperm reaching the egg to fertilise it and prevents the egg from reaching the uterus; *see also* vasectomy diagram (p. 283); and *see* diagram below

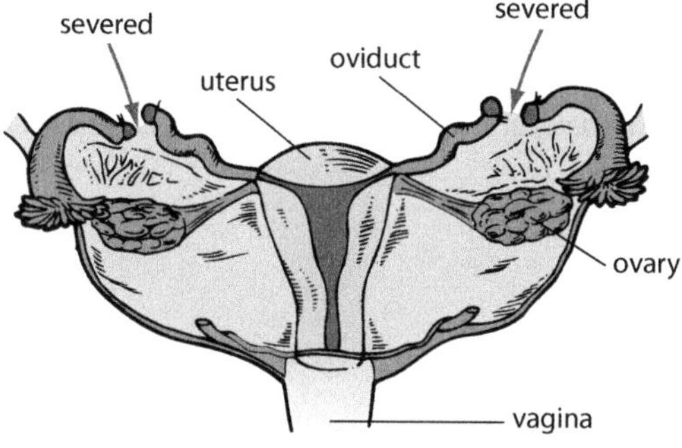

Tubal ligation

tuberculosis an inflammation of the lungs and pleurae caused by infection by the bacterium *Mycobacterium tuberculosis*; results in destruction of the lung tissue and its replacement by fibrous connective tissue; the bacterium may be carried in the blood to bone from an infection in the lungs, resulting in tuberculosis of the bone; tuberculosis of the lungs was once commonly referred to as *consumption*

tubular reabsorption REABSORPTION of substances from the tubules of the kidney nephrons

tubular secretion the process whereby ions and drugs are secreted from the blood into the kidney tubule in response to the body's specific needs

tumour an abnormal growth of tissue; *see* neoplasm

Turner's syndrome a condition caused by the presence of only one *X*-chromosome in the cells of an individual; such persons are female and tend to be short in stature, lack secondary sexual characteristics and are infertile; with appropriate medical treatment these women are able to lead normal lives

twins a pair of children who develop at the same time, in the same mother **1.** *fraternal twins* are produced as a result of two separate eggs being fertilised at the same time by two different sperm; such twins are *non-identical*, can be of different sexes and are no more alike than ordinary brothers and sisters; because they develop from two separate zygotes they are also known as *dizygotic twins* **2.** *identical twins* are produced when a single fertilised egg divides completely during early embryonic development, resulting in two separate individuals being formed; the two individuals are genetically identical and therefore always of the same sex; because they develop from one zygote they may be known as *monozygotic twins*, and because they develop from one egg they may be called *uniovular twins*

twitch **1.** the complete cycle of stimulus, contraction and relaxation in a muscle or a muscle cell **2.** a sudden involuntary or spasmodic muscle spasm (e.g. a twitch of the eye)

two-egg twins an alternative name for *fraternal twins*; *see* twins

tympanic membrane a thin layer of fibrous tissue between the auditory canal and the middle ear; sound waves striking the tympanic membrane cause it to vibrate and the vibration is passed via the bones in the middle ear to the liquid in the cochlea, where sensory cells are stimulated; also called the *eardrum*; *see also* diagram of ear (p. 82)

type 1 diabetes an alternative name for *insulin-dependent diabetes*; *see* diabetes mellitus

type 2 diabetes an alternative name for *adult-onset diabetes*; *see* diabetes mellitus

typhoid fever an acute disease resulting from infection by the bacterium *Salmonella typhi*; transmitted by water contaminated with faeces or by carriers directly handling food; causes high temperature, abdominal pain and skin spots

typhus a disease caused by one of the rickettsia bacteria; symptoms are high temperature, headache and skin rashes; transmitted by the bites of fleas and lice; intermediate hosts of the rickettsias are mice and rats

tyrosine one of the 20 amino acids that are common in proteins; not essential in the human diet; *see also* list of amino acids (p. 11)

U

ubiquitous proteins proteins that appear to be present in all species, from bacteria to humans; the small protein *ubiquitin* was so-named because it is present in all types of cells

ulcer an open sore on the surface of an organ or tissue due to inflamed tissue being shed many times; may occur when there is a prolonged inflammatory response due to tissue being continuously injured; most ulcers occur in the first part of the duodenum and are called *duodenal ulcers*—these are usually caused by infection with the bacterium *Helicobacter pylori*; *peptic ulcers* occur when acid from the stomach causes the mucous membrane lining to erode; a peptic ulcer in the stomach is called a *gastric ulcer*

ulna the lower arm bone that forms the point of the elbow; the larger of the two bones in the forearm; articulates with the humerus at the elbow and with the carpals at the wrist; *see also* diagram of skeleton (p. 249)

ultra- a prefix meaning beyond or excessive (e.g. ultracentrifuge—a centrifuge that spins at a speed beyond that of a normal centrifuge)

ultracentrifuge a very fast spinning CENTRIFUGE

ultrafiltration the process whereby particles pass through a semipermeable membrane under pressure or suction; in the kidney, ultrafiltration occurs when water and dissolved blood components are forced from the capillaries of the glomerulus into the glomerular capsule of the nephron to become the filtrate

ultrasound very high frequency, inaudible soundwaves (frequency greater than 20 000 hertz); can be directed into the body and the echoes they produce as they strike various tissues can be used to define the positions of organs; used in particular to produce an image of an unborn foetus; also used to break up kidney stones and in physiotherapy to treat inflammation

ultrastructure the structure of a cell that is too small to be seen with the light microscope (e.g. microtubules and ribosomes are part of the ultrastructure of a cell—they can only be seen with an electron microscope)

ultraviolet light an alternative term for ULTRA-VIOLET RADIATION

ultraviolet radiation (UV) radiation with a wavelength shorter than normal light; beyond the violet end of the visible spectrum; causes sunburn, tanning of the skin, skin cancers and cataracts of the eye; a certain amount is needed for production of vitamin D in the skin; also called *ultraviolet light*

umbilical arteries the two ARTERIES that pass through the UMBILICAL CORD and carry DEOXYGENATED BLOOD from the FOETUS to the PLACENTA

umbilical cord the cord that attaches the foetus to the placenta; contains two UMBILICAL ARTERIES and one UMBILICAL VEIN supported by a jelly-like connective tissue; after birth, with the cutting of the umbilical cord, blood no longer flows through the umbilical vessels; the stump of the umbilical cord shrivels and drops off, leaving a small scar—the *umbilicus* (or navel)—on the abdomen, marking the cord's former attachment to the foetus; *see also* diagram of foetal circulation (p. 212)

umbilical vein the single VEIN that passes through the UMBILICAL CORD, carrying oxygen and nutrients from the PLACENTA to the FOETUS

umbilicus the scar on the abdomen that marks the former attachment of the UMBILICAL CORD; also called *navel* or 'belly button'

uncontrolled variable a factor not controlled in an experiment; *see* variable

underweight the condition of having less body fat than the optimum for good health; defined as a BODY MASS INDEX of less than 18.5

unipolar neuron a NEURON with a single extension, an AXON; the CELL BODY is to one side of the axon; in humans such neurons are SENSORY NEURONS and carry messages to the SPINAL CORD; *see* diagram (p. 187)

universal donor an individual of O blood group; *see* ABO blood group system

universal recipient (receiver) an individual of AB blood group; *see* ABO blood group system

unmyelinated fibre a nerve fibre (axon or dendron) that has no MYELIN SHEATH; such fibres are enclosed by Schwann cells but the cell is not wrapped around the fibre many times as it is with a myelinated fibre

unsaturated fat FAT made up of a high proportion of UNSATURATED FATTY ACIDS

unsaturated fatty acids FATTY ACIDS with molecules that are not completely saturated with hydrogen atoms

upper motor neuron a NEURON that starts in the motor cortex of the brain (*see* figure on page 174) and whose processes terminate within the MEDULLA OBLONGATA or SPINAL CORD; upper motor neurons relay information to the LOWER MOTOR NEURONS

uracil one of the four bases that form part of the RNA molecule; *see* ribonucleic acid

uranium-thorium dating an ABSOLUTE DATING technique that uses the properties of the radioactive HALF-LIFE of Uranium-238 and Thorium-230; used to determine the age of materials such as marine sediment, bone, wood, coral, stone and soil; uranium-thorium dating has an upper age limit of about 600 000 years

urea CON_2H_4; the main nitrogen-containing waste excreted in the urine of mammals; formed from AMMONIA in the liver as a result of the breakdown of proteins; soluble in water

ureter the tube that leaves each kidney and carries urine to the urinary bladder; *see also* diagram of urinary system, next page

urethra the tube that empties the bladder to the outside; in males it also carries sperm; *see also* diagram of urinary system, opposite

uric acid $C_5H_4O_3N_4$; the nitrogen-containing waste of insects, reptiles and birds; insoluble in water; found in small quantities in the urine of humans; the product of the breakdown of nucleic acids

urinary bladder a hollow, muscular organ near the base of the abdominal cavity; collects urine from the kidneys via the two ureters; urine is passed to the outside of the body through the urethra; has a capacity of about 700 mL, but when the volume of urine exceeds 200–400 mL, a reflex initiates the conscious desire to expel urine; commonly referred to as the *bladder*; *see* diagram of urinary system, below

urinary system the organs involved in the removal of metabolic waste, toxic materials and excess of essential materials from the blood; controls the composition and volume of the blood; consists of the kidneys, ureters, bladder and urethra; *see* diagram below

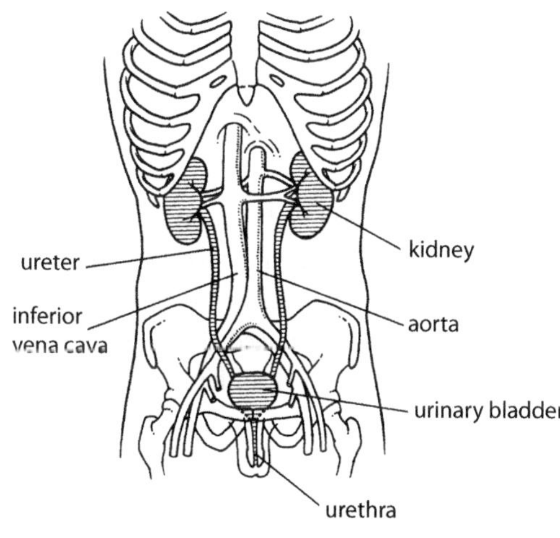

Urinary system

urination the act of expelling urine from the urinary bladder; also called *micturition*

urine the fluid produced by the kidneys that contains wastes or excess materials; passed out of the body through the urethra

urine analysis examining the URINE to determine the general health of the body and, in particular, kidney function; used to detect diseases or disorders of the kidneys or urinary tract and to monitor patients with DIABETES; usually includes measuring the urine's acidity (PH), and testing for the presence of blood, protein, sugar and KETONE BODIES; frequently referred to as *urinalysis*

urogenital system the organs making up the excretory and reproductive systems; also called the *urinogenital system* or *urogenital tract*

uterine tube the tube that carries eggs from the ovary to the uterus; lined with mucous membranes that have cells with cilia on the inside of the tube; movement of an egg along the tube is assisted by beating of the cilia and by contractions of muscles in the wall of the tube; also called *Fallopian tube* or *oviduct*; *see also* diagram of the female reproductive system (p. 235)

uterus a hollow, pear-shaped organ situated between the urinary bladder and the rectum of females; has a lining called the endometrium in which the embryo implants and develops if fertilisation occurs; if fertilisation does not occur, the surface layer of the endometrium breaks down and is shed at menstruation; also called the *womb*; *see also* diagram of female reproductive system (p. 235)

utricle the part of the membranous labyrinth of the inner ear into which the semicircular canals open; contains a receptor organ that gives information about the position of the head; also called the *utriculus*; *see also* saccule, diagram of structures of the inner ear (p. 140)

utriculus an alternative name for UTRICLE

uvula a soft, fleshy, V-shaped structure that hangs from the soft palate at the back of the throat

V

vaccination introducing a VACCINE into the body to provide IMMUNITY; commonly used as an alternative name for IMMUNISATION, however immunisation can also occur naturally

vaccine an antigen preparation used in immunisation; injected to cause an immune response and thus create active IMMUNITY; a vaccine may be a weakened (attenuated) form of a pathogen (e.g. the measles vaccine), dead bacteria (e.g. typhoid fever vaccine), a less virulent strain of the organism or an inactivated toxin (e.g. tetanus vaccine)

vacuole a membrane-bound cavity within the cytoplasm of a cell; vacuoles in animal cells are usually very small and are often called *vesicles*; *see also* diagram of cell structure (p. 44)

vagina the canal leading from the uterus to the exterior of the body; capable of considerable stretching to receive the penis during sexual intercourse or to allow passage of the foetus during childbirth; *see also* diagram of female reproductive system (p. 235)

validity **1.** the accuracy of a measurement; the degree to which an instrument, statistical technique, or test measures what it is supposed to measure; differs from RELIABILITY, which is the consistency of a measurement **2.** the strength of a conclusion, inference or proposition

valine one of the 20 amino acids that are common in proteins; essential in the human diet; *see also* list of amino acids (p. 11)

variable any factor that may change during an experiment; *controlled variables* are those that are kept the same for the control and experimental set-ups in an experiment; *uncontrolled variables* are those that could not be kept the same for the experimental and control set-ups, or those that the experimenter neglected to control; in an experiment the *independent variable* is the factor that is being investigated, the factor that is deliberately changed to determine its effect and is deliberately different between the control and the experimental set-ups—also called the *experimental variable* or the *manipulated variable*; the *dependent variable* in an experiment is the factor that changes in response to changes in the independent variable; it is also known as the *responding variable*

variation the differences that exist between individuals or populations; anthropologists study both cultural and biological variation; natural selection depends on biological variation because individuals with the most favourable variations have a better chance of survival and reproduction than those with unfavourable variations; that is, the favourable variations are selected

varicose vein an enlarged vein; may be due to an inherited weakness of the walls and valves of the veins; may be caused by pressure on the veins such as caused by pregnancy or prolonged standing; may also be a result of ageing; usually occurs in the legs, in veins near the surface

vas deferens (plural *vasa deferentia*) the tube that carries the sperm away from the testis; extends from the epididymis in the scrotum, upwards into the abdominal cavity, over the bladder and joins the urethra near the prostate gland; is able to store sperm for several months and has muscular walls that contract to propel the sperm towards the urethra during ejaculation; also called the *sperm duct*, *seminal duct* or *ductus deferens*; *see also* diagram of male reproductive system (p. 235)

vascular having many blood vessels or relating to blood vessels (e.g. a highly vascular tissue contains many blood vessels, vascular disease is disease of the blood vessels)

vascular system an alternative name for the CIRCULATORY SYSTEM

vasectomy male sterilisation; the removal of a small piece of each vas deferens; a minor operation performed through a small incision in the scrotum; following the operation semen lacks sperm but ejaculations are normal and sexual behaviour is not affected; sperm produced are reabsorbed in to the body; *see also* tubal ligation; diagram top right

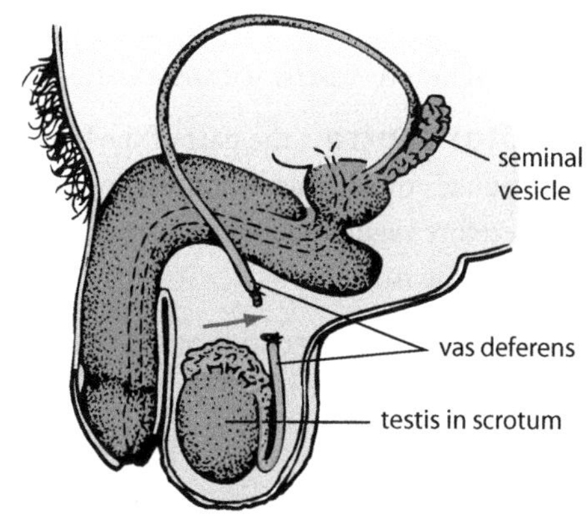

Vasectomy

vasoconstriction a decrease in the diameter of arteries and arterioles that reduces flow of blood through the capillaries supplied by those vessels; caused by contraction of smooth muscle in the wall of the vessels (e.g. cold causes vasoconstriction of blood vessels in the skin); *vasoconstrictors* are substances that produce a local constriction of blood vessels (e.g. adrenaline and noradrenaline cause vasoconstriction of arterioles in the skin and abdomen, but they also cause dilation of arterioles in the heart and skeletal muscles); *see also* vasodilation

vasoconstrictors substances that cause constriction of blood vessels; *see* vasoconstriction

vasodilation an increase in the diameter of arteries and arterioles that increases flow of blood through the capillaries supplied by those vessels; brought about by relaxation of smooth muscle in the wall of the vessels (e.g. hot conditions cause vasodilation of blood vessels in the skin); *vasodilators* are substances that produce a local widening, or dilation, of blood vessels (e.g. histamine and kinins produce vasodilation as part of the inflammatory response); *see also* vasoconstriction

vasodilators substances that cause dilation of blood vessels; *see* vasodilation

vasomotor centre the part of the brain that regulates the diameter of blood vessels and therefore regulates blood pressure; located in the medulla

vasopressin an alternative name for ANTIDIURETIC HORMONE

vector **1.** an agent, such as an insect, capable of transferring a disease-causing organism from one person to another (e.g. the *Anopheles* mosquito is the vector for the malarial parasite) **2.** in RECOMBINANT DNA technology, the term refers to a bacterial plasmid, a viral phage or another agent used to transfer genetic material from one cell to another

vegan a person who does not use or consume animal products of any kind, eating plant products only, and who uses no products derived from animals, such as silk, fur or leather

vein a blood vessel that carries blood towards the heart; has thinner and less muscular walls than an artery and may contain valves to prevent backflow of blood; *see* diagram of major veins, right

vena cava one of two large veins that open into the right atrium; carry blood from the systemic circulation into the heart; the *inferior vena cava* returns blood to the heart from the lower part of the body; the *superior vena cava* returns blood to the heart from parts of the body above the level of the heart; *see also* diagram of the heart (p. 121), diagram of major veins, right

venereal disease (VD) an alternative name for SEXUALLY TRANSMITTED DISEASE

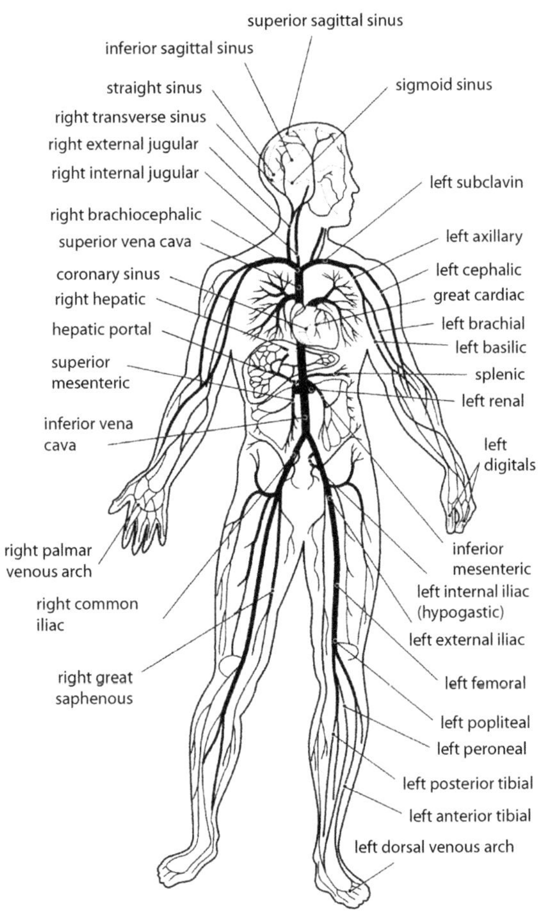

Major veins of the body

venous return the return of blood to the right ATRIUM of the heart via the SUPERIOR VENA CAVA and the INFERIOR VENA CAVA

ventilation moving air into and out of the lungs; breathing in and out

ventouse delivery the birth of a baby that is aided by an instrument consisting of a suction cap fitted to the head of the baby; pulling on the suction cap helps to move the baby through the birth canal

ventral towards the lower side (the belly) of an animal; in animals that stand erect, like humans, the front of the animal

ventral root one of the two roots that link a spinal nerve to the spinal cord; *see* root

ventricle **1.** one of the two pumping chambers of the heart; has a much thicker muscle wall than the atria (the receiving chambers); the right ventricle pumps blood to the lungs; the left ventricle pumps blood to the remainder of the body; *see also* diagram of heart (p. 121) **2.** one of the cavities in the brain; the ventricles of the brain are connected to one another and to the central canal of the spinal cord; they contain cerebrospinal fluid

ventricular folds a pair of membranes above the VOCAL FOLDS in the larynx

ventricular systole contraction of the ventricles of the heart; *see* cardiac cycle

venule a small vein; collects blood from capillaries and carries it to a vein; differs from a capillary in that it has connective tissue in the wall; *see also* arteriole

verification establishing or confirming the truth or accuracy of a fact, hypothesis or theory; provides additional evidence in support of something that was thought to be true

vermiform appendix an alternative term for APPENDIX

vernix waxy material covering a baby at birth

vertebra one of a series of 33 bones that make up the backbone; *see* vertebral column, diagram top right

vertebral canal the opening in the vertebra through which the spinal cord passes; made up of the *vertebral foramina* of all the vertebrae

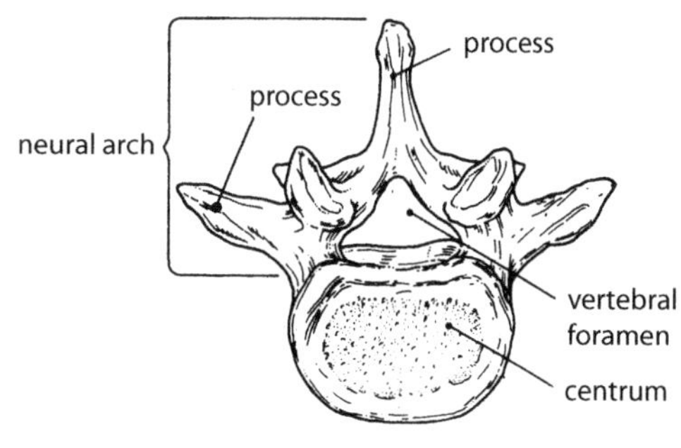

Vertebra

vertebral column the backbone; made up of 33 bones called *vertebrae* (singular *vertebra*), some of which are fused together; encloses and protects the spinal cord and provides points of attachment for the ribs and back muscles; also called the *spinal column* or *spine*; the 7 vertebrae in the neck region are the *cervical vertebrae*; there are 12 *thoracic vertebrae* in the chest region, each articulating with a pair of ribs; the 5 bones that support the lower back, the *lumbar vertebrae*, are the largest and strongest of the vertebrae; the 5 *sacral vertebrae* are fused together to form a single bone, the *sacrum*, that is joined to the hip bones of the pelvic girdle; the fused vertebrae at the end of the vertebral column form the *coccyx* (there are usually 4 in the coccyx but the number can vary from 3 to 5); *see* diagram of vertebral column next page, diagram of skeleton (p. 249)

vertebral foramen the opening in a single vertebra through which the spinal cord passes; the segment of bone that encloses the opening is called the NEURAL ARCH; collectively, all the vertebral foramina make up the vertebral canal; *see also* diagram of vertebra, left

Vertebrata a subphylum of the phylum Chordata; defined by the presence of an internal, segmented spinal column and bilateral symmetry; commonly referred to as *vertebrates*; those animals possessing a backbone; includes fish, amphibians, reptiles, birds and mammals

vesicle a small membrane-bound cavity in the cytoplasm of a cell; smaller than a vacuole; *see also* diagram of cell structure (p. 44)

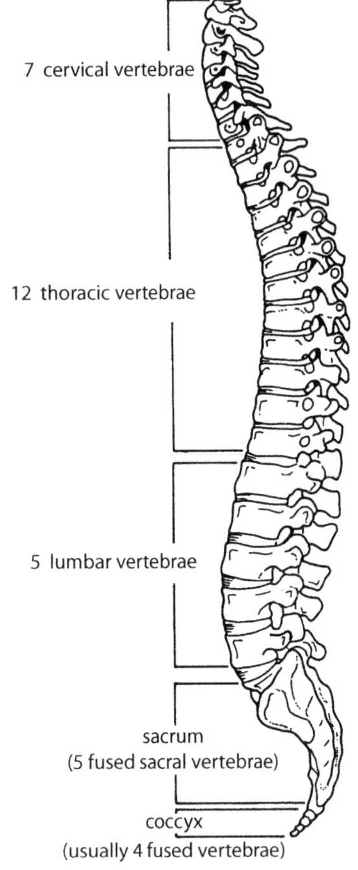

Vertebral column

vesicula seminalis an alternative name for SEMINAL VESICLES

vesicular transport the transport of materials in or out of a cell in membrane-bound sacs called VESICLES; also known as *bulk transport*

vestibule a general name for a cavity that forms an entrance to an organ; in particular **1.** the oval, central portion of the bony labyrinth of the inner ear; contains the utricle and the saccule; forms the entrance to the cochlea; *see also* diagram of structure of the inner ear (p. 140) **2.** the opening between the labia (in females); contains the openings to the vagina and the urethra

vestibulocochlear nerve the eighth cranial nerve; has two parts, the *cochlear* and *vestibular* nerves; both of these nerves relay sensory information from receptors in the inner ear to the brain; the cochlear nerve carries impulses related to hearing; the vestibular nerve carries impulses concerned with balance; *see* diagram of ear (p. 82)

vestigial organ a structure that, during the course of evolution, has been reduced in size or has lost its function (e.g. the ear muscles of humans that may once have been able to move the external part of the ear); vestigial organs bear a strong resemblance to structures in probable ancestors

vibrio a bacterial cell that is curved, often like a comma; *see* bacteria

villi (singular *villus*) **1.** projections from the internal lining of the small intestine; increase the surface area for absorption of digested food; each is covered by a single layer of epithelial cells that have microvilli extending from the surface to further increase surface area; contain a network of capillaries and a single lymph capillary (lacteal); *see also* diagram of villus, opposite **2.** finger-like projections of the chorion that grow into the lining of the uterus; *see* chorionic villi

viral load the number of viral particles in a sample of blood PLASMA; frequently used to determine the progression of a disease such as HIV

virion a mature VIRUS

virulence the disease-producing power of a micro-organism (e.g. bacteria with a capsule are more virulent because the capsule makes it easier for the bacterium to evade the body's defences)

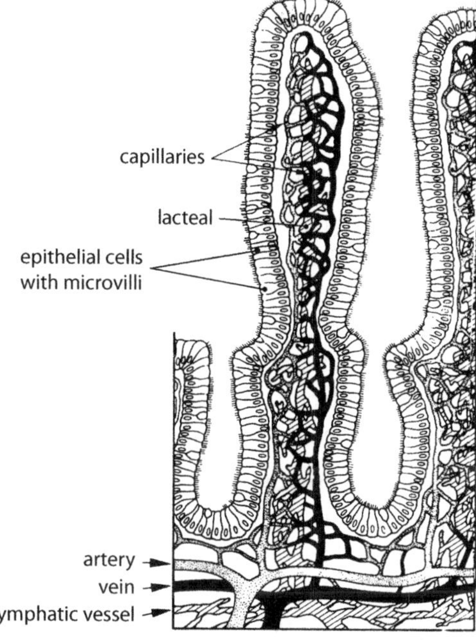

Villus from the small intestine

virus an infectious agent consisting of a protein sheath surrounding a core of nucleic acid (either RNA or DNA); can only be seen with an electron microscope; unable to reproduce on its own and incapable of carrying out any of the usual activities of living organisms; infects a living cell by injecting the nucleic acid strand into the cell; the nucleic acid forms new viruses, which are released when the cell membrane ruptures; a mature virus, with its protein coat, is called a *virion*; when the nucleic acid strand is injected into the cell, the coat is left behind; a *retrovirus* (or *RNA virus*) is a virus that contains RNA instead of the usual DNA; retroviruses must make DNA copies of their RNA before they begin to replicate inside a cell; *see also* diagram of virus, below

viscera the internal organs, especially the organs in the ABDOMINAL CAVITY

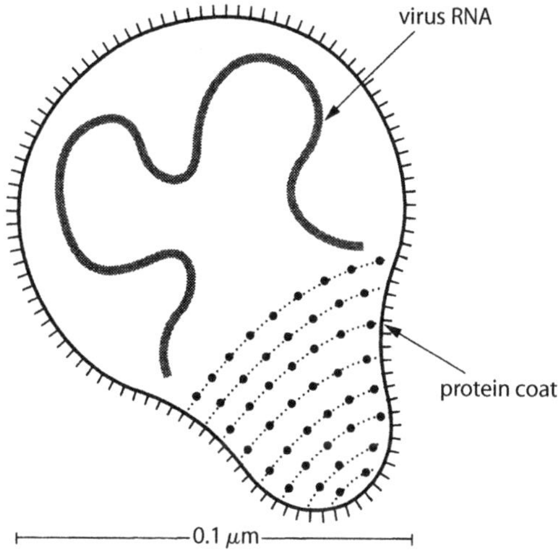

Influenza virus

visceral reflex a REFLEX that causes secretion by glands or involuntary muscle

visceral sensory neuron a NEURON that takes impulses from the internal organs to the brain and spinal cord; monitors the levels of carbon dioxide, oxygen and sugar in the blood, the pressure in the arteries, and the chemical composition of the contents of the stomach and intestines

visual purple an alternative name for RHODOPSIN

vital capacity the maximum volume of air that can be forcibly breathed out after taking as much air as possible into the lungs; *see* lung volumes

vitamin an organic compound (with relatively small molecules) necessary, in very small amounts, for the normal functioning of the body; insufficient intake of a vitamin can lead to a DEFICIENCY DISEASE; *see also* individual vitamins by name, table of vitamins (pp. 289–290)

vitamin A a fat-soluble vitamin that occurs in liver, yellow and green vegetables, milk and butter; essential for health of epithelial cells and growth of bones and teeth; also essential for the formation of *rhodopsin*, the light-sensitive chemical in the rods of the eye; some results of deficiency include dry skin and hair, drying of the cornea of the eye (xerophthalmia) and poor night vision (night blindness); also known as *retinol*

vitamin B complex a group of eight water-soluble vitamins; most function as co-enzymes in metabolic reactions; the vitamins are THIAMINE (B_1), RIBOFLAVIN (B_2), PANTOTHENIC ACID (B_5), PYRIDOXINE (B_6), BIOTIN, NICOTINIC ACID, FOLIC ACID and COBALAMIN (B_{12})

vitamin B_1 also known as THIAMINE

vitamin B_2 also known as RIBOFLAVIN

vitamin B_3 also known as NICOTINIC ACID

vitamin B_5 also called PANTOTHENIC ACID

vitamin B_6 also called PYRIDOXINE

vitamin B_9 also known as FOLIC ACID

vitamin B_{12} also called COBALAMIN

vitamin C also known as ASCORBIC ACID

vitamin D a small group of fat-soluble vitamins; also called *cholecalciferol*; good sources are fish-liver oils and egg yolk, but it can also be made by the skin when exposed to ultraviolet radiation from sunlight; needed in order to absorb and use calcium and phosphorus from the small intestine; deficiency in children causes rickets—soft bones and teeth; in adults, a deficiency causes calcium and phosphorus to be removed from the bones, especially of the legs, pelvis and spine, a condition known as OSTEOMALACIA

vitamin E a small group of fat-soluble vitamins; good sources are nuts, wheat germ, green leafy vegetables; inhibits breakdown of certain fatty acids and therefore preserves cell membranes; deficiency may cause sterility; also called *tocopherol*

vitamin H usually known as BIOTIN

vitamin K a fat-soluble vitamin; good sources are liver, cabbage, cauliflower and spinach; stored in the liver; essential for the production of prothrombin, an important substance in blood clotting; deficiency causes slow blood clotting; also called *phylloquinone*

vitreous humour the jelly-like material filling the rear cavity of the eyeball, between the lens and the retina; maintains the shape of the eyeball; *see also* aqueous humour, diagram of eye (p. 94)

vocal cords membranes in the larynx that vibrate to produce sound; *see* vocal folds

vocal folds the lower of two pairs of horizontal mucous membranes in the larynx (sometimes called the *true vocal cords*); air passing through the opening between the folds (the *glottis*) causes them to vibrate and produce sound; the pitch of the sound depends on the tension on the vocal folds; the upper of the two pairs of membranes is the *ventricular folds* (false vocal cords), which are able to come together to close off the larynx and hold air in the lungs; *see* diagram of vocal folds and glottis (p. 112)

vocalisation the type of sound made by an animal; its vocal behaviour

Table 9 *The principal vitamins, examples of their functions and some results of deficiency*

Vitamin	Examples of important functions	Some effects of deficiency
Water soluble		
Vitamin B_1 (Thiamine)	A co-enzyme involved in carbohydrate metabolism and in metabolism of many amino acids; essential for acetylcholine synthesis	Deficiency leads to beri-beri—partial paralysis of muscles of the alimentary canal, skeletal muscle paralysis and possible cardiac muscle weakness and heart failure; also polyneuritis—degeneration of the myelin sheaths of nerve fibres resulting in muscle atrophy and possible paralysis
Vitamin B_2 (Riboflavin)	Part of co-enzymes such as FAD which are involved in the reactions of the Krebs cycle and in protein metabolism in some cells	Deficiency may lead to cracking of the skin, especially at the corners of the mouth, blurred vision and cataracts
Niacin (Nicotinic acid or nicotinamide)	Part of the co-enzyme NAD that is involved in many energy releasing reactions	Deficiency results in pellagra which is characterised by dermatitis, diarrhoea and mental disturbances
Vitamin B_6 (Pyridoxine)	A co-enzyme involved in protein and amino acid metabolism; also involved in production of circulating antibodies	Most common symptom is dermatitis but can also cause nausea and retarded growth in children
Vitamin B_{12} (Cobalamin or cyano-cobalamin)	A co-enzyme essential for formation of red blood cells and of the amino acid methionine; also involved in nucleic acid metabolism	Pernicious anaemia—insufficient production of red blood cells; malfunction of the nervous system resulting from degeneration of axons in the spinal cord
Pantothenic acid	A part of co-enzyme A that is essential for the transfer of pyruvic acid into the Krebs cycle and for breakdown of fatty acids	Uncertain, but possibly fatigue, muscle spasms and degeneration of nerve and muscle
Folic acid (Folate or folacin)	Essential for the synthesis of the purines and thymine that are necessary for DNA formation; necessary for normal red and white blood cell production	Macrocytic anaemia—abnormally large red blood cells; neural tube defects in the foetus if there is a deficiency before and during pregnancy
Biotin	Involved in amino acid and protein metabolism	Muscle pain, fatigue, poor appetite, nausea
Vitamin C (Ascorbic acid)	Important for collagen formation; works with antibodies; promotes wound healing	Scurvy—tender, swollen gums, loose teeth, defective bone formation, bleeding due to fragile blood vessel walls; also anaemia

(continued)

Table 9 *(continued)*

Vitamin	Examples of important functions	Some effects of deficiency
Fat soluble		
Vitamin A	Important for maintaining epithelial cells; also essential for formation of light sensitive pigments; important for normal growth of bones and teeth	Drying of skin and hair; increased incidence of digestive, respiratory and urinary tract infections; xerophthalmia—drying and opaqueness of cornea; night blindness—impaired vision in low light intensity; slow and faulty growth of bones and teeth
Vitamin D	Needed for absorption of calcium and phosphorus from the alimentary canal	In children, rickets—soft bones and teeth because of defective absorption of calcium and phosphorus; in adults bones are demineralised, a condition known as osteomalacia
Vitamin E (Tocopherol)	Inhibits breakdown of some fatty acids; also involved in formation of RNA, DNA and red blood cells	Uncertain, but may cause abnormalities of mitochondria, lysosomes and cell membranes
Vitamin K	A co-enzyme essential for synthesis of prothrombin and other blood clotting factors in the liver	Excessive bleeding following injury—due to delayed blood clotting time

See individual entries for good food sources of vitamins

voluntary muscle an alternative name for SKELETAL MUSCLE; *see also* muscular tissue

vomeronasal organ (VNO) a pair of minute dents on either side of the tissue between the two nostrils and on top of the vomer bone; a CHEMORECEPTOR that in many animals is able to detect small amounts of PHEROMONES; experts are divided on whether the organ has a function in humans

vomiting the forcible ejection of the STOMACH contents through the mouth, usually as a series of involuntary spasmic movements

vulva the region comprising the external genitalia of the female; includes the labia, clitoris, vestibule and the opening of the vagina

W

Wallace's line an imaginary line dividing the animal populations of Borneo (which resemble Asian fauna) from those of Sulawesi and Lombok (which resemble the fauna of Australia); proposed by the British naturalist Alfred Russel Wallace in his book *The Malay Archipelago* (1869); it was during his travels in the Malay Archipelago that Wallace formulated his theory of natural selection, which (together with the work of Charles Darwin) changed scientific thinking about the evolution of life on earth; *see* map below

warts small, usually painless growths, caused by infection with certain of the human papilloma viruses; they occur on the skin or mucous membranes, such as the tissue that lines the nose, throat, digestive tract, and other body openings; can be passed from one person to another and can travel from one part of the body to another; *see also* genital warts

wastes includes the products of METABOLISM passed out in the URINE, some of the substances lost in SWEAT and when breathing

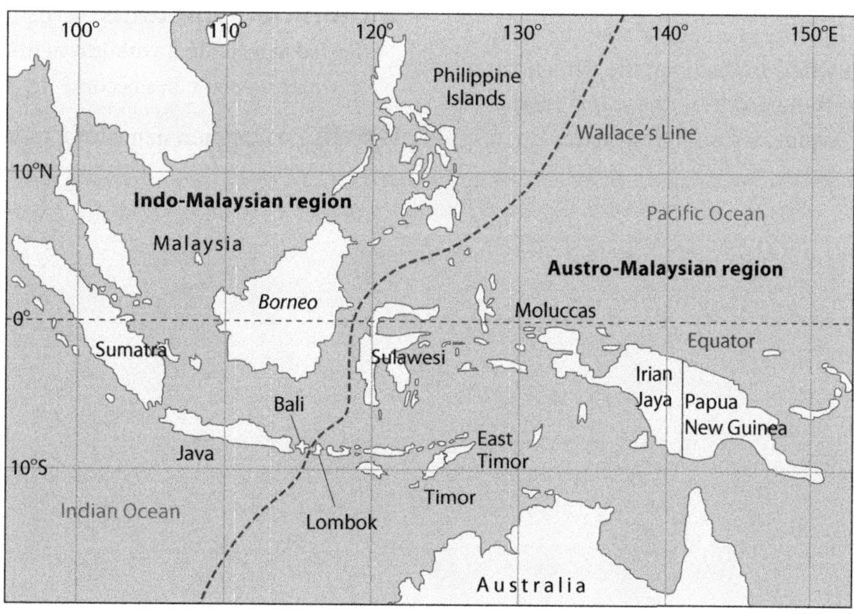

Wallace's line

out, and the undigested residue of food eliminated from the body (the FAECES)

water intoxication a potentially life-threatening condition caused by drinking too much water when the amount of salt (and other ELECTROLYTES) in the body is low; commonly caused by long bouts of intensive exercise during which electrolytes are not replenished and large amounts of water are consumed

Wernicke's area an area of the brain that is important in language development; occupies a large region of the PARIETAL and TEMPORAL LOBES on the left side of the CEREBRUM; responsible for the comprehension of speech (*see also* Broca's area, which is involved in *producing* speech); damage to the area can result in impaired language development and use

white blood cells the common name for LEUCOCYTES

white matter the part of the brain and spinal cord made up of myelinated fibres; myelin is white in colour so that the tissue appears white; in the brain the white matter is towards the inside, in the spinal cord it is on the outside; *see also* grey matter

whole blood blood as taken from a blood vessel; includes plasma and all the different cell types

whooping cough an infection of the mucous membranes of the respiratory system caused by the bacterium *Bordetella pertussis*; initial symptoms are similar to those of the common cold but develop into a severe cough with a 'whooping' sound as air is taken into the lungs; more common in children and most are immunised in infancy, although outbreaks do occur because people have become complacent about the need for immunisation

wisdom teeth the third pair of molar TEETH in each jaw

withdrawal **1.** stopping, or reducing, intake of a drug on which one has become dependent; *see also* withdrawal symptoms **2.** during sexual intercourse, prevention of fertilisation by removal of the penis from the vagina before ejaculation

withdrawal symptoms symptoms experienced as a result of withdrawal from a drug on which a person has become dependent

womb a common name for UTERUS

X

***X*-chromosome** one of the SEX CHROMOSOMES

***X*-linkage** the control of characteristics by genes in the *X*-chromosome; *see* sex linkage

***X*-linked** a term frequently used to describe a characteristic that is controlled by a gene that occurs in the *X*-chromosome; *see also* sex linkage

***X*-linked characteristics** characteristics controlled by genes located in the *X*-chromosome; *see* sex linkage

X-ray a form of electromagnetic radiation that is able to penetrate soft body tissues but is stopped by more solid material such as bone; used to diagnose medical conditions such as tuberculosis, lung and breast cancers and broken bones; also used to determine the structure of large, complex molecules by a technique called X-ray diffraction

xenotransplantation transplantation of body parts from one species into another (e.g. the possible transplantation of pigs' livers into humans)

Y

***Y*-chromosome** one of the SEX CHROMOSOMES

***Y*-linked characteristics** characteristics controlled by genes located in the *Y*-chromosome; *see* sex linkage

ya an abbreviation meaning 'years ago' (e.g. 10 000 ya means 10 000 years ago)

yellow bone marrow the region of a bone where fat is stored; *see* marrow

yellow fever a serious viral disease of tropical and subtropical regions; causes destruction of cells in the liver, spleen, bone marrow, lymph nodes and kidneys; the virus is spread by a particular species of mosquito

yellow spot an area at the centre of the RETINA of the eye, directly opposite the centre of the lens; also called the *macula lutea*; the yellow spot has no rods, only cones; in the centre of the yellow spot is a small depression, the *fovea* (or *fovea centralis*), the area of sharpest vision; when we look directly at an object, the light rays reflected from that object are focused onto the fovea; *see also* diagram of eye (p. 94)

yolk sac one of the FOETAL MEMBRANES

Z

Z line part of the structure of skeletal muscle; *see* sarcomere

Z membrane an alternative name for *Z line*; *see* sarcomere

zidovudine (AZT) a drug used to slow the progress of infection by human immunodeficiency virus (HIV)

zona pellucida the jelly-like layer that surrounds the egg; is broken down by the action of chemicals contained in the heads of sperm to allow fertilisation to take place; *see also* acrosome reaction

zoology the science dealing with the study of the structure, functions, behaviour, history, classification and distribution of animals

zoonosis (plural *zoonoses*) any disease of animals that can be passed on to humans (e.g. rabies, typhus and bubonic plague)

zygomatic arch an arch of bone at the side of the skull; connects the zygomatic and temporal bones; anchors muscles used in chewing; particularly prominent in the fossil skulls of *Australopithecus boisei*

zygomatic bone a bone on each side of the skull that forms the protruding part of the cheek and the lower part of the eye socket; also called the *cheekbone*; a projection of the zygomatic bone joins a projection of the temporal bone to form the zygomatic arch, which is just above the joint of the lower jaw

zygote the fertilised egg from which a new individual develops; a diploid cell resulting from the union of a haploid sperm and a haploid egg; forms a new individual by repeated mitotic divisions

zygotene the second stage of prophase I of MEIOSIS; during zygotene the like chromosomes pair and lie alongside one another